D0214829

DNA MICROARRAYS AND RELATED GENOMICS TECHNIQUES

Design, Analysis, and Interpretation of Experiments

Biostatistics: A Series of References and Textbooks

Series Editor
Shein-Chung Chow
Duke Clinical Research Institute
Duke University
Durham, NC, USA

1. *Design and Analysis of Animal Studies in Pharmaceutical Development,* Shein-Chung Chow and Jen-pei Liu
2. *Basic Statistics and Pharmaceutical Statistical Applications,* James E. De Muth
3. *Design and Analysis of Bioavailability and Bioequivalence Studies, Second Edition, Revised and Expanded,* Shein-Chung Chow and Jen-pei Liu
4. *Meta-Analysis in Medicine and Health Policy,* Dalene K. Stangl and Donald A. Berry
5. *Generalized Linear Models: A Bayesian Perspective,* Dipak K. Dey, Sujit K. Ghosh, and Bani K. Mallick
6. *Difference Equations with Public Health Applications,* Lemuel A. Moyé and Asha Seth Kapadia
7. *Medical Biostatistics,* Abhaya Indrayan and Sanjeev B. Sarmukaddam
8. *Statistical Methods for Clinical Trials,* Mark X. Norleans
9. *Causal Analysis in Biomedicine and Epidemiology: Based on Minimal Sufficient Causation,* Mikel Aickin
10. *Statistics in Drug Research: Methodologies and Recent Developments,* Shein-Chung Chow and Jun Shao
11. *Sample Size Calculations in Clinical Research,* Shein-Chung Chow, Jun Shao, and Hansheng Wang
12. *Applied Statistical Design for the Researcher,* Daryl S. Paulson
13. *Advances in Clinical Trial Biostatistics,* Nancy L. Geller
14. *Statistics in the Pharmaceutical Industry, 3rd Edition,* Ralph Buncher and Jia-Yeong Tsay
15. *DNA Microarrays and Related Genomics Techniques: Design, Analysis, and Interpretation of Experiments,* David B. Allsion, Grier P. Page, T. Mark Beasley, and Jode W. Edwards

DNA MICROARRAYS AND RELATED GENOMICS TECHNIQUES

Design, Analysis, and Interpretation of Experiments

EDITED BY

DAVID B. ALLISON

GRIER P. PAGE

T. MARK BEASLEY

JODE W. EDWARDS

Chapman & Hall/CRC
Taylor & Francis Group
Boca Raton London New York

Published in 2006 by
Chapman & Hall/CRC
Taylor & Francis Group
6000 Broken Sound Parkway NW, Suite 300
Boca Raton, FL 33487-2742

Printed in the United States of America on acid-free paper
10 9 8 7 6 5 4 3 2 1

International Standard Book Number-10: 0-8247-5461-1 (Hardcover)
International Standard Book Number-13: 978-0-8247-5461-7 (Hardcover)
Library of Congress Card Number 2005050488

Library of Congress Cataloging-in-Publication Data

DNA microarrays and related genomics techniques : design, analysis, and interpretation of experiments / editors, David B. Allison ... [et al.].
p. cm. -- (Biostatistics)
Includes bibliographical references and index.
ISBN 0-8247-5461-1 (alk. paper)
1. DNA microarrays. I. Allison, David B. (David Bradley), 1963- II. Biostatistics (New York, N.Y.)

QP624.5.D726.D636 2005
572.8'636--dc22 2005050488

Visit the Taylor & Francis Web site at
http://www.taylorandfrancis.com

and the CRC Press Web site at
http://www.crcpress.com

Series Introduction

The primary objectives of the *Biostatistics Book Series* are to provide useful reference books for researchers and scientists in academia, industry, and government, and also to offer textbooks for undergraduate and graduate courses in the area of biostatistics and bioinformatics. This book series will provide comprehensive and unified presentations of statistical designs and analyses of important applications in biostatistics and bioinformatics, such as those in biological and biomedical research. It gives a well-balanced summary of current and recently developed statistical methods and interpretations for both statisticians and researchers/scientists with minimal statistical knowledge who are engaged in the field of applied biostatistics and bioinformatics. The series is committed to providing easy-to-understand, state-of-the-art references and textbooks. In each volume, statistical concepts and methodologies will be illustrated through real world examples whenever possible.

In recent years, the screening of thousands of genes using the technique of expression microarrays has become a very popular topic in biological and biomedical research. The purpose is to identify those genes that may have an impact on clinical outcomes of a subject who receives a test treatment under investigation and consequently establish a medical predictive model. Under a well-established predictive model, we will be able not only to identify subjects with certain genes who are most likely to respond to the test treatment, but also to identify subjects with certain genes who are most likely to experience (serious) adverse events. This concept plays an important role in the so-called personalized medicine research. This volume summarizes various useful experimental designs and statistical methods that are commonly employed in microarray studies. It covers important topics in DNA microarrays and related genomics research such as normalization of microarray data, microarray quality control, statistical methods for screening of high-dimensional biology, and power and sample size calculation. In addition, this volume provides useful approaches to microarray studies such as clustering approaches to gene microarray data, parametric linear models, nonparametric procedures, and Bayesian analysis of microarray data. It would be beneficial to biostatisticians, biological and biomedical researchers, and pharmaceutical scientists who are engaged in the areas of DNA microarrays and related genomics research.

Shein-Chung Chow

Preface

WHAT ARE MICROARRAYS?

Microarrays have become a central tool used in modern biological and biomedical research. This book concerns expression microarrays, which for the remainder of the book, we simply refer to as microarrays. They are tools that permit quantification of the amount of all mRNA transcripts within a particular biological specimen. There are several different technologies for producing microarrays that have different strengths and weaknesses. These platforms and alternatives are discussed in Chapter 1 by Gaffney et al.

Viewed as "hot" and highly exotic tools as recently as the late 1990s, they are now ubiquitous in biological research and the modern biological researcher can no more be unaware and unexposed to microarray research and its results than one can remain ignorant of clinical trials, questionnaire studies, genome scans, animal models, or any of the other tools that have become standard parts of our armamentarium. Although much development in microarray research methodology is still needed, it is clear that microarrays are here to stay.

WHY THIS BOOK?

In one sense, microarrays are simply measurement assays. Just as one can measure, for example, the amount of insulin (which is the product of a gene) in blood, we can measure the products of genes with microarrays in any tissue. What distinguishes microarrays from traditional approaches is their "omic" nature. That is, they have capacity to measure *all* gene transcripts at once. This ushered in the subfield of *transcriptomics*. A particular challenge is that because of the expense of microarray research and the fact that it is often directed at basic discovery and hypothesis generation/exploration missions, the number of variables (transcripts) available in microarray studies tends to exceed the number of cases (subjects) by several orders of magnitude. Traditional statistical approaches to design and analysis were not developed in the context of such high dimensional and small sample problems. We and many others now find that our training in traditional statistical methods is not especially well-suited to such situations.

We (the editors) were first introduced to the analysis of microarray data ca. 1999. At that time, there were almost no statistical papers providing approaches to analyze microarray data or design microarray studies from a statistical perspective. By 2003, this situation had changed dramatically and we estimate that there were hundreds of papers thereon (Mehta et al., 2004). This overwhelming deluge of methods from these papers is quite daunting to either the applied investigator looking for methodologies to utilize or the methodologist trying to keep up with the field.

As part of the research efforts funded by the National Science Foundation, we have hosted an annual retreat for scientists interested in analytic methods for microarray research for the last five years. The impetus for this book came in part from discussions held at those retreats. We felt there was a need for a book that consolidated many of the existing methodologic advances and compiled many of the issues and methods into a single volume. This book is aimed at both the investigator who will conduct analyses of microarray data and at the methodologists who will evaluate existing and develop future methodologies.

WHAT IS HERE?

We have structured this book in a manner that we believe parallels the steps that an investigator or an analyst will go through while conducting and analyzing a microarray experiment from conception to interpretation. We begin with the most foundational issues: ensuring the quality and integrity of the data and assessing the validity of the statistical methods we employ. We then move on to the often neglected, but critical aspects of designing a microarray experiment. Gadbury et al. (Chapter 5) address issues such as power and sample size, where only very recently have developments allowed such calculations in a high dimensional context. The third section of the book is the largest, addressing issues of the analysis of microarray data. The size of this section reflects both the variety of topics and the amount of effort investigators have devoted to developing new methodologies. Finally, we move on to the intellectual frontier — interpretation of microarray data. New methods for facilitating and affecting formalization of the interpretation process are discussed. The movement to make large high dimensional datasets public for further analysis and methods for doing so are also addressed.

WHAT IS NOT HERE?

This book is not a detailed exposition of software packages (although some are mentioned in specific chapters), biochemistry, or the mechanics of the physical production of microarrays or biological specimens for analysis via microarrays. Interested readers should consult other more topical books in these areas (Jordan, 2001; Grigorenko, 2002; Ye and Day, 2003) Many closely related disciplines such as proteomics and metabolomics are not discussed in any depth although the astute reader will readily see the commonalities among the statistical and design approaches that can be applied to such data.

THE FUTURE

There is no question that this field will continue to advance rapidly and some of the specific methodologies we discuss herein will be replaced by new advances in the near future. Nevertheless, we believe the field is now at a point where a foundation of key categories of methods has been laid and begun to settle. Although the details may change, we believe that the majority of the key principles described herein and the foundational categories are likely to stand the test of time and serve as a useful guide to the reader. We look forward to new biological knowledge that we anticipate will emerge from the evermore sophisticated technologies and analysis as well as the exciting new statistical advances sure to come.

REFERENCES

Girgorenko E.V. (2002) *DNA Arrays: Technologies and Experimental Strategies*. CRC Press, Boca Raton, FL.

Jordan B.R. (ed.) (2001) *DNA Microarray: Gene Expression Applications*. Springer-Verlag, Berlin.

Mehta T., Tanik M., and Allison D.B. (2004) Towards sound epistemological foundation of statistical methods for high-dimensional biology. *Nature Genetics* 36: 943–947.

Ye S. and Day I.N.M. (2003) *Microarray and Microplates*. Bios Press, Oxford.

Editors

David B. Allison received his Ph.D. from Hofstra University in 1990. He then completed a postdoctoral fellowship at the Johns Hopkins University School of Medicine and a second postdoctoral fellowship at the NIH-funded New York Obesity Research Center at St. Luke's/Roosevelt Hospital Center. He was a research scientist at the New York Obesity Research Center and Associate Professor of Medical Psychology at Columbia University College of Physicians and Surgeons until 2001. In 2001, he joined the faculty of the University of Alabama at Birmingham where he is currently Professor of Biostatistics, Head of the Section on Statistical Genetics, and Director of the NIH-funded Clinical Nutrition Research Center. He has authored over 300 scientific publications and edited three books. He has won several awards, including the 2002 Lilly Scientific Achievement Award from the North American Association for the Study of Obesity and the 2002 Andre Mayer Award from the International Association for the Study of Obesity, holds several NIH and NSF grants, served on the Council of the North American Association for the Study of Obesity from 1995 to 2001, and has been a member of the Board of Trustees for the International Life Science Institute, North America, since January 2002. He serves on the editorial boards of *Obesity Reviews; Nutrition Today; Public Library of Science (PLOS) Genetics; International Journal of Obesity; Behavior Genetics; Computational Statistics and Data Analysis; and Human Heredity*.

Dr. Allison's research interests include obesity, quantitative genetics, clinical trials, and statistical and research methodology.

Grier P. Page, Ph.D. was born in Cleveland, Ohio in 1970. He received his B.S. in Zoology and Molecular Biology from the University of Texas, Austin. Then he received his M.S. and Ph.D. in Biomedical Sciences from the University of Texas–Health Sciences Center—Houston under the mentorship of Drs. Eric Boerwinkle and Christopher Amos. Dr. Page has been involved in the use and analysis of microarrays since 1998 for expression, genomics, and genotyping. He is very active in the development of new methods for the analysis of microarray data as well as methods and techniques for the generation on the highest quality microarray data. He uses microarrays in his research in the mechanisms of cancer development, nutrient production, and nutrient gene interactions especially in cancer and plants. He is currently a member of the Section on Statistical Genetics, Department of Biostatistics the University of Alabama, Birmingham.

T. Mark Beasley, Ph.D. is Associate Professor of Biostatics and a member of the Section on Statistical Genetics at the University of Alabama at Birmingham. He is the leader of the measurement and inferences teams for a funded National Science Foundation (NSF) grant to further the development of microarray analysis methods. He has a Ph.D. in Statistics and Measurement from Southern Illinois University and a strong research record in the area of statistical methodology, focused in methodological problems in statistical genetics; nonparametric statistics; simulation studies; and the use of linear models. He also has a strong background in measurement theory and the multivariate methods (e.g., factor analysis, structural equation models; regression models). Dr. Beasley teaches courses on Applied Multivariate Analysis and General Linear Models at UAB and is currently Editor of *Multiple Linear Regression Viewpoints*, a journal focused on applications of general linear models and multivariate analysis. He has published articles in applied statistics journals such as the *Journal of Educational & Behavioral Statistics, Journal of the Royal Statistical Society, Computational Statistics & Data Analysis, Multivariate Behavioral Research*, and *Communications in Statistics*. He has also published articles on methodological problems in statistical genetics in leading journals such as the *American Journal of Human Genetics; Behavior Genetics; Genetic Epidemiology; Genetics, Selection, and Evolution* and *Human Heredity*.

Jode W. Edwards received a Ph.D. in plant breeding and genetics with a minor in statistics from Iowa State University in 1999. He then spent 3 years with Monsanto Company as a statistical geneticist working in the areas of marker-assisted plant breeding and QTL mapping. Dr. Edwards joined the Section on Statistical Genetics as a Postdoctoral Fellow in 2002. His research involved application of Empirical Bayes methods to microarray analysis and development of software for microarray data analysis. Using SAS as a prototyping platform, he designed experimental versions of the HDBStat! software that is now distributed by the Section on Statistical Genetics. Additionally, Dr. Edwards helped initiate efforts to build the microarray Power Atlas, a tool to assist investigators in designing microarray experiments. In 2004, he completed his postdoctoral studies and assumed a position as a Research Geneticist with the Agricultural Research Service of the United States Department of Agriculture, in Ames, IA. His research is focused on quantitative genetics of maize, application of Bayesian methods in plant breeding, and breeding for amino acid balance in maize protein.

Contributors

David B. Allison
Department of Biostatistics
University of Alabama at Birmingham
Birmingham, Alabama

T. Mark Beasley
Department of Biostatistics
University of Alabama at Birmingham
Birmingham, Alabama

Jacob P.L. Brand
Department of Biostatistics
University of Alabama at Birmingham
Birmingham, Alabama

Jane Y. Chang
Department of Applied Statistics and Operational Research
Bowling Green State University
Bowling Green, Ohio

Kei-Hoi Cheung
Department of Genetics
Center for Medical Informatics
Yale University School of Medicine
New Haven, Connecticut

Tzu-Ming Chu
SAS Institute
Cary, North Carolina

Christopher S. Coffey
University of Alabama at Birmingham
Birmingham, Alabama

Stacey S. Cofield
University of Alabama at Birmingham
Birmingham, Alabama

Robert R. Delongchamp
Division of Biometry and Risk Management
National Center for Toxicological Research
Jefferson, Arizona

Shibing Deng
SAS Institute
Cary, North Carolina

Jode W. Edwards
Department of Biostatistics
University of Alabama at Birmingham
Birmingham, Alabama

David Finkelstein
Hartwell Center
St. Jude Children's Research Hospital
Memphis, Tennessee

Gary L. Gadbury
Department of Mathematics and Statistics
University of Missouri-Rolla
Rolla, Missouri

Patrick M. Gaffney
University of Minnesota
Minneapolis, Minnesota

Elizabeth Garrett-Mayer
Division of Oncology Biostatistics
Sidney Kimmel Comprehensive Cancer Center
Baltimore, Maryland

Pulak Ghosh
Department of Mathematics and Statistics
Georgia State University
Atlanta, Georgia

Bernard S. Gorman
Nassau Community College and
Hofstia University
Garden City, New York

Jason C. Hsu
Department of Statistics
Ohio State University
Columbus, Ohio

Michael Janis
Department of Chemistry,
Biochemistry, and
Molecular Biology
University of California
at Los Angeles
Los Angeles, California

Christina M. Kendziorski
Department of Biostatistics and
Medical Informatics
University of Wisconsin-Madison
Madison, Wisconsin

Jeanne Kowalski
Division of Oncology Biostatistics
Johns Hopkins University
Baltimore, Maryland

Jeffrey D. Long
Department of Biostatistics
University of Alabama at Birmingham
Birmingham, Alabama

Tapan Mehta
Department of Biostatistics
University of Alabama at Birmingham
Birmingham, Alabama

Kathy L. Moser
Department of Medicine
Institute of Human Genetics and
Center for Immunology
University of Minnesota
Medical School
Minneapolis, Minnesota

Michael V. Osier
Yale Center for Medical Informatics
Yale University School of Medicine
New Haven, Connecticut

Grier P. Page
Department of Biostatistics
University of Alabama at Birmingham
Birmingham, Alabama

Rudolph S. Parrish
Department of Bioinformatics and
Biostatistics School of Public
Health and Information Sciences
University of Louisville
Louisville, Kentucky

Jacques Retief
Iconix Pharmaceuticals
Mountain View, California

Douglas M. Ruden
Department of Environmental Health
Sciences
University of Alabama at Birmingham
Birmingham, Alabama

Chiara Sabatti
Department of Human Genetics
University of California at Los Angeles
Los Angeles, California

Kathryn Steiger
Division of Biostatistics
University of California at Berkeley
Berkeley, California

Murat Tanik
Department of Biostatistics
University of Alabama at Birmingham
Birmingham, Alabama

Alan Williams
Affymetrix
Santa Clara, California

Russell D. Wolfinger
SAS Institute
Cary, North Carolina

Qinfang Xiang
Department of Mathematics and Statistics
University of Missouri-Rolla
Rolla, Missouri

Stanislav O. Zakharkin
Department of Biostatistics
University of Alabama at Birmingham
Birmingham, Alabama

Kui Zhang
Department of Biostatistics
University of Alabama at Birmingham
Birmingham, Alabama

Zhen Zhang
Department of Pathology School of Medicine
Johns Hopkins University
Baltimore, Maryland

Contents

Chapter 1
Microarray Platforms 1
Patrick M. Gaffney and Kathy L. Moser

Chapter 2
Normalization of Microarray Data 9
Rudolph S. Parrish and Robert R. Delongchamp

Chapter 3
Microarray Quality Control and Assessment 29
David Finkelstein, Michael Janis, Alan Williams, Kathryn Steiger, and Jacques Retief

Chapter 4
Epistemological Foundations of Statistical Methods for High-Dimensional Biology 57
Stanislav O. Zakharkin, Tapan Mehta, Murat Tanik, and David B. Allison

Chapter 5
The Role of Sample Size on Measures of Uncertainty and Power 77
Gary L. Gadbury, Qinfang Xiang, Jode W. Edwards, Grier P. Page, and David B. Allison

Chapter 6
Pooling Biological Samples in Microarray Experiments 95
Christina M. Kendziorski

Chapter 7
Designing Microarrays for the Analysis of Gene Expressions 111
Jane Y. Chang and Jason C. Hsu

Chapter 8
Overview of Standard Clustering Approaches for Gene Microarray Data Analysis 131
Elizabeth Garrett-Mayer

Chapter 9
Cluster Stability 159
Bernard S. Gorman and Kui Zhang

Chapter 10
Dimensionality Reduction and Discrimination 177
Jeanne Kowalski and Zhen Zhang

Chapter 11
Modeling Affymetrix Data at the Probe Level 197
Tzu-Ming Chu, Shibing Deng, and Russell D. Wolfinger

Chapter 12
Parametric Linear Models 223
Christopher S. Coffey and Stacey S. Cofield

Chapter 13
The Use of Nonparametric Procedures in the Statistical Analysis of Microarray Data 245
T. Mark Beasley, Jacob P.L. Brand, and Jeffrey D. Long

Chapter 14
Bayesian Analysis of Microarray Data 267
Jode W. Edwards and Pulak Ghosh

Chapter 15
False Discovery Rate and Multiple Comparison Procedures 289
Chiara Sabatti

Chapter 16
Using Standards to Facilitate Interoperation of Heterogeneous Microarray Databases and Analytic Tools 305
Kei-Hoi Cheung

Chapter 17
Postanalysis Interpretation: "What Do I Do with This Gene List?" 321
Michael V. Osier

Chapter 18
Combining High Dimensional Biological Data to Study Complex Diseases and Quantitative Traits 335
Grier P. Page and Douglas M. Ruden

Index 361

1 Microarray Platforms

Patrick M. Gaffney and Kathy L. Moser

CONTENTS

1.1 Introduction 1
1.2 Microarray Technology 1
1.3 Autoantigen and Cytokine Microarrays 2
1.4 DNA and Oligonucleotide Microarrays 4
1.5 Tiling Arrays 4
1.6 Data Analysis 5
1.7 Future Directions 5
References 6

1.1 INTRODUCTION

As with the development of any novel and potentially powerful technology, the prospect of revealing new information that may dramatically change our understanding of biological processes can generate much excitement. Such is true for the emerging genomic approaches that make possible high-density assays using microarray platforms. Indeed, it is difficult, if not impossible, to imagine any area of biology that could not be affected by the wide range of potential applications of microarray technology. Numerous examples, such as those from the field of oncology, provide striking evidence of the power of microarrays to bring about extraordinary advances in molecularly defining important disease phenotypes that were otherwise unrecognized using conventional approaches such as histology.

In this chapter, we present a general overview of microarray platforms currently in use with particular emphasis on high-density DNA arrays. We touch briefly on approaches to data analysis leaving most of the details for the ensuing chapters. For those just entering the microarray arena or interested in more details, a series of particularly useful reviews have recently been published that take stock of the latest developments and discuss the most pressing challenges of this technology [1].

1.2 MICROARRAY TECHNOLOGY

Microarray technology provides an unprecedented and uniquely comprehensive probe into the coordinated workings of entire biological pathways and genomic-level

TABLE 1.1
Potential Objectives of Studies Utilizing Microarray Technology

1. Distinguish patients from normal controls
2. Identify subsets of patients
3. Characterize host responses
4. Examine cellular pathways
5. Compare alternative experimental conditions
6. Examine drug response
7. Follow temporal changes in gene expression
8. Identify candidate genes for genetic studies

processes. In general terms, microarrays refer to a variety of platforms in which high density assays are performed in parallel on a solid support. Thousands to tens of thousands of datapoints may be generated in each experiment. The growth of scientific literature since the mid-1990s may provide some indication for the potential impact of this technology in biomedical sciences. A majority of applications have been in oncology, although many examples from other fields are rapidly emerging and include examination of host response to pathogens, examination of drug responses, identification of temporal changes in gene expression, and comparisons of various experimental conditions.

Three major types of microarrays exist — tissue, protein, and DNA. Tissue microarrays immobilize small amounts of tissue from biopsies of multiple subjects on glass slides for immunohistochemical processing, while protein arrays immobilize peptides or intact proteins for detection by antibodies or other means (see Section 1.3). For the last several years, much excitement and attention has focused on DNA microarrays and most of this book will concentrate on DNA microarray analysis. Regardless of the specific platform used, these approaches offer new opportunities to address biologic questions in a way never possible before. Table 1.1 provides just a few examples of the potential ways in which microarray technology can be utilized.

1.3 AUTOANTIGEN AND CYTOKINE MICROARRAYS

Applications of protein microarrays include assessment of enzyme–substrate, protein–protein, and DNA–protein interactions. Although efforts to develop these proteomic tools predate the first descriptions of DNA microarrays [2], progress has been relatively slower — in part due to challenges posed by natural inherent differences in proteins compared with DNA. As examples, proteins consist of highly diverse conformational structures that result from 20 amino acids vs. the 4 nucleic acid building blocks that generate a relatively uniform structure in DNA. Proteins may exist as large complexes, can be hydrophilic or hydrophobic, acidic or basic, and contain posttranslational modifications such as acetylation, glycosylation, or phosphorylation. Functional and conformational properties of proteins must often remain intact when immobilized onto a microarray in order to retain the desired binding properties for detection of target ligands.

The development of protein microarrays to detect immunologic targets such as cytokines or autoantibodies has enormous potential for research and diagnostic applications in autoimmune diseases. Several groups, including Joos and colleagues in Germany [3], and Robinson and colleagues at Stanford University [4], have made important strides in developing autoantigen microarrays for multiplex characterization of autoimmune serum. Joos and colleagues spotted 18 common autoantigens onto silane-treated glass slides and nitrocellulose at serial dilutions. Bound antibodies from minimal amounts of 25 characterized autoimmune serum samples and ten normal blood donors were titered by using variable amounts of autoantigen. The autoimmune serum samples were obtained from patients with autoimmune thyroiditis (Hashimoto's thyroiditis and Graves' disease), systemic lupus erythematosus (SLE), Sjogren's syndrome (SS), mixed connective tissue disease (MCTD), scleroderma, polymyositis, systemic vasculitis, and antiphospholipid syndrome. These assays proved to be highly specific and similar in sensitivity when compared to a standard ELISA format. Further developments will include optimizing the nature of the autoantigen material to minimize possible loss of antigenicity and expanding the representation of autoantigens on the array.

Similarly, Robinson and colleagues have developed a 1152-feature array containing 196 distinct biomolecules representing major autoantigens targeted by antibodies produced by rheumatic autoimmune disease patients [4]. The autoantigens included hundreds of proteins, peptides, DNA, enzymatic complexes, and ribonucleoprotein complexes. Examples of autoantigens spotted include Ro52, Ro60, La, jo-1, Sm-B/B′, U1-70 kD, U1 snRNP-C, topoisomerase 1, pyruvate dehydrogenase (PDH), and histone H2A. The arrays were characterized using multiple sera from eight human autoimmune diseases and included SLE, SS, MCTD, polymyositis, primary biliary cirrhosis, rheumatoid arthritis (RA), and both limited and diffuse forms of scleroderma. This work demonstrates the feasibility of using large-scale, fluorescence-based autoantigen microarrays to detect human autoantibodies with simple protocols and widely available equipment in a low-cost and low-sample volume format. Some of the potential applications for this technology include (1) rapid screening for autoantibody specificities to facilitate diagnosis and treatment, (2) characterization of the specificity, diversity, and epitope spreading of autoantibody responses, (3) determination of isotype subclass of specific autoantibodies, (4) guiding development and selection of antigen-specific therapies, and (5) use as a discovery tool to identify novel autoantigens or epitopes.

Microarrays that simultaneously detect multiple cytokines have been developed by Huang and colleagues at Emory University [5]. Their method utilizes capture antibodies spotted onto membranes, incubation with biological samples such as patient serum, and detection by biotin-conjugated antibodies and enzymatic-coupled enhanced chemiluminescence. Twenty-eight cytokines were detected using this method, including interleukins-1α, 2, 3, 5, 6, 7, 8, 10, 13, and 15; tumor necrosis factors α, and β; interferon-γ, and others. In addition to detecting multiple cytokines simultaneously, these assays were shown to be more sensitive than conventional ELISAs, with broader detection ranges. The ability to readily scale up this approach to include much larger numbers of cytokines and other proteins will undoubtedly fuel further development of this powerful tool for studying complex and dynamic cellular processes such as immune reactions, apoptosis, cell proliferation, and differentiation.

1.4 DNA AND OLIGONUCLEOTIDE MICROARRAYS

DNA microarrays were first introduced in the mid-1990s [6] and have been the most widely utilized application of microarray technology. There are two commonly available DNA microarray systems. First are the cDNA microarrays fabricated by robotic spotting of PCR products, derived primarily from the 3′ end of genes and expressed sequence tags (ESTs), onto glass slides — this is the method popularized by, among others, Dr. Patrick Brown at Stanford and Dr. Louis Staudt at the NIH [7,8]. The second method uses *in situ* synthesized oligonucleotide arrays that are fabricated using photolithographic chemistry on silicon chips — this is the method used in the proprietary Affymetrix™ system [9] and recently by NimbleGen™. A third method involves spotting previously synthesized longer (40 to 70mer) oligonucleotides on either glass (Amersham™ and Agilent™) or nylon and plastic (clonetech™ and SuperArray™). The data generated using these systems are highly concordant, as demonstrated in parallel studies of the yeast cell cycle [10,11]. In the spotted cDNA and long oligo microarray systems, two probes with different fluorescent tags are hybridized to the same array, one serving as the experimental condition and the other as a control. The ratio of hybridization between the two probes is calculated, allowing a quantization of the hybridization signal for each spot on the array. In this system, the probe is 1st strand cDNA generated by oligo-dT primed reverse transcription from an RNA sample (for additional details see http://cmgm.stanford.edu/pbrown/). In the Affymetrix™ system, only a single labeled probe is used and each gene on the chip is represented by 8 to 10 wild-type 25-mer oligonucletides and the same number of single base mutant 25-mer oligonucleotides synthesized next to one another on the array. Signal intensity and the ratio of specific to nonspecific hybridization allows the generation of quantitative data regarding gene expression in the sample (for more details see http://www.affymetrix.com/technology/tech_probe.html).

1.5 TILING ARRAYS

Recently several groups have developed arrays with long stretches of chromosomes or whole-genomic sequences probed onto arrays. Potential uses for such whole-genome arrays include empirical annotation of the transcriptome [12], identification of novel transcripts [13,14], analysis of alternative and cryptic splicing, characterization of the methylation state of the genome, polymorphism discovery and genotyping, comparative genome hybridization, and genome resequencing [15]. These arrays have great future potential for studying new aspects of the genome and providing greater insights into the function of living organisms.

1.6 DATA ANALYSIS

Microarray analysis is often considered a discovery-based rather than hypothesis-driven approach [16,17], largely due to the potential for discovering altered expression of novel genes for which little or no prior information was available to suggest a role in the disease or experimental condition examined. However, high quality experiments are driven by addressing a scientific question (even if it is simply — "are there genes that are differentially expressed between a group of patients and controls?"), consistency in execution of experimental protocols, use of sample sizes with as many replicates as is feasible, and a plan for statistical analysis and interpretation of the data. Including statistical expertise during the early phase of experimental design (i.e., prior to any data collection) is critical, particularly in the setting of microarray analysis where each experiment can carry significant cost.

1.7 FUTURE DIRECTIONS

The majority of human diseases undoubtedly involves the complex interplay of many genes. Although the number and type of genes are not yet known, global assessment of gene expression is a very powerful approach for gaining insight into these processes. Identification of these genes will certainly contribute to advancing our understanding of the molecular basis for human diseases and identifying novel therapeutic targets. Within a relatively short period of time, the information learned from the application of microarray technology to address complicated biological questions has not only met, but often exceeded expectations. Despite their success, microarray studies are not without their challenges. Continued refinement of these techniques, including development of improved statistical methods for extracting information from large datasets and software tools for data processing, management, and storage as described in the following chapters of this book, will likely increase the applicability and general use of these technologies. Additionally, establishing common standards for the publishing and sharing of microarray generated data will be important. The applicability of this technology in translational medicine is only beginning to be appreciated and it is likely that microarray technologies will have a substantial impact on our understanding of human disease now and into the future.

REFERENCES

1. J.M. Trent and A.D. Baxevanis. Chipping away at genomic medicine. *Nat. Genet.* (Suppl): 462, 2002.
2. G. MacBeath. Protein microarrays and proteomics. *Nat. Genet.* 32 (Suppl): 526–532, 2002.
3. T.O. Joos, M. Schrenk, P. Hopfl, K. Kroger, U. Chowdhury, D. Stoll, D. Schorner, M. Durr, K. Herick, S. Rupp, K. Sohn, and H. Hammerle. A microarray enzyme-linked immunosorbent assay for autoimmune diagnostics. *Electrophoresis* 21: 2641–2650, 2000.
4. W.H. Robinson, C. DiGennaro, W. Hueber, B.B. Haab, M. Kamachi, E.J. Dean, S. Fournel, D. Fong, M.C. Genovese, H.E. de Vegvar, K. Skriner, D.L. Hirschberg, R.I. Morris, S. Muller, G.J. Pruijn, W.J. van Venrooij, J.S. Smolen, P.O. Brown, L. Steinman, and P.J. Utz. Autoantigen microarrays for multiplex characterization of autoantibody responses. *Nat. Med.* 8: 295–301, 2002.
5. R.P. Huang. Simultaneous detection of multiple proteins with an array-based enzyme-linked immunosorbent assay (ELISA) and enhanced chemiluminescence (ECL). *Clin. Chem. Lab. Med.* 39: 209–214, 2001.
6. M. Schena, D. Shalon, R.W. Davis, and P.O. Brown. Quantitative monitoring of gene expression patterns with a complementary DNA microarray. *Science* 270: 467–470, 1995.
7. A. Alizadeh, M. Eisen, D. Botstein, P.O. Brown, and L.M. Staudt. Probing lymphocyte biology by genomic-scale gene expression analysis. *J. Clin. Immunol.* 18: 373–379, 1998.
8. J.L. DeRisi, V.R. Iyer, and P.O. Brown. Exploring the metabolic and genetic control of gene expression on a genomic scale. *Science* 278: 680–686, 1997.
9. A.C. Pease, D. Solas, E.J. Sullivan, M.T. Cronin, C.P. Holmes, and S.P. Fodor. Light-generated oligonucleotide arrays for rapid DNA sequence analysis. *Proc. Natl Acad. Sci., USA* 91: 5022–5026, 1994.
10. R.J. Cho, M.J. Campbell, E.A. Winzeler, L. Steinmetz, A. Conway, L. Wodicka, T.G. Wolfsberg, A.E. Gabrielian, D. Landsman, D.J. Lockhart, and R.W. Davis. A genome-wide transcriptional analysis of the mitotic cell cycle. *Mol. Cell.* 2: 65–73, 1998.
11. P.T. Spellman, G. Sherlock, M.Q. Zhang, V.R. Iyer, K. Anders, M.B. Eisen, P.O. Brown, D. Botstein, and B. Futcher. Comprehensive identification of cell cycle-regulated genes of the yeast *Saccharomyces cerevisiae* by microarray hybridization. *Mol. Biol. Cell* 9: 3273–3297, 1998.
12. E.E. Schadt, S.W. Edwards, D. GuhaThakurta, D. Holder, L. Ying, V. Svetnik, A. Leonardson, K.W. Hart, A. Russell, G. Li, G. Cavet, J. Castle, P. McDonagh, Z. Kan, R. Chen, A. Kasarskis, M. Margarint, R.M. Caceres, J.M. Johnson, C.D. Armour, P.W. Garrett-Engele, N.F. Tsinoremas, and D.D. Shoemaker. A comprehensive transcript index of the human genome generated using microarrays and computational approaches. *Genome Biol.* 5: R73, 2004.
13. V. Stolc, M.P. Samanta, W. Tongprasit, H. Sethi, S. Liang, D.C. Nelson, A. Hegeman, C. Nelson, D. Rancour, S. Bednarek, E.L. Ulrich, Q. Zhao, R.L. Wrobel, C.S. Newman, B.G. Fox, G.N. Phillips Jr., J.L. Markley, and M.R. Sussman. *Arabidopsis thaliana* by using high-resolution genome tiling arrays. *Proc. Natl. Acad. Sci., USA* 102: 4453–4458, 2005.
14. P. Bertone, V. Stolc, T.E. Royce, J.S. Rozowsky, A.E. Urban, X. Zhu, J.L. Rinn, W. Tongprasit, M. Samanta, S. Weissman, M. Gerstein, and M. Snyder. Global

identification of human transcribed sequences with genome tiling arrays. *Science* 306: 2242–2246, 2004.

15. T.C. Mockler and J.R. Ecker. Applications of DNA tiling arrays for whole-genome analysis. *Genomics* 85: 1–15, 2005.
16. L.M. Staudt and P.O. Brown. Genomic views of the immune system. *Annu. Rev. Immunol.* 18: 829–859, 2000.
17. S.M. Albelda and D. Sheppard. Functional genomics and expression profiling: be there or be square. *Am. J. Respir. Cell Mol. Biol.* 23: 265–269, 2000.

2 Normalization of Microarray Data

Rudolph S. Parrish and Robert R. Delongchamp

CONTENTS

2.1 Objectives of Normalization 10
2.1.1 What Is Normalization? 10
2.1.2 Sources of Variation 10
2.1.3 Background Correction 10
2.1.4 Platforms 11
2.2 Statistical Basis of Normalization 11
2.2.1 Microarray Data 11
2.2.2 Transformations 11
2.2.3 Analysis of Variance Models 12
2.2.4 Variance Components 12
2.2.5 Significance Testing 13
2.2.6 Bias and Variance Reduction 13
2.2.7 Variance within Arrays 15
2.2.8 Analysis of Covariance Models 15
2.3 Normalization Algorithms 17
2.3.1 Reference Genes 17
2.3.1.1 Housekeeping Genes 17
2.3.2 Global and Local Methods 18
2.3.2.1 Linear Scaling to a Common Mean and Range 18
2.3.2.2 Nonlinear Scaling 18
2.3.2.3 Overall Mean or Median 18
2.3.2.4 Scaling for Heterogeneity of Variance 19
2.3.2.5 Loess on Two-Channel or Paired Arrays 19
2.3.2.6 Cyclic Loess on Single-Channel Arrays 20
2.3.2.7 Loess on an Orthonormal Basis 20
2.3.2.8 Quantile Normalization 20
2.3.3 Linear Model Based Methods 20
2.3.3.1 ANOVA-Based Model 21
2.3.3.2 Mixed-Effects Model 21
2.3.3.3 Split-Plot Design 22
2.3.3.4 Subset Normalization 22
2.3.4 Local Methods 22

2.4 Evaluating Normalization Methods 24
References 25

2.1 OBJECTIVES OF NORMALIZATION

2.1.1 WHAT IS NORMALIZATION?

Normalization of microarray data is any procedure meant to reduce or account for systematic variation among or within arrays. This variation is a component common to all the genes that are measured on an array or, more generally, to a subset of genes on the array. Normalization methods are often applied prior to the application of statistical analysis methods, which are usually designed to detect differential expression. However, normalization includes procedures that adjust for known effects as part of the statistical analysis as well as those that replace the actual data with modified values prior to the statistical analysis. In either case, normalization represents an effort to obtain more powerful tests by reducing variation in the data or otherwise accounting for it mathematically.

2.1.2 SOURCES OF VARIATION

The raw data from gene microarrays involve variation due to several sources [1,2]. The intent of a typical experiment is to determine whether treatment groups of experimental units (e.g., subjects, patients, mice, etc.) exhibit differential gene expression patterns. Such comparisons are based on an assessment of the variation among the experimental units within groups, which is the experimental error variance. In addition to variation in mRNA levels from unit to unit, there is variation arising from the measurement process. Normalization methods attempt to remove or reduce the influence of these additional sources of variation.

Some writers distinguish between "biological" and "technical" variation. Variation that is inherent to the characteristics of the experimental units is considered as biological variation. Variation that derives from the characteristics of the arrays themselves (due to manufacturing issues), the processing of the samples applied to arrays (e.g., sample preparation, mRNA extraction, labeling), hybridization of sample material onto the arrays, and measurement of intensities (e.g., optical properties, label intensity, scanner settings) all are considered as technical variation; see also [3]. If all technical variation could be eliminated, there would be some sense of purity in the data that should reflect group effects and involve only natural unit-to-unit variation. An ideal normalization method would remove all effects of technical variation.

Although some procedures may result in transformed data that are approximately normally distributed (i.e., Gaussian distribution), achieving normality is not the primary objective of normalization.

2.1.3 BACKGROUND CORRECTION

Many proposed methods incorporate the use of a background correction procedure in which measured intensities are adjusted according to some level of background noise. Although these methods also result in a modification to the data, in this chapter background correction algorithms are regarded as attempts to reduce bias, whereas normalization methods are regarded as attempts to reduce variance due to technical sources. Background corrections mainly affect the low expressions. Thus, background correction will not be considered in this treatment of normalization.

2.1.4 Platforms

Normalization methods generally apply to any platform, although some methods obviously are designed for one- or two-color systems or are specific to a platform. These platforms include high-density oligonucleotide arrays [4], cDNA spotted arrays using two labels per spot, and spotted arrays using one label per spot [5].

2.2 STATISTICAL BASIS OF NORMALIZATION

2.2.1 Microarray Data

Microarray data from an individual array basically form just a high-dimension multivariate observation of gene expressions. The array corresponds to an experimental unit or a sampling unit within the experimental unit, and genes correspond to the variables measured on the unit. In two-color systems, the red and green dyes often correspond to paired specimens from the same or different experimental units. As part of multivariate observations, it is natural to assume that correlations exist among the variables. Obviously, genes may be correlated through biological relationships. Variables may be correlated also by virtue of being associated with the same spots on two-color arrays.

Typically, there is only a single array for each experimental unit. That is, there usually is no replication of multiple arrays per experimental unit. In classical experimental designs, such replication forms the basis for obtaining purer estimates of the experimental error variance, and this principle can be applied to microarray experiments [6,7]. With microarrays, a different technique is employed that is based on assumed relationships involving hundreds or thousands of genes on each array. Basically, normalization methods are developed under the assumption that the average gene does not change significantly among the experimental units even under the various experimental conditions.

2.2.2 Transformations

Nearly all investigators employ a logarithmic (usually base 2) transformation on expression values prior to analysis or normalization. For 16-bit images, this means that $\log_2$-transformed expression values will be real values between 0 and 16. The purpose of transforming the data is mostly to address potential multiplicative error structures that give rise to instability of variances, but also to achieve normality so that subsequent statistical inferences will be valid. However, considering all choices, the logarithmic transformation may not be the one that most nearly produces a normal distribution, and the most appropriate choice of transformation is likely to be different for various genes when seeking normality [8,9]. Nonetheless, it is usually assumed that a logarithmic transformation stabilizes the variance, at least approximately [10].

2.2.3 Analysis of Variance Models

It is useful to consider statistical models for microarray data in order to characterize variability mathematically and to assess normalization methods. Several models, as described below, have been proposed which provide a framework for understanding the variances that are present in microarray expression data. Among these are analysis of variance type models with additive errors as given by several authors [3,11–16]. Models involving multiplicative errors have also been introduced [17,18]. Most utilize a logarithmic transformation of the expression data. Normalization methods that involve modeling probe intensity levels have also been proposed [19].

Various normalization methods make use of presumed relationships with other genes (or their overall characteristics) to modify the data so that the among-arrays variance is reduced. In a linear model context, the data are not modified directly but rather other effects in the model are adjusted for in order to reduce estimates of standard errors of treatment differences.

Most papers consider normalization as an adjustment that precedes the analysis for treatment effects. Examples include the mean (median) subtraction or locally-weighted regression (loess) adjustments. In these cases, the data are normalized and then the normalized values are analyzed for treatment effects. However, normalization can be directly incorporated into the analysis, as with the analysis of variance models. The analysis of variance approach is attractive because it explicitly accounts for sources of variation that impact inferences about treatments including the "array effect," which invariably is a major source of variation.

2.2.4 Variance Components

A simple and typical experimental design for single-channel arrays involves two treatment groups ($k = 2$), multiple subjects per treatment (n subjects), and a single array per subject ($r = 1$). There are two components of random variation: one associated with variation among subjects treated alike (subjects within treatment groups) and the other associated with arrays within subjects. The first of these is the experimental error variance, denoted by σ_s^2. The second is an array-specific variance, denoted by σ_a^2. The analysis of variance involves three sources of variation in the indicated expected mean squares. Because the number of degrees of freedom for arrays is zero when $r = 1$, the corresponding variance component is not estimable. In this model, subjects and arrays are considered as random effects.

Source	DF	Expected mean square
Treatment (Trt)	$k-1$	$Q_{\text{Trt}} + r\sigma_s^2 + \sigma_a^2$
Subjects (Trt)	$k(n-1)$	$r\sigma_s^2 + \sigma_a^2$
Arrays (Subjects Trt)	$kn(r-1)$	σ_a^2

Normalization may be thought of as an attempt to reduce the magnitude of the "among-arrays" variance component σ_a^2. The "among-subjects" component σ_s^2 is the experimental error component and should not be modified by normalization. Q_{Trt} is a quadratic form based on treatment means.

2.2.5 Significance Testing

In view of the ultimate objectives in microarray analysis, significance testing is conducted in one form or another in order to discover genes that exhibit differential expression. As such, normalization methods should seek to improve power of such tests and to reduce false discovery rates. Thus, in a real sense, the impact on significance testing is one of the most important characteristics of a normalization procedure. Methods that replace the data with adjusted values, by definition, attempt to reduce variability and are likely to produce a higher frequency of significant results. Consideration of the F test, which is formed by the ratio of the mean square among treatments, $\text{MS}_{(\text{Trt})}$, to the mean square among subjects within treatments, $\text{MS}_{\text{Subjects(Trt)}}$, gives rise to the following ratios of variances estimated by these mean squares in the presence or absence of variance associated with arrays

$$\frac{E(\text{MS}_{\text{Trt}})}{E(\text{MS}_{\text{Subjects(Trt)}})} = \frac{Q_{\text{Trt}} + r\sigma_s^2 + \sigma_a^2}{r\sigma_s^2 + \sigma_a^2} < \frac{Q_{\text{Trt}} + r\sigma_s^2}{r\sigma_s^2}$$

Thus, a normalization procedure that is effective in reducing σ_a^2 will increase the F statistics (or, equivalently, t statistics for the case of two treatments), and therefore reduce p values, assuming σ_s^2 remains constant. A problem arises if the normalization method reduces variability associated not only with the array effects but also that associated with the experimental error (i.e., among-subjects variance). This issue has been discussed for two normalization methods applied to prostate cancer data [20].

2.2.6 Bias and Variance Reduction

Normalization procedures attempt to reduce variance among arrays, which improves the resolution of treatment differences. However, normalization also may bias the estimated treatment effects, which impairs the resolution of treatment differences. The bias and variance can be examined in a simple case, which allows a heuristic evaluation of the properties to be expected for less tractable cases. Let y_{ga} denote the logarithm of the observed intensity for gene "g" on array "a." A simple normalization method is to subtract the global mean for each array. This formula can be written in matrix form as $(\mathbf{I} - \mathbf{J}/m)\mathbf{y}_a$ where $\mathbf{I}$ denotes an $(m \times m)$ identity matrix, $\mathbf{J}$ denotes an $(m \times m)$ matrix of 1s, and $\mathbf{y}_a = (y_{1a}, y_{2a}, \ldots, y_{ma})'$. The gth element of $(\mathbf{I} - \mathbf{J}/m)\mathbf{y}_a$ is simply $y_{ga} - \bar{y}_{\cdot a}$. If μ_a and Σ denote the mean and covariance of y_{ga}, then the mean and covariance of the normalized data are, respectively, $(\mathbf{I} - \mathbf{J}/m)\mu_a$ and $(\mathbf{I} - \mathbf{J}/m)\Sigma(\mathbf{I} - \mathbf{J}/m)'$.

The underlying logic for normalization is an assumption that the m genes measured on an array share a component of variation, the array effect. The aim of normalization is to eliminate this effect. To see this, denote the variance of the array effect by σ_a^2, then the variance Σ can be partitioned into two parts such that $\Sigma = \mathbf{D} + \mathbf{J}\sigma_a^2$, where $\mathbf{D}$ is a diagonal matrix with elements equal to the gene-specific variances (over subjects). Then, the variance of the normalized values is $(\mathbf{I} - \mathbf{J}/m)\mathbf{D}(\mathbf{I} - \mathbf{J}/m)'$, which does not depend on σ_a^2. This matrix has diagonal elements equal to $\sigma_g^2 - (2/m)\sigma_g^2 + (1/m^2)\sum_{g=1}^{m}\sigma_g^2$, representing variances of the normalized variables. The covariances are given by $-(1/m)\sigma_i^2 - (1/m)\sigma_j^2 + (1/m^2)\sum_{g=1}^{m}\sigma_g^2$. In effect, sources of variation shared by all intensities on the array are eliminated by this normalization. Because the array variance can be quite large relative to the variance among subjects for specific genes, the data may have substantially lower variance after normalization. For large m, the variance of the normalized data will be approximately $\mathbf{D}$.

Global normalization of this form also introduces bias. Let Δ denote the logarithm of the true fold changes between two arrays, that is, the expectation of $\mathbf{y}_a - \mathbf{y}_b$. Then the expected difference after subtracting the respective means is $(\mathbf{I} - \mathbf{J}/m)\Delta \neq \Delta$. In general, estimated fold changes, which are based on normalized data, are biased in the opposite direction of the average logarithm of the fold change. If most of the interrogated genes are unaffected by treatment and m is large, this bias is negligible and the accompanying reduction in variance far outweighs any detriment from bias. However, the potential to seriously corrupt inferences about treatment effects through normalization should be a concern whenever large numbers of genes exhibit differential expression and/or whenever the data are highly smoothed as this conceptually corresponds to normalizing within subsets of the interrogated genes (i.e., effectively small m).

There are two generic directions in which the global means normalization can be modified. These modifications encompass a large percentage of the normalization procedures that have been proposed. One direction is to apply the mean normalization within subsets of the interrogated genes. For example, genes can be placed into subsets based on their physical location or the magnitude of their intensity. Procedures that adjust for each print pin or those that regress on the magnitude can be viewed this way. The other direction of modification is to replace the mean with alternative estimators. The bias can be mitigated somewhat by the use of 'outlier' resistant methods (e.g., using the median rather than the mean). An outlier in the context of normalization is any intensity that is affected by factors in addition to the array effect. The model outlined above only accounts for treatment effects and array effects, so "outliers" in the context of this model are the genes affected by treatment. Finessing in both directions of modification leads to more complex procedures such as loess regression.

Two other procedures also can be interpreted in the context of means normalization. If one replaces the rows of $\mathbf{J}$ with an indicator of genes for which there are no treatment effects, then the variance is reduced and there is no bias. However, this requires a priori knowledge of a subset of interrogated genes which are not affected by treatment. This is the basis for normalizations based upon housekeeping genes, although the assertion that they are unaffected by treatment is problematic. Another approach is to use the normalizing value (global mean) as a covariate in an analysis of the observed log-intensities. In essence, $\mathbf{J}$ is replaced by $\mathbf{BJ}$ where $\mathbf{B}$ is a diagonal matrix of gene-specific coefficients estimated by the analysis of covariance.

Conceptually, this accounts better for different correlations between the observed intensities and the normalizing value.

2.2.7 Variance within Arrays

Normalization by subtraction of the global means does not adjust for possible different array-specific variability among genes that might exist. In this situation, individual gene expressions will have variances across subjects (σ_g^2) that may be very large, even with respect to σ_a^2. By considering features of the array-specific distributions of gene expressions or probe intensities (e.g., the interquartile range or particular quantiles), additional adjustments may be in order. The method based on the median and interquartile range is an example of a combined approach.

2.2.8 Analysis of Covariance Models

It is not straightforward to test for treatment effects in analysis of variance (ANOVA) models unless one makes the assumptions that all genes have the same residual variance and that all genes are independent, which are unlikely to be satisfied in practice. From a strict statistical perspective, a better approach is to analyze the intensity data by individual genes and to extract the "normalizing" information residing in the other interrogated genes by enlisting a summary statistic as a covariate. For example, instead of subtracting the median from each log-intensity and then analyzing for treatment effects, one could analyze the log-intensities for treatment effects incorporating the median as a covariate. Conceptually, such an analysis makes a better adjustment because it estimates the attenuation associated with measurement error that is implicit in using the median (or other summary estimate) as a surrogate measure of the array effect.

To elaborate, consider a simplified hypothetical setting where the analysis involves two genes per array with several arrays receiving a treatment and several arrays serving as controls. We make the additional assumption that the second gene is unaffected by treatment (essentially, the basis for normalizing by housekeeping genes). Then $\mathbf{y}_a$ as previously defined can be written

$$\begin{pmatrix} Y_1 \\ Y_2 \end{pmatrix}_a \sim \left[\begin{pmatrix} \mu_1 + \tau \\ \mu_2 \end{pmatrix}, \begin{pmatrix} \sigma_1^2 + \sigma_a^2 & \sigma_{12} + \sigma_a^2 \\ \sigma_{12} + \sigma_a^2 & \sigma_2^2 + \sigma_a^2 \end{pmatrix} \right]$$

where $\tau = 0$ if the array is from a control sample. In the analysis of covariance, we are interested in testing the first gene for a treatment effect using the distribution of Y_1, given an observed value of Y_2. In general, the expectation of this distribution depends upon the distributional assumptions in addition to the means and variances. For a bivariate normal distribution, it is known that $E(Y_1|y_2) = \mu_1 + \tau + \beta(y_2 - \mu_2)$ where

$$\beta = \frac{\sigma_{12} + \sigma_a^2}{\sigma_2^2 + \sigma_a^2}$$

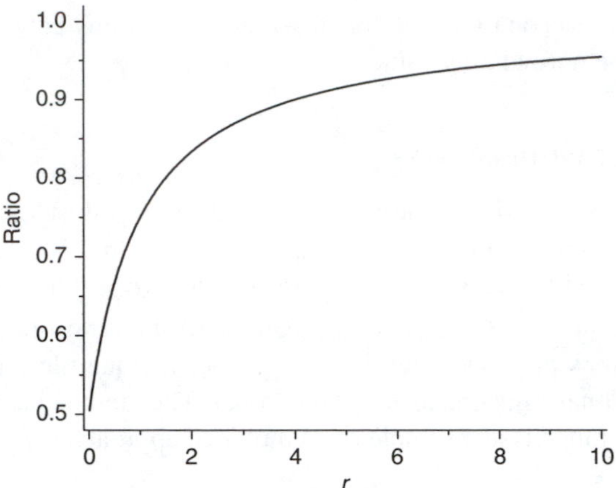

FIGURE 2.1 Ratio of the variances of treatment effect: variance using analysis of covariance divided by variance using the difference.

In particular, the variance is $\text{var}(Y_1|Y_2) = \sigma_1^2 + \sigma_a^2 - \beta(\sigma_{12} + \sigma_a^2)$ giving the variance of the estimated treatment effect, $\text{var}(\hat{\tau}) = (2/n)(\sigma_1^2 + \sigma_a^2 - \beta(\sigma_{12} + \sigma_a^2))$, assuming there are n treated and n control arrays. The variance of the treatment effect estimated from the differences, $y_1 - y_2$, is $(2/n)(\sigma_1^2 + \sigma_2^2 - 2\sigma_{12})$. Arguably, the better of these methods is the one that has the smallest variance, which is straightforward to evaluate if all the component variances are specified. As an example, suppose

$$\boldsymbol{\sigma} = \sigma^2 \begin{pmatrix} 1+r & r \\ r & 1+r \end{pmatrix}$$

This represents cases where $\sigma_1^2 = \sigma_2^2 \equiv \sigma^2$, $\sigma_{12} = 0$, and $\sigma_a^2 = r\sigma^2$, that is, r is the relative magnitude of the variance of the array effect. In this example, the ratio of these variances only depends on r. In Figure 2.1, the ratio is formed with $\text{var}(\hat{\tau})$ in the numerator and plotted as a function of r. Since the ratio is less than 1, the analysis of covariance produces more precise estimates of the treatment effect. At least under the assumed $\boldsymbol{\sigma}$, an analysis of covariance would be preferable to analyzing the differences. This preference would likely apply whenever normalization is based upon the mean of a few "housekeeping" genes.

In practice, when the median of all the interrogated genes is employed as a covariate, the mathematics becomes intractable. However, the median should behave similarly as the mean where the ratio is essentially 1, so there is little advantage in regard to the precision of estimated effects. Simulations using the median confirm that there is little if any increase in the precision of estimated treatment effects when a covariance analysis is compared to an analysis based upon the differences. Like the difference, the analysis of covariance also suffers from bias, which justifies use of the median rather than the mean. Hence, our preference for the analysis of covariance is largely aesthetic. An analysis of covariance seems better because it renders an assessment of the variation explained by normalization.

2.3 NORMALIZATION ALGORITHMS

A large number of normalization methods have been proposed, several of which are identified in Table 2.1; this list is not exhaustive. Bolstad et al. [19] distinguished between "complete data" methods, in which data from all arrays are used to normalize, and methods that use a baseline array to establish the normalization relation. The global normalization methods generally can be applied to all arrays together or separately to arrays within treatment groups. Nonetheless, there is no consensus on classification of these methods.

In the following sections, some methods that have been frequently reported are described in detail; however, this does not imply that these methods are necessarily more appropriate or more effective than others. Comparisons among normalization methods have not established which ones have the most desirable statistical performance characteristics.

2.3.1 Reference Genes

2.3.1.1 Housekeeping Genes

A set of housekeeping genes may be used as a group of reference genes for adjusting array values [21,22], provided it can be assumed that true expression values for these genes are unaffected by experimental conditions and do not vary across subjects or samples. Such genes are selected in advance of the experiment. Methods for selection

TABLE 2.1
Selected Normalization Methods

Method	Reference
Reference genes	
Housekeeping and control genes	[21,44]
Global and local methods	
Global mean or median	
Linear scaling	[4]
Nonlinear scaling	[25]
Invariant set	[17]
Consistent set	[45]
Median-Interquartile range	[28]
Signal dependent q-spline	[46]
Variance stabilization	[8,10,47]
Quantile	[19,32,33]
Local regression	[29,30,36,48]
Variance regularization	[49]
Spatial normalization	[46]
Linear models	
Analysis of variance	[11]
Mixed-effects models	[12,35]
Split-plot design	[13]
Subset/global intensity	[14]

of control genes have been described [23]. Some arrays (e.g., HG-U133) are designed with probe sets that are not expected to vary across different tissue types. Typically, an adjustment factor would be computed for each array that makes the means of the reference genes all equal. Some investigators have reported difficulty in selecting suitable housekeeping genes [24,25].

2.3.2 Global and Local Methods

2.3.2.1 Linear Scaling to a Common Mean and Range

The Affymetrix algorithm [4] involves simple scaling according to the expression

$$y_{ij}^{*} = y_{ij} \times \left(y_{\text{baseline}}^{(m)} / y_{i}^{(m)} \right)$$

where $y_{\text{baseline}}^{(m)}$ is the (trimmed) mean of the expression values on the reference array, and similarly $y_i^{(m)}$ is that corresponding to the ith array. This may be applied at the probe intensity level. Under a linearity assumption, this is effectively fitting a line through the origin for array i values vs. baseline array values, paired according to individual genes or probes. This produces for array i the same mean and same range of variation as for the baseline array. This method does not correct for situations where the low-level expressions or intensities have a slope different from that for the larger values (i.e., if the expression distributions are very different).

2.3.2.2 Nonlinear Scaling

Instead of assuming linearity between each array and the baseline array, one can employ nonlinear relationships. Schadt et al. [25] described use of a nonlinear normalizing relation based on a subset of genes considered to be invariant relative to the ordering of expressions on an array compared to the baseline array. They proposed an algorithm based on ranks that finds an approximately invariant set of genes, which then are used with the generalized cross-validation smoothing spline algorithm (GCVSS) given by Wahba [26]. This algorithm has been implemented in dChip software [25]. This data transformation may be represented generally as

$$y_{ij}^{*} = f_i(y_{ij})$$

where f_i represents the nonlinear scaling function.

Li and Wong [17] employed a piecewise running median line instead of the GCVSS approach, and Bolstad et al. [19] utilized a loess smoothing approach [27] on probe intensities.

2.3.2.3 Overall Mean or Median

An additive adjustment factor can be computed for each array in order to make the array means or medians all equal to one another. That is, it makes the total intensity for all arrays equal or nearly so. This can be implemented by choosing a reference (i.e., baseline) array arbitrarily and adjusting the other arrays to that total intensity value, or it can be implemented by computing the mean of the means or median of the medians as the target value. This type of normalization might be valid if log expression values are affected as a direct result of differing quantities of mRNA. Mathematically, this is represented as

$$y_{ij}^{*} = y_{ij} + (y.^{(m)} - y_i^{(m)})$$

where y_{ij} is the prenormalized log-expression value for array i and gene j, $y_i^{(m)}$ is the mean or median value for array i based on values of all genes on the array or it may be the value from the baseline array, $y.^{(m)}$ is the mean of means or median of medians over all arrays, and y_{ij}^{*} is the normalized log-expression value. An arithmetic mean or a trimmed mean can be used. In a distribution sense, this method equates the central tendencies for all arrays. A global mean or median adjustment can be applied across all arrays together or within treatment groups separately.

2.3.2.4 Scaling for Heterogeneity of Variance

The global mean or median adjustment provides shift corrections but does not alter the variance of gene expression values within arrays. A simple adjustment for differing variances can be accomplished by selecting a scaling constant for each array. The scaling constants can be based on a measure of dispersion using either a baseline array or the mean or median of that measure over all arrays. Commonly, the interquartile range (IQR) is used because it is not affected by outliers as is the case with the standard deviation [28]. Incorporation of both global mean or median adjustment and an IQR-based dispersion adjustment is given mathematically by

$$y_{ij}^{*} = (y_{ij} + y.^{(m)} - y_i^{(m)}) \times (D./D_i)$$

where D_i is the measure of dispersion for the ith array and $D.$ is the mean or median or maximum of the D_i values over all arrays. The IQR-based method can be generalized for use of any two other quantiles, such as the normal range based on the 2.5th and 97.5th percentiles. When the IQR criterion is used, all arrays will have the same IQR after normalization.

2.3.2.5 Loess on Two-Channel or Paired Arrays

Dudoit et al. [29] described a loess-based method based on M vs. A (MvA) plots. This method plots

$$M_j = \log_2(y_{ij}^{(1)}/y_{ij}^{(2)}) \quad \text{vs.} \quad A_j = 0.5 \log_2(y_{ij}^{(1)} \times y_{ij}^{(2)})$$

or, equivalently,

$$M_j = \log_2(y_{ij}^{(1)}) - \log_2(y_{ij}^{(2)}) \quad \text{vs.} \quad A_j = [\log_2(y_{ij}^{(1)}) + \log_2(y_{ij}^{(2)})]/2$$

for each array i over all genes (or probes) j, and then fits a loess smoothing function. Here, $y_{ij}^{(1)}$ represents the expression (or probe) values for one channel and $y_{ij}^{(2)}$ similarly for the other channel.

Normalized values are a function of the deviations, denoted by M_j', from the fitted regression line; particularly,

$$\log_2(y_{ij}^{(1)*}) = A_j + 0.5M_j' \quad \text{and} \quad \log_2(y_{ij}^{(2)*}) = A_j - 0.5M_j'$$

where $y_{ij}^{(1)*}$ and $y_{ij}^{(2)*}$ represent the normalized values for the two channels.

2.3.2.6 Cyclic Loess on Single-Channel Arrays

The method of Dudoit et al. [29] was adapted by Bolstad et al. [19] for application to single-channel arrays by considering all pairs of arrays when constructing *MvA* plots. Their algorithm iteratively finds adjustments that ultimately result in normalized values.

2.3.2.7 Loess on an Orthonormal Basis

A version of the *MvA* method was introduced by Astrand [30] in which he first transforms the log probe intensity vector for each array using an orthonormal contrast matrix with dimensions equal to the number of probes, in order to create an alternative basis of the data. This is followed by applying a loess method to the *MvA* plots where a fixed reference vector in the alternative basis is paired with each of the other arrays in the alternative basis. This algorithm is implemented in the *R* software *maffy* [2,31].

2.3.2.8 Quantile Normalization

Bolstad and coworkers [19,32–34] introduced a method based on quantiles of the underlying distribution of probe intensities. That method creates identical distributions of probe intensities for all arrays by replacing the intensity values in order to attain a straight-line relationship on quantile–quantile plots for any two arrays. This is accomplished by projecting the points of an n-dimensional quantile plot onto a unit diagonal vector, according to the following steps (a) Form a data matrix $\mathbf{X}$ of dimension $p \times n$ where p is the number of probes and n is the number of arrays; (b) Sort each column of values from low to high to produce order statistics for each column; (c) Replace all values in each row of the sorted matrix by the mean of that row's values (i.e., the mean of the ith order statistics from all columns); (d) Rearrange the elements of each column back to the original ordering. The resulting matrix is the normalized data matrix from which expression values then are calculated.

2.3.3 Linear Model Based Methods

In the following statistical models, the response variable is generally taken to be the $\log_2$ of expression.

2.3.3.1 ANOVA-Based Model

Kerr et al. [11] introduced the use of ANOVA models that accounted for array, dye, and treatment effects for cDNA arrays. In this fashion, normalization was accomplished intrinsically without preliminary data manipulation. The model they proposed may be written as

$$y_{ijkg} = \mu + A_i + T_j + D_k + G_g + AG_{ig} + TG_{jg} + e_{ijkg}$$

where μ is the mean expression, A_i is the effect of the ith array, T_j is the effect of the jth treatment, D_k is the effect of the kth dye, G_g is the gth gene effect, and AG_{ig} and TG_{jg} represent interaction effects. Of interest for testing differential expression are the interaction effects, TG_{jg}, for which appropriate contrasts can be estimated for each gene. In this model, all effects were considered as fixed effects. Other terms could be incorporated into this model.

2.3.3.2 Mixed-Effects Model

Wolfinger et al. [12] utilized a mixed-effects model for cDNA data where the array effect and related interaction terms are considered as random effects. A model involving all effects simultaneously can be written as

$$y_{ijg} = \mu + A_i + T_j + AT_{ij} + G_g + AG_{ig} + TG_{jg} + e_{ijg}$$

Like the ANOVA model, this also intrinsically adjusts for the effects of arrays without modifying the data directly, although they recommend first fitting the model

$$y_{ijg} = \mu + A_i + T_j + AT_{ij} + e_{g(ij)}$$

and calculating residuals, denoted by r_{ijg}. Then the residuals are used as the dependent variables in the gene-specific models given by

$$r_{ijg} = G_g + AG_{ig} + TG_{jg} + e_{ijg}$$

or, equivalently, for each gene

$$r_{ij}^{(g)} = \mu^{(g)} + A_i^{(g)} + T_j^{(g)} + e_{ij}^{(g)}$$

The residuals are the normalized values. The effect of interest for testing differential expression is the $T_j^{(g)}$ term.

For single-channel arrays, in which different arrays are used within treatments, this approach can be represented with the overall model

$$y_{ijg} = \mu + T_j + A(T)_{i(j)} + G_g + TG_{jg} + e_{gi(j)}$$

with the analogous normalization and gene-specific models. The error term is the equivalent of the $GA(T)_{gi(j)}$ term. Chu et al. [35] described a mixed-effects model for probe-level data.

2.3.3.3 Split-Plot Design

Emptage et al. [13] suggested that microarray experiments should be viewed as split-plot designs in which the arrays are whole plots and the probe sets or spots are subplots. For two-channel arrays, this would take the general form

$$y_{ijkg} = \mu + A_i + T_j + D_k + d_{ijk} + G_g + AG_{ig} + TG_{jg} + DG_{kg} + e_{ijkg}$$

where d_{ijk} is the whole-plot error term and e_{ijkg} is the subplot error term. Normalization is based on adjustment for array and dye effects and array $\times$ gene and dye $\times$ gene interactions.

For single-channel arrays, the model is

$$y_{ijg} = \mu + T_j + d_{ij} + G_g + TG_{jg} + e_{ijg}$$

where $d_{ij} \equiv A(T)_{i(j)}$ and $e_{ijg} \equiv GA(T)_{gi(j)}$.

2.3.3.4 Subset Normalization

Chen et al. [14] considered normalization of cDNA intensity ratios based on a subset of genes to adjust for location biases combined with global normalization for intensity biases. Subset normalization involves partitioning the genes on an array into disjoint subsets and then adjusting the values using a mean, median, or loess estimate. Partitions can be based on either location or intensity. The model proposed is given by

$$y_{ijl} = \mu + L_{il} + I_i + e_{ijl}$$

where μ is the mean expression, L_{il} is the effect of location l on array i, and I_i is the effect of intensity on array i. Residuals from this model are the normalized values.

Chen also considered subset normalization of individual intensities. The model is

$$y_{ijkg} = \mu + L_{ijkl} + I_{ijk} + T_j + D_k + G_g + e_{ijkg}$$

where μ is the mean expression, T_j is the effect of the jth treatment, D_k is the effect of the kth dye, and G_g is the gth gene effect.

2.3.4 Local Methods

Any normalization method that is defined globally (i.e., defined for all of the spots on an array) can usually be applied locally (i.e., to a specified subset of the array's spots). Often arrays have natural subdivisions (e.g., blocks of genes that are replicated within the array or blocks of genes that were printed with the same print pin), and it is a straightforward extension to apply normalization methods within these subdivisions. There can also be systematic intensity differences observed across an array's surface, which do not correspond to natural subdivisions of the array. In such cases the spots on arrays can be arbitrarily subdivided into groups and normalized within these subdivisions.

The spots on an array can also be arbitrarily grouped based upon the magnitude of their intensity. A clear-cut application is when arrays or samples within arrays are naturally paired. For example, this occurs when two samples are hybridized to a single array, or when before/after treatment samples are assayed within a subject. In these settings, the *M*v*A* plot is well defined, and the spots can be arbitrarily grouped based upon the average log-intensities (A). While loess normalization is widely applied in these situations [36], many global methods can also be applied. For example, one can compute the average log-intensity for each array or hybridization and simply subtract the corresponding group mean from all spots in that group. While this simple procedure is not recommended as a replacement for loess, the latter can be viewed as finessing this procedure. In essence, loess adopts a moving window (subset based upon A) and a robust estimate of central tendency replaces the average. The importance of the means within subset normalization is that its statistical properties are easily tractable while those of loess are not. Because loess is conceptually a refinement of the means within subset normalization, expected properties of loess can be inferred.

The underlying ideas behind the *M*v*A* plot and grouping by values of A can be extended to multiple arrays by defining A as the average over arrays. Then for array 'a', M_a is the difference between the each spot's log-intensity and A. Again, it is straightforward to apply loess regression or to apply most global procedures to A-based subsets of spots.

The spots on an array can be grouped based upon their location within the array and they can be grouped based upon their magnitude. If the magnitude groups are nested within the location groups or vice versa, then each spot belongs to one group and global methods can be applied as previously discussed. Such nesting can result in groups with only a few spots and normalizing within small groups can be very biasing. One doesn't have to nest the groupings in which case each spot is a member of two groups, the location group and the magnitude group. Using indicators of group membership, it is straightforward to implement a means normalization using analysis of variance methods [11,12], or a median normalization using an iteratively reweighted least squares algorithm to minimize the absolute deviations [37,38]. Likewise, local means, local medians, or loess predicted values, which correspond to the spot, can be entered as a pair of covariates into a "by spot" statistical analysis.

It is our experience that an aggressive subdivision of the spots on each array based upon their location and magnitude can correct systematic effects induced by uneven hybridization, differential background, and saturation. However, the uniformity across arrays that is imposed by aggressive normalization increases the potential for biases that can seriously misrepresent the effects of treatments. As a rule, one should opt for as little "normalizing" as possible. Because the intensities on an array are arbitrarily scaled, a global normalization is reasonable and unlikely to introduce serious bias under many scenarios. In the analysis of variance models that use group indicators, it is easy to partition the sums-of-squares into the reduction associated with the global normalization and the additional reduction from the local normalization. With reasonably clean arrays, the global correction frequently captures the bulk of the variance that can be removed by normalization.

2.4 EVALUATING NORMALIZATION METHODS

The ideal microarray data set would be one obtained under an appropriate experimental design that completely eliminated or otherwise dealt with the problems associated with poor-quality data. Experimental design methods involve blocking, replication, and randomization [14]. Blocking groups together experimental units that are similar in some respect, and it thereby serves to account for variation that otherwise would reduce the precision of effect estimates. Replication of arrays within experimental subjects and replication of spots within arrays provide a level of protection against ill effects of array-specific or spot-specific biases. This generally reduces variance through intrinsic use of means computed over replicates. Randomization reduces the likelihood of system-specific and selection biases and provides a basis for valid estimation of variance components. In one sense, normalization attempts to overcome the problems resulting from shortcomings in the experimental design (e.g., small sample size, inadequate replication, no blocking). Whether normalization methods are effective in doing this is an open question.

Some of the statistical issues that should be addressed for normalization methods include impacts that the methods have on bias in estimates of effects, variance components, robustness relative to test assumptions, and significance tests or other analysis methods. Proposed methods need to be evaluated with respect to the statistical properties that they possess considering the type of data to which they are likely to be applied and the underlying purpose of the corresponding statistical analysis. For example, it is important to know how different variance components are affected. Tsodikov et al. [39] considered various normalization procedures and their underlying assumptions. The techniques that have been used when considering normalization algorithms include pairwise differences between arrays, *M*v*A* (or *R* vs. *I*) plots, distributions of probe intensities [19], and significance tests [20,40].

Several data set types have been used for assessing proposed normalization methods. These include spike-in experiments [41], dilution-mixture experiments [33], real data sets, and simulated data sets. The first two approaches present scenarios in which the "truth" (i.e., which genes actually are differentially expressed and by how much) is known but they may not adequately represent real-world data. The third choice is realistic for a selected genre of studies, but the "truth" is never known. While the fourth option is difficult to model, the "truth" would be known even if not realistic itself. Except for simulation studies, most test data sets involve small numbers and, by definition, are not capable of producing much insight into the statistical properties of the normalization methods.

Comparison of different normalization methods with one another has been undertaken by several investigators [19,20,42]. In many cases, it is difficult to evaluate the normalization methods independently of the expression estimation methods.

Simulation of microarray data, although complex and computationally demanding, may be the best way to deal with ascertaining the statistical properties of proposed normalization methods. The statistical and computational methodology for accomplishing this has not been sufficiently exploited. The challenges include modeling realistic situations with appropriate (multivariate) distribution models and developing new computational algorithms and computer code to handle the high dimensionality of the problem. Given such a simulation system, the statistical properties of proposed estimation (and expression estimation algorithms) could be examined effectively and objectively. An on-line system known as *Gene Expression Data Simulator* [43] has been developed (*http://bioinformatics.upmc.edu/GE2/index.html*). Application of computationally intensive methods may be fruitful also.

REFERENCES

1. G.A. Churchill. Fundamentals of experimental design for cDNA microarrays. *Nature Genetics* 32 (Suppl): 490–495, 2002.
2. G. Parmigiani, E.S. Garrett, and S. Ziegler (eds.), *The Analysis of Gene Expression Data: Methods and Software*. Springer-Verlag, Berlin, 2003.
3. S. Draghici, A. Kuklin, B. Hoff, and S. Shams. Experimental design, analysis of variance and slide quality assessment in gene expression arrays. *Current Opinion in Drug Discovery and Development* 4: 332–337, 2001.
4. Affymetrix. *Statistical Algorithms Description Document*. Santa Clara, CA; Affymetrix, Inc. 2002. http://www.affymetrix.com/support/technical/whitepapers/sadd-whitepaper.pdf.
5. D.V. Nguyen, A.B. Arpat, N. Wang, and R.J. Carroll. DNA microarray experiments: biological and technological aspects. *Biometrics* 58: 701–717, 2002.
6. M.T. Lee, F.C. Kuo, G.A. Whitmore, and J. Sklar. Importance of replication in microarray gene expression studies: statistical methods and evidence from repetitive cDNA hybridizations. *Proceedings of the National Academy of Sciences, USA* 97: 9834–9839, 2000.
7. J.K. Lee. Analysis issues for gene expression array data. *Clinical Chemistry* 47: 1350–1352, 2001.
8. W. Huber, A. von Heydebreck, H. Sultmann, A. Poustka, and M. Vingron. Variance stabilization applied to microarray data calibration and to the quantification of differential expression. *Bioinformatics* 18 (Suppl 1): S96–S104, 2002.
9. R.S. Parrish and H.J. Spencer. Distribution modeling of gene microarray data. In: *Proceedings of the First Annual Conference of the Mid-South Computational Biology and Bioinformatics Society*, Little Rock, AR, 2003.
10. S.C. Geller, J.P. Gregg, P. Hagerman, and D.M. Rocke. Transformation and normalization of oligonucleotide microarray data. *Bioinformatics* 19: 1817–1823, 2003.
11. M.K. Kerr, M. Martin, and G.A. Churchill. Analysis of variance for gene expression microarray data. *Journal of Computational Biology* 7: 819–838, 2000.
12. R.D. Wolfinger, G. Gibson, E.D. Wolfinger, L. Bennett, H. Hamadeh, P. Bushel, C. Afshari, and R.S. Paules. Assessing gene significance from cDNA microarray expression data via mixed models. *Journal of Computational Biology* 8: 625–637, 2001.

13. M.R. Emptage, B. Hudson-Curtis, and K. Sen. Treatment of microarray experiments as split-plot designs. *Journal of Biopharmaceutical Statistics* 13: 159–178, 2003.
14. J.J. Chen, R.R. Delongchamp, T. Chen-An, V. Desai, and J. Fuscoe. Sources of variation in microarray gene expression data. *Bioinformatics* 20: 1436–1446, 2004.
15. J.J. Chen and C.-H. Chen. Microarray gene expression. In: *Encyclopedia of Biopharmaceutical Statistics*, 2nd ed., Shein-Chung Chow (ed.), Marcel-Dekker, New York, 2003.
16. X. Cui and G.A. Churchill. Statistical tests for differential expression in cDNA microarray experiments. *Genome Biology* 4: 210.1–210.10, 2003.
17. C. Li and W.H. Wong. Model-based analysis of oligonucleotide arrays: expression index computation and outlier detection. *Proceedings of the National Academy of Sciences* USA 98: 31–36, 2001.
18. Y.-J. Chen, R. Kodell, F. Sistare, K.L. Thompson, S. Morris, and J.J. Chen. Normalization methods for analysis of microarray gene-expression data. *Journal of Biopharmaceutical Statistics* 13: 57–74, 2003.
19. B.M. Bolstad, R.A. Irizarry, M. Astrand, and T.P. Speed. A comparison of normalization methods for high-density oligonucleotide array data based on variance and bias. *Bioinformatics* 19: 185–193, 2003.
20. R.S. Parrish and H.J. Spencer. Effect of normalization on significance testing for oligonucleotide microarrays. *Journal of Biopharmaceutical Statistics* 14: 575–589, 2004.
21. H.K. Hamalainen, J.C. Tubman, S. Vikman, T. Kyrola, E. Ylikoski, J.A. Warrington, and R. Lahesmaa. Identification and validation of endogenous reference genes for expression profiling of T helper cell differentiation by quantitative real-time RT-PCR. *Analytical Biochemistry* 299: 63–70, 2001.
22. P. Gieser, G.C. Bloom, and E.N. Lazaridis. Introduction to microarray experimentation and analysis. In: *Methods in Molecular Biology 184: Biostatistical Methods*, S.W. Looney (ed.), pp. 29–49. Humana Press, Totowa, NJ, 2001.
23. J. Vandesompele, K. De Preter, F. Pattyn, B. Poppe, N. Van Roy, A. De Paepe, and F. Speleman. Accurate normalization of real-time quantitative RT-PCR data by geometric averaging of multiple internal control genes. *Genome Biology* 3: 0034.1–0034.11, 2002.
24. X. Wang, M.J. Hessner, Y. Wu, N. Pati, and S. Ghosh. Quantitative quality control in microarray experiments and the application in data filtering, normalization and false positive rate prediction. *Bioinformatics* 19: 1341–1347, 2003.
25. E.E. Schadt, C. Li, B. Ellis, and W.H. Wong. Feature extraction and normalization algorithms for high-density oligonucleotide gene expression array data. *Journal of Cellular Biochemistry* 37 (Suppl): 120–125, 2001.
26. G. Wahba. Spline methods for observational data. In: *Proceedings of the CBMS-NSF Regional Conference Series in Applied Mathematics*, SIAM, Philadelphia, 1990.
27. W.S. Cleveland and S.J. Devlin. Locally-weighted regression: an approach to regression analysis by local fitting. *Journal of the American Statistical Association* 83: 596–610, 1988.
28. Insightful. *S+ArrayAnalyzer Version 1.1 software*. Insightful Corporation, Seattle, WA, 2003.
29. S. Dudoit, Y.H. Yang, M.J. Callow, and T.P. Speed. Statistical methods for identifying differentially expressed genes in replicated cDNA microarray experiments. *Statistica Sinica* 12: 111–139, 2002.

30. M. Astrand. Contrast normalization of oligonucleotide arrays. *Journal of Computational Biology* 10: 95–102, 2003.
31. Bioconductor. *Methods for Affymetrix Oligonucleotide Arrays (affy)*. The Bioconductor Project, Version 1.2, 2003. http://www.bioconductor.org/.
32. R.A. Irizarry, B.M. Bolstad, F. Collin, L.M. Cope, B. Hobbs, and T.P. Speed. Summaries of Affymetrix GeneChip® probe level data. *Nucleic Acids Research* 31: e15, 2003.
33. R.A. Irizarry, B. Hobbs, F. Collin, Y.D. Beazer-Barclay, K.J. Antonellis, U. Scherf, and T.P. Speed. Exploration, normalization, and summaries of high-density oligonucleotide array probe level data. *Biostatistics* 4: 249–264, 2003.
34. R.A. Irizarry, L. Gautier, and L. Cope. An R package for analysis of Affymetrix oligonucleotide arrays. In: R.I.G. Parmigiani, E.S. Garrett, and S. Ziegler (eds.), *The Analysis of Gene Expression Data: Methods and Software*, pp. 102–119. Springer-Verlag, Berlin, 2003.
35. T.M. Chu, B. Weir, and R.D. Wolfinger. A systematic statistical linear modeling approach to oligonucleotide array experiments. *Mathematical Biosciences* 176: 35–51, 2002.
36. Y.H. Yang, S. Dudoit, P. Luu, D.M. Lin, V. Peng, J. Ngai, and T.P. Speed. Normalization for cDNA microarray data: a robust composite method addressing single and multiple slide systematic variation. *Nucleic Acids Research* 30: e15, 2002.
37. R.R. Delongchamp, C. Velasco, M. Razzaghi, A. Harris, and D. Casciano. Median-of-subsets normalization of intensities for cDNA array data. *DNA and Cell Biology* 23: 653–659, 2004.
38. R.A. Thisted. *Elements of Statistical Computing*. Chapman and Hall, New York, 1988.
39. A. Tsodikov, A. Szabo, and D. Jones. Adjustments and measures of differential expression for microarray data. *Bioinformatics* 18: 251–260, 2002.
40. R. Hoffmann, T. Seidl, and M. Dugas. Profound effect of normalization on detection of differentially expressed genes in oligonucleotide microarray data analysis. *Genome Biology* 3: 0033.1–0033.11, 2002.
41. A.A. Hill, E.L. Brown, M.Z. Whitley, G. Tucker-Kellogg, C.P. Hunter, and D.K. Slonim. Evaluation of normalization procedures for oligonucleotide array data based on spiked cRNA controls. *Genome Biology* 2: 0055.1–0055.13, 2001.
42. O. Hartmann, B. Samans, D. Hoffman, J. WeBels, M. Klingenspor, and H. Schäfer. Comparison of normalization methods and signal intensity measures for Affymetrix GeneChips®, Marburg, Germany. In: *The Genetic and Molecular Basis of Human Disease. Proceedings of the Symposium of the German National Genome Research Network (NGFN) and the German Human Genome Project (DHGP)*, Berlin, November 2002.
43. J. Lyons-Weiler. Experimental design and analysis bottlenecks in the discovery of cancer biomarkers from microarray experiments. In: *Proceedings of the Conference on Advancing Practice, Instruction, and Innovation Through Informatics*, October 2003. http://apiii.upmc.edu/abstracts/posterarchive/2002/lyons_weiler.html.
44. J. Schuchhardt, D. Beule, A. Malik, E. Wolski, H. Eickhoff, H. Lehrach, and H. Herzel. Normalization strategies for cDNA microarrays. *Nucleic Acids Research* 28: e47, 2000.
45. T.B. Kepler, L. Crosby, and K.T. Morgan. Normalization and analysis of DNA microarray data by self-consistency and local regression. *Genome Biology* 3: 0037.1–0037.12, 2002.
46. C. Workman, L.J. Jensen, H. Jarmer, R. Berka, L. Gautier, H.B. Nielsen, H.H. Saxild, C. Nielsen, S. Brunak, and S. Knudsen. A new nonlinear normalization method

for reducing variability in DNA microarray experiments. *Genome Biology* 3: 0048.1–0048.16, 2002.

47. B.P. Durbin, J.S. Hardin, D.M. Hawkins, and D.M. Rocke. A variance-stabilizing transformation for gene-expression microarray data. *Bioinformatics* 18 (Suppl 1): S105–S110, 2002.
48. C. Colantuoni, G. Henry, S. Zeger, and J. Pevsner. Local mean normalization of microarray element signal intensities across an array surface: quality control and correction of spatially systematic artifacts. *BioTechniques* 32: 1316–1320, 2002.
49. J. Quackenbush. Microarray data normalization and transformation. *Nature Genetics* 32 (Suppl): 496–501, 2002.

3 Microarray Quality Control and Assessment

David Finkelstein, Michael Janis, Alan Williams, Kathryn Steiger, and Jacques Retief

CONTENTS

3.1 Introduction 29
3.2 Array Quality and Design 30
3.3 Bioinformatic Quality 30
3.4 Manufacturing Quality 32
3.5 Experimental Design Quality 33
3.6 Experimenatal Execution 34
3.7 Quality Control Metrics 36
3.7.1 RNA Quality 36
3.7.2 Probe-Based Quality Metrics 36
3.7.3 Image Analysis 37
3.8 Data Analysis Quality 37
3.8.1 Transformations and Normalization 38
3.8.2 Statistical Approaches 39
3.8.3 Multiple Test Correction 39
3.9 Quality of Interpretation 40
3.10 Quality of Validation 41
3.11 Making Decisions Based on Quality 41
3.11.1 Biological Effects 41
3.11.2 Global or Probe Set Level Metrics 45
3.11.3 Censor, Impute, or Repair 46
3.11.4 Other Methods for Quality Analysis 47
3.11.5 Quality Control for Core Facilities 49
3.12 Conclusions 50
References 51

3.1 INTRODUCTION

There are numerous successful microarray experiments that have advanced our understanding of biology [1–5]. The important role of quality control (QC) in such successes is often overlooked. Successfully completing microarray experiments includes assessing the quality of the array design, the experimental design, the experimental execution, the data analysis, and the biological interpretation. At each step data quality and data integrity should be maintained by minimizing both systematic and random measurement errors. Data integrity is defined as the ability to faithfully identify each data point with its transcript sequence, annotation, and the experimental conditions that generated it. Data integrity is maintained through skilled design, collection, analysis, and annotation. Data integrity is maintained by ensuring the precise classification of transcripts, the application of appropriate analytical methods and the accurate interpretation of biological results. Data quality refers to the familiar technical and biological sources of error in the data. In this chapter we provide advice and a strategy for assessing and maintaining data quality and integrity at each stage of a microarray experiment. Special emphasis is given to the Affymetrix® GeneChip® microarray platform. However, many QC issues are universal and the implication to other platforms will be discussed.

This chapter provides a broad overview of data quality. It is intended to heighten the awareness of the analyst to bioinformatics and array design issues that may affect data quality. The quality of experimental designs, data analysis, interpretation, and validation are also discussed. Although there is a greater emphasis on data from Affymetrix® GeneChip® arrays, some of the specific concerns of other platforms are addressed. Furthermore this chapter describes quality control strategies that are platform independent.

3.2 ARRAY QUALITY AND DESIGN

Array quality is usually considered as the accurate synthesis or spotting of the intended transcript in the proper location on the array. This is highly platform dependent and usually addressed by the manufacturers. Array design involves universal bioinformatic quality issues, such as the selection of sequences from the millions available in the public databases, whereas choosing the number and length of probes or clones to represent those sequences are platform specific. The identification of authentically expressed transcripts and the accurate annotation of those transcripts are all data integrity issues. Commercial providers monitor their own design process and provide varying levels of annotation. (http://www.affymetrix.com/support/technical/whitepapers/probeset_annotations.pdf, http://www/affymetrix.com/analysis/index.affx)

3.3 BIOINFORMATIC QUALITY

The quality of the sequence information used in array design greatly influences the analysis, interpretation, and validation of the experiment. In practice no knowledge is absolute and no data perfect, so an array design is a series of compromises. The data analyst inherits all these compromises and should be aware of their consequences.

The thermodynamic hybridization properties of DNA, called probe affinity, are sequence specific and affects the linearity and variance of intensity measurements [6].

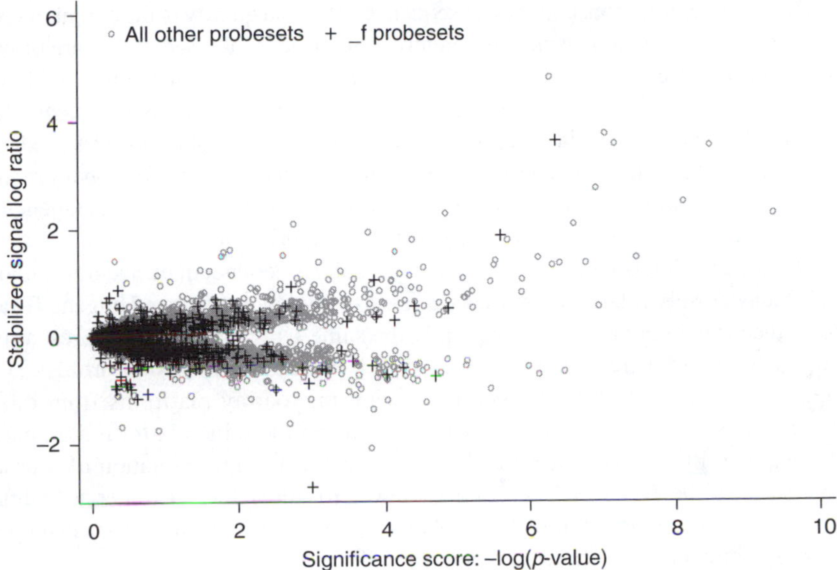

FIGURE 3.1 The effect of array design on data analysis. Probesets representing multiple family members (+) are more variable than other probesets (o). The y-axis is stabilized signal log ratio and the x-axis is the significance score denoting the significance of a signal change between control and experiment. The stabilized signal log ratio was calculated as follows:

$$\log \exp\left[\sum \log(\text{Signal}_{\text{control replicates}} + 50)/n\right] / \exp\left[\sum \log(\text{Signal}_{\text{treated replicates}} + 50)/n\right] / \log(2)$$

where signal from MAS 5.0 from replicate treatment arrays were stabilized by the addition of a small constant (50 in this case) log transformed averaged, exponentiated to get the geometric mean. A second geometric stabilized mean was found for control array signal and the ratio of these summaries was $\log_2$ transformed to produce a stabilized signal log ratio. On the x-axis is the $-\log_{10}(p\text{-value})$ where the p-value is derived from a two sample t-test of log(signal + 50) values from all replicates.

In a good array design, probes are selected to have similar probe affinities and optimal linear responses [7].

Many genes are highly similar and it may be impossible to select unique probes to represent the gene. In these cases we select probe sets to represent the family of genes as uniquely as possible. These probesets may represent the behavior of the multiple genes and denoted by an _for _x postscript on the probeset name. As a class, _f_at probesets may have lower statistical significance (Figure 3.1). However, in many biological systems some _f_at probesets will measure authentic expression changes. To address this problem a stratified approach may be used where poor-quality probesets are weighed differently from high-quality probe sets or analyzed separately. Ultimately, the best performing _f_at probesets will require more extensive bench validation to distinguish closely related transcripts.

The most obvious consequence of sequence selection quality is the extent of cross hybridization. If a genome is not completely sequenced, or the sequences are of poor quality, all possible cross hybridization cannot be excluded. While many reliable validated sequences exist some transcripts are selected from expressed sequence tags (ESTs) (which may be partial or rearranged) or from theoretical transcripts predicted from genomic sequences. For predicted gene transcripts, even professional organizations such as Celera and Ensembl frequently do not agree [8]. As a consequence all large data sets can be assumed to have some cross hybridization.

The accurate interpretation of microarray results is dependent on accurate annotations. Many closely related expressed sequences may have the same Unigene ID and inherit the same annotation. Consequently the same Unigene ID may represent two completely different transcripts with completely different expression profiles. This problem is particularly severe when comparing microarray platforms from different manufacturers, or even different array designs on the same platform. Ultimately only mapping by sequence truly defines the identity and interpretation of a probe. Tools such as the NetAffx™ sequence viewer to allow the customer to validate the sequence and annotation of the probesets (http://www.affymetrix.com/support/technical/whitepapers/probeset_annotations.pdf).

3.4 MANUFACTURING QUALITY

The type of error produced during manufacture varies between different platforms. For example, in mechanically spotted systems sequences we may expect differences in solution deposition and spot topography, which may affect only some of the spots. *In situ* synthesis, where all probes are synthesized in parallel, may be subject to incomplete synthesis, which will affect all the probes. The two manufacturing approaches will produce two completely different error models.

Affymetrix® GeneChip® arrays use fused silica as an array substrate. This high quality material has a more consistently flat surface and does not fluoresce to the same degree as boro-silicate glass. Background fluorescence is a special concern with two color arrays because inconsistencies across the surface of a slide and between dyes contribute to measurement errors [9,10]. Some of these effects can be detected [11,12] and addressed with special experimental designs [10,13,14], normalization methods [15–17], a third dye for cDNA arrays [18], or specialized equipment [9]. Other problems such as the ozone sensitivity of cyanine dyes are best handled through alternative chemistries [19].

Another distinction in platform is the spatial arrangement of probes. On Affymetrix® GeneChip® arrays each probe pair is arrayed by sequence and are distributed across the array effectively randomizing spatial effects. By contrast, cDNA arrays and long oligo 2 color arrays generally use a single probe per transcript. For these arrays spatial effects may influence results. The placement of replicated sets of control probes to assess this problem is recommended. Alternatively, normalization by printing pin may reduce the effect of systematic spatial bias on the data.

For *in situ* synthesized probes the placement and sequence of every probe is known. With mechanically spotted arrays the identity of the sequence itself needs to be tracked. This tracking should include maintaining records on materials, manufacturing conditions, and sequence verification of clones and primers [20]. When the Arabidopsis Functional Genomic Consortium validated their cDNA arrays by re-sequencing 700 clones at random, approximately 1% of the plasmid inserts were found to be misidentified [11].

3.5 EXPERIMENTAL DESIGN QUALITY

Experimental designs are rarely assessed for quality in a technical sense, but do contribute greatly to the success of the experiment. There are special design considerations that are universal to all micro array experiments, and some design considerations that are platform dependent. Here we focus on the practical measures of quality such as cost, complexity, and context of oligonucleotide array experimental designs.

Cost is a direct product of the complexity of the experimental design and number of replicates. Estimating the appropriate number of replicates for a microarray experiment is complex and linked to the data analysis method [21,22]. For example to achieve the same power, parametric methods require fewer replicates than non-parametric methods. After a log transformation and variance stabilizing transforms, microarray data behaves well in parametric ANOVA models and they are usually the method of choice [23–25]. The population of probe sets on an array will range in variance and power estimates. It is helpful to arrange the data in quartiles to show the proportion of probesets that are sufficiently tested. Fortunately there are several methods in the literature that detail the special concerns of microarray power estimates from the simple [26] to the complex [21]. The choice of algorithm used to calculate signal will also influence the number of replicates required. Fewer arrays are required for variance minimized methods than bias minimized methods as power estimates are based on variance.

Cost is a consideration in most experimental designs and pooling strategies may be employed to reduce microarray costs. However, pooling introduces a bias [27,28] which may increase the frequency of false positives. False positives in turn may be costly to validate. Another hidden cost of pooling is the increased complexity of design and interpretation. In some cases pooling is never appropriate. In toxicology, for example, the frequency of uncommon toxic responses may be of interest. Pooling across such samples will dilute and mask rare toxic events. In addition, if sample-specific covariates are known, they should be included in the model. In these cases, pooling is not appropriate.

Other experimental design issues are related to number of parameters to be measured. Each array measures many thousands of probesets simultaneously while only a few arrays are used as replicates. As a consequence, microarray experimental designs are statistically underpowered. This is compounded if too many questions are asked of the data. Ranking the importance of variables allows you to maximize replication where it provides the most value. Another hidden cost to a highly parameterized design is complexity of execution. If time, dose, drug, and tissue are all varied in the same experiment the organizational burden and opportunity for mistakes increase.

The design of a high quality microarray experiment proceeds in stages. First, the biologist sets goals and the analyst defines and explains statistical assumptions. Next a small-scale pilot study is performed and evaluated for quality and relevance to goals. Pilot studies allow biological controls, assays, time points, and analytic methods to be tested and refined. To improve the likelihood of success, include extra replicates based on the estimated failure rate from the pilot study or current laboratory history. From this pilot we capture an estimate of power and gain a sense of the complexity of the treatments. Adjustments to experimental conditions or time points can be made and the fullscale experiment planned. Finally, by planning for validation and biological interpretation the experimental design becomes a complete experimental plan.

3.6 EXPERIMENATAL EXECUTION

A key part of the execution of the experiment is the capture of critical metrics. The most useful metrics are provided by manufacturer's controls that are designed to test assay and array performance. Some QCs and measures, such as the optical density ratio of 260/280 nm wavelength for RNA (OD ratio) must be recorded in the laboratory. In a core facility or long term professional laboratory these metrics should be captured for each RNA sample and each array.

In addition to the expression changes of interest, the data from genome-wide microarrays will also reflect other biological changes such as changes in temperature [29,30], time of day [31–34], diet [35–37], mechanical stress [38,39], and even humidity [40,41]. Some responses are unpredictable and good laboratory practice requires that as many variables as possible be controlled and recorded. If these measures are not controlled there is an increased risk of false positives or increased biological variability. For example, Figure 3.2 depicts the expression profile of one transcript from a

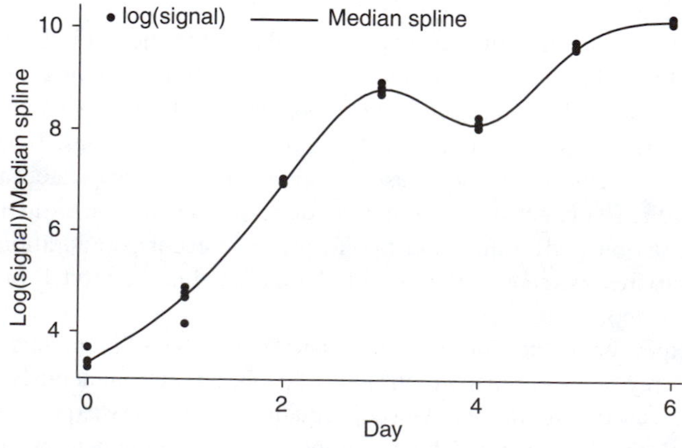

FIGURE 3.2 Effect of an unexpected event. The line represents an expression profile of a single probeset over a time series with four measurements per day. The line is a median spline and the dots are the original measurements. The change in the trend from day 3 to day 4 correlates with media replenishment on day 3.

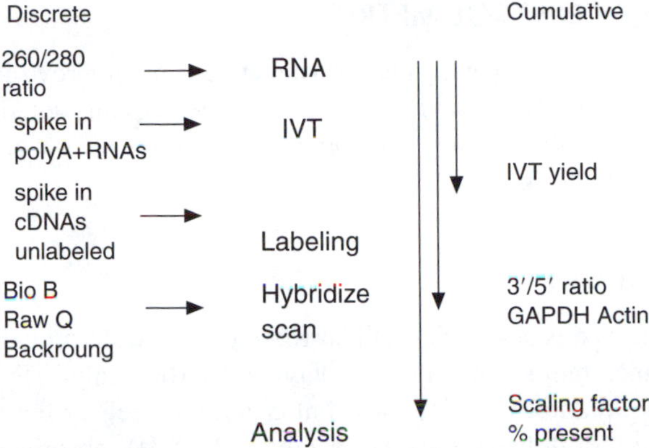

FIGURE 3.3 Quality control metrics are either discrete or cumulative. The best metrics for diagnostic purposes are discrete.

cell line over the course of one week. Note the break in linearity from day 3 to day 4. This same disjointed pattern was seen in hundreds of genes and correlated with a replenishment of cell media. Such responses can confound biological interpretation.

In clinical samples many variables cannot be controlled. These variables should be recorded and included in the analysis. Variables such as treatment, age, and sex are obvious, but microarrays are sensitive to very subtle effects. For example, in a blood study where microarray analyses were performed on donor samples collected at different times of day a small set of genes had distinctly time dependent patterns [34]. This environmental sensitivity highlights the value of a pilot study where unexpected problems can be detected at a reduced risk.

The execution of microarray experiments is a multistage serial process and error can be introduced at any point. Many of these errors are also cumulative. This results in dependence and covariance amongst the different quality metrics (Figure 3.3). Spike-in controls are partially processed standardized RNAs or DNAs of known concentration. These spike-in controls are independent of all prior processing steps and are critical for the unambiguous detection of variability in a given experiment. QC throughout the process allows the diagnosis of problems and assures the analyst and experimenter that the measured gene expression changes are accurate and authentic.

If a source of error cannot be eliminated through experimental control, it is possible to minimize its influence through randomization or orthogonalization. If the effect due to a given systematic factor is expected to be low relative to the biological test then randomization is a reasonable approach. For example, randomizing the manufacturing lot number between control and treated samples eliminates any confounding effect from array lot.

If the systematic effect is large it may be necessary to employ an orthogonalization method. Returning to the blood drawing example, we may plan our study to collect blood from control and treated patients at regular intervals throughout the day. Then the predictable effects of time of day can be readily modeled and eliminated from the analysis by using residuals.

3.7 QUALITY CONTROL METRICS

Measures of quality are frequently platform dependent and improve and develop with experience. A complete listing and metric by metric description would be too platform and error specific for this chapter. The following is a brief review of quality metrics for Affymetrix® Probe Arrays.

3.7.1 RNA Quality

Direct quality measures begin with cellular RNA yield and RNA quality as measured by UV absorbance ratios and gel electrophoresis or BioAnalyzer™ (Agilent Technologies, Inc, Palo Alto, CA). Pass and fail criteria are set for these measures and they can also be useful diagnostics. For example RNA UV absorbance ratios of the 260/280 nm range of light is characteristically between 1.9 and 2.0 for high quality RNA. If this ratio is 1.7 or lower, degradation or contamination is likely and we can justifiably exclude this sample. For the BioAnalyzer™ quality RNA provides a distinct profile. The ideal profile includes sharply defined peak in the expected place and in the expected proportion. Further the ideal profile has no indication of peaks derived from degradation.

The measure of amplification or *in vitro* transcription (IVT) reflects RNA quality. In practice, this measure only indicates that a reaction was successful or failed and we have rarely observed a strong correlation between this metric and the variability of signal when cRNA yield is adequate. However, if yield is low then poor synthesis is indicated and data quality will be eroded. The extraction solution lot numbers and RNA storage method should be recorded in the event that an IVT failure needs to be diagnosed. After the IVT and labeling reactions, the samples are hybridized to the array and all subsequent quality measures are derived from the array. Once hybridized, the discovery of a low quality sample will require running additional arrays or analytical repair. For Affymetrix arrays, a test array is available, which can detect labeling and background issues prior to the use of full-scale array.

3.7.2 Probe-Based Quality Metrics

Probe-based quality metrics can be broadly divided into two types: array-wide and probe-level. Global metrics, meaning metrics that average across all probes on an array, are indicators of the sample quality. Linking probe-level information to array-wide metrics requires summary descriptive stats of the behavior of the array population for example total discrimination and percent present. Indicators of the labeling reaction are background levels, raw q (for raw probe quality score: a measure of pixel noise), low discrimination between perfect and mismatch probes, the CV of pixels, and the number of gridding outliers (CEL file). For long term sustained quality, such as in a core facility, subtle variability can be detected by tracking which operator, hybridization chamber, wash station and scanner was associated with each array. Detailed descriptions of these metrics and the equations that generate them are available at http://www.affymetrix.com/support/technical/whitepapers/sadd_whitepaper.pdf.

3.7.3 Image Analysis

Once a process is complete and hybridized scanned image recorded, all quality control is retrospective. However, the quality measures are useful in decision-making and in the detection of the causes of quality control failures.

Imperfections in arrays may be due to sample contamination, scratches, precipitation or imperfect gridding. Image quality is usually assessed by visual inspection. Such inspection is subjective, depends on operator experience and not suitable for the analysis of large numbers of arrays. However, there are methods available to assess image quality from a downstream intensity file as well as from the image file itself. Exploring the quality of the information in an image file has advantages. Imperfections in arrays may be due to speckling, scratches, precipitation or imperfect gridding, or spot detection. Automated means for detecting, censoring or flagging specific probes or spots exist in commercial and academic software packages. These packages generally rely on relative measures based on replicates or known limitations in signal [17,42,43].

These software packages and the measures of error are frequently platform specific. For example, spot shape estimates are cDNA array specific metrics [44–46]. These arrays require specialized gridding automated software that locally detects and qualifies spots. Even commercially manufactured Affymetrix array images are improved by an algorithm that locally detects the feature. In GeneChip® Operating System (GCOS) we provide automated local feature extraction software that minimizes the coefficient of variance of pixels.

Once identified by automated means, a visual inspection of outlier arrays is useful for diagnosing problems as specific visual patterns may indicate specific QC failings. For example, unusually bright pixels in the margins of features on an Affymetrix array are signs of incomplete fragmentation of labeled target. A poorly labeled center commonly called a "crop circle," denotes insufficient hydration of the array and may indicate low buffer volume or failed wash station circulation. For two color arrays hollow spots or "donuts" arise when printing pins tap the glass slide with excessive speed or force. Irregular spotting may be due to changes in humidity during printing or in DNA concentration due to uneven PCR amplification or evaporation. As a rule once an image anomaly is detected mathematically the analyst should enlist the aid of the technical expert to assess the causes of failure.

3.8 DATA ANALYSIS QUALITY

Microarray data analysis can range from the basic to extremely complex. A basic analysis may ignore some dependencies for the sake of simplicity, while complex methods run the risk of biasing the data and removing authentic effects. To maintain data integrity it is useful to assess the quality of the analysis itself.

Microarray data analysis stretches the limits of existing statistical methods by simultaneously testing thousands of parameters with orders of magnitude fewer replicates. In addition the sources and distributions of measurement errors are complex and nonnormal. To make the situation worse, dependencies of unknown frequency and complexity exist between measurements. It is therefore no accident that microarray topics are commonplace at statistical meetings. For example, there were 66 microarray talks at the 2003 Joint Statistical meeting in San Francisco (www.amstat.org). This reflects that the field is not standardized and that there is little consensus over the best methods to employ. The MIAME standard is a move toward standardized descriptions of methods and is now required by many scientific journals [47–49]. The guiding principle for all data analysis should be transparency. In other words, the analysis should be reproducible and the assumptions explicit. This is most easily achieved by using peer reviewed methods (e.g., ANOVA [50], mixed models [10,17], singular value decomposition (SVD) [51], principal component analysis (PCA) [52], permutation [53], variance stabilizing transformations [23,24], and False discovery rates [54]).

The ultimate goal for any microarray experiment is to answer a biological question. The best quality analysis is the analysis, which offers the most accurate and appropriate biological answer. Most of the successful data analysis methods have been developed in a partnership between statisticians and biologists. When developing new methods it is critical to validate the new methods by testing them against existing methods. It is also a good practice to be certain that the experimentalists you are working with are comfortable with an unproven approach and the assumptions used in the approach. Each data transformation, filter or normalization method should be justified both statistically and biologically. Taking the time to lay out the process and to detail what form the results will take will illicit responses from your biologist partner. This is the time to plan for replicates and to discuss what to do if a sample or array fails. If the experimental failure rate is known, alter the experimental plan by adding replicates. Discuss the methods for identifying outlier arrays or data. Plan which variables, such as time of day, solution lot, and operator identity should be identified. Discuss which quality checks that will be performed in the laboratory and what remedial steps will be taken. For example, will poor quality RNA be discarded rather than re-isolated? Now is the time to discuss expectations and differentiate between exploratory analysis, serial analysis, and principled multiple testing. Discuss validation strategies and the relative cost of false positives vs. false negatives. Ideally, include a pilot study to test the system and the assumptions. The final data analysis plan should describe the normalization method, statistical approach, and any downstream clustering or biological function analysis. Such defined goals and specific endpoints will maintain the biological focus and avoid over-analysis.

3.8.1 Transformations and Normalization

Transformations and normalizations have substantial impact on the result of an analysis. Transformations make data more statistically tractable. For example, a variance stabilizing transformation is needed to satisfy the assumption of normality required by parametric tests. Normalization, on the other hand, is used to remove systematic errors due physical causes such as labeling efficiency or scanner setting. Both these techniques should be applied carefully. Incomplete normalizations can lead to false positives while excessive normalization may mask authentic biological results or technical errors. In all cases strive to keep methods invertible, transparent, and ensure that the underlying assumptions are valid.

Transformations and normalization methods change parameters, such as outliers and variance, used for QC and assessment. It therefore follows that quality assessments should be made before transformations. The extent to which the data has been altered in the normalization process can in itself be a useful quality parameter. The scaling factor is such a parameter that is usually tracked for simple linear normalization methods such as those employed in microarray analysis suite (MAS) 5.0. Tracking these changes for complex methods such as quantile normalization must be planned as a separate test. For example, the sum of the absolute differences in expression before and after normalization can be tracked per array and treated as measure of the degree of normalization.

3.8.2 Statistical Approaches

The quality of a given statistical approach depends on its application and expeditiousness. As a rule of thumb we recommend performing a rapid, standard analysis with strict false positive protection as a reference. The initial analysis will validate the biological relevance of the results and serve as a reference to compare any novel analysis against. Beyond this philosophical approach there are some fundamental guidelines for a quality analysis detailed by the ethical standards produced by the American Statistical Association. These guidelines include common sense advice about stating assumptions clearly and detailing all methodological steps, especially those that result in censoring data (www.amstat.org). For microarray analysts there are special considerations, for example the simultaneous testing of thousands of entities of unknown dependence creates novel multiple comparison problems where dependency must be modeled [54].

3.8.3 Multiple Test Correction

The quality of microarrays results is often assessed by the number of probesets that pass a threshold, such as a *p*-value (Figure 3.4). However, each microarray experiment measures many thousands of transcripts simultaneously and is subject to extremely high likelihoods of false positives and negatives. The Bonferroni correction is often used to correct for this effect but is generally regarded as excessively strict for microarray applications. Furthermore genes are coherently expressed and the Bonferroni assumption of independence is almost certainly false. Several other correction methods have been published based on resampling [55] step down methods, and false discovery rate [53,54,56,57].

Despite the risks of multiple testing, there is a strong temptation to reanalyze microarray data with each newly developed technique. This exploratory approach lowers quality of the experiment in subtle ways. First, the answers are cast into doubt, as any discrepancy between methods must be investigated. Second, serial reanalysis delays bench validation and follow up on the biological points of interest. Third, the return on a reanalysis is low when compared to the return more replicates, a new time point, or dose may produce. Finally, serial reanalysis effectively converts a designed experiment with clear goals into an open ended exploratory analysis.

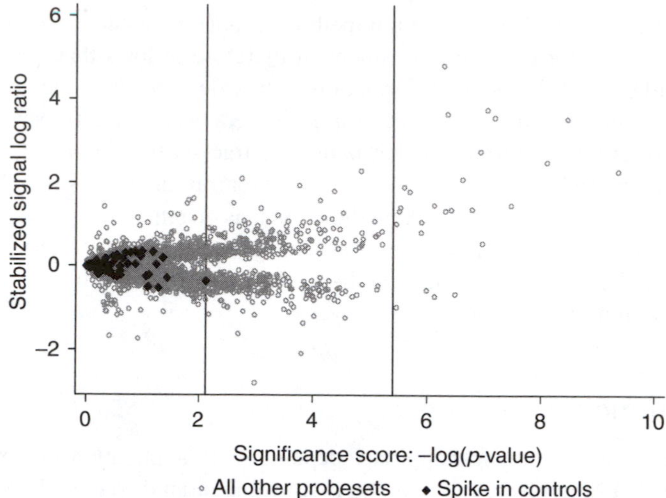

FIGURE 3.4 The severity of the Bonferroni correction. Probesets representing spike in controls (diamonds) are negative controls. Other probe sets are designated as circles (o). The y-axis is stabilized signal log ratio and the x-axis is the significance score denoting the significance of a signal change between control and experiment. The stabilized signal log ratio was calculated as follows:

$$\log\left[\exp\left[\sum \log(\text{Signal}_{\text{control replicates}}+50)/n\right]/\exp\left[\sum \log(\text{Signal}_{\text{treated replicates}}+50)/n\right]\right]/\log(2)$$

where signal from MAS 5.0 from replicate treatment arrays were stabilized by the addition of a small constant (50 in this case) log transformed averaged, exponentiated to get the geometric mean. A second geometric stabilized mean was found for control array signal and the ratio of these summaries was $\log_2$ transformed to produce a stabilized signal log ratio. On the x-axis is the $-\log_{10}(p\text{-value})$ where the p-value is derived from a two sample t-test of log(signal + 50) values from all replicates. Note that the most significant negative control is well to the left (first vertical line) of the Bonferroni cut-off ($\alpha = .05$, sec vertical line), indicating the severity of this correction in this example.

3.9 QUALITY OF INTERPRETATION

To confirm the quality of an analysis an analyst must check that trivial confounding reasons do not explain the observed phenomenon. To do this the data should be compared to the control parameters captured during the execution of the experiment (see Section 3.1.5). By verifying that pattern of gene expression is not coincident with the technician, labeling solution or time of day data integrity is maintained and valid interpretation is possible. Interpretation is primarily a function of the biologist. However, it is responsibility of the analyst to design experiments that prevent well-known confounding factors and to verify that the experiment followed those guidelines.

Interpretation quality is also dependent on the Bioinformatic quality of the arrays used (see Section 3.3). Annotation quality in public databases is uneven, sparse, and the linkage to the transcript may be imprecise or not biologically appropriate. For example, the cytochrome P450 genes are annotated as reduction–oxidation genes. This does not reflect their important function as drug degradation enzymes in toxicological studies. Gene ontology™ (GO) evidence codes provide some basis of the quality of each node and statistic methods based on likelihood ratios can be applied to asses where a given code appeared in this data set by chance [58,59]. More advanced techniques that combine GO with expression to predict gene functions [60,61] are best used to generate hypotheses rather than draw conclusions as broad biological conclusion require bench validation.

3.10 QUALITY OF VALIDATION

Validation and verification of the biology must ultimately be determined by alternative biological means [62]. Repeating the expression pattern by RT-PCR, Northern or other means has limited utility because such techniques suffer from technical issues of their own and do not reflect the biological reality any more accurately than a well-executed microarray experiment. All of these techniques measure transcript abundance, which is a proxy for the proteins that actually produce the biological effect. Ideally, the protein activity should be validated. As this kind of validation is slow and laborious, false positives are potentially very expensive. Once again discuss validation with the biologist, as analytical choices may influence the likelihood of success in validation.

3.11 MAKING DECISIONS BASED ON QUALITY

The first step in data analysis is to evaluate the data and to decide whether to exclude or repair low quality data. There are several basic methods for deciding what constitutes poor quality. First, some quality metrics are array independent and have universal thresholds based on experience. For example, poor RNA UV absorbance ratios are sufficient basis to exclude a sample. Where quality parameters are ambiguous and there are more than three replicates, outlier detection methods based on standard deviation or mean absolute deviations are a principled way to distinguish the relative quality of the arrays in a set. In general, the rare catastrophic failure of an entire array is easily detected. Rather it is the subtle erosion of quality that we seek to detect. These subtle effects require consideration before a correction action is taken.

3.11.1 Biological Effects

As a general rule array-wide quality metrics should not track with treatment. For example, if a drug induces severe necrosis, the necrotic tissue cannot be expected to yield the same quality and amount of RNA as healthy tissue. In a typical experiment studying dose and time, the ANOVA model of dose and time should also be tested against the background intensity, percent present, or IVT yield or other quality metrics captured during the execution of the experiment (see Section 3.1.5). Another common problem is that different types of tissues yield different amounts and quality of RNA. If background or RNA yield tracks with tissue or treatment then this may be due to authentic biological differences. For example elevated CO_2 can induce elevated carbohydrate storage in *Arabidopsis* [63]. As nucleic acid purification is sensitive to oligosaccharides and secondary compounds [64] the net result is that poor RNA yields covaries with treatment from high CO_2 exposure. To mitigate this problem, specialized methods of RNA purification may be required. If such a confounding result occurs it may bias all valid results and remedial steps should be considered. A useful strategy is to use smaller doses, shorter times, or less severely affected tissues. This shifts the focus to the more informative early biological processes and uses the sensitivity of micro arrays to better advantage.

To test whether quality tracks with treatment we recommend applying the same statistical model used to validate gene expression to each quality metric. So if ANOVA was used, test whether global background is significantly influenced by treatment. By testing each QC metric, the quality of interpretation is verified. We expect that if global quality tracks with treatment we have insufficiently controlled for confounding systematic errors and some positive results may be false.

Figure 3.5 demonstrates that the bimodal distribution of the average signal (Figure 3.5[a]) is due to a difference in tissue type where the hepatocytes have unique behavior (Figure 3.5[b]). This can be trivially attributed to the vast number of probesets, which are uniquely expressed in liver cells in response to drug treatments. However this global effect may be influencing the normalization of expression profiles and thereby introducing an interpretation problem. This theory is testable by examining the behavior of expression signal against each quality metric. If a given metric does not influence our results, it can usually be discounted. For example, variable background values may not influence the quantization of highly expressed transcripts.

The following strategy uses relationship of Signal values to create quality metrics:

1. First fit the statistical model to the expression data and collect the residuals.
2. These residuals are then standardized by the variance for each probeset.
3. Apply a rule to define outliers observations based on these residuals. For example, residuals that were + or −2 standard deviations from 0 can be defined as outliers.
4. These outliers were then counted on a per array basis.

This outlier count is a performance based quality metric, which will be compared with the process quality metrics, collected during the execution of the experiment. This comparison will help discern the relevance of each process quality metric to performance.

Next each array level quality metric was regressed to predict the level of outlier counts per array. A grouped logit regression was used to predict the non-normally distributed always positive outlier count. This per array outlier count is the *y*-axis of Figure 3.5, Panel A. From this figure we observe that outlier counts are roughly similar in all tissues by tissue with the exception of one extreme array with an exceptionally high outlier count. Methods of handling this array are discussed in section below on choices. Once regressions of per array outlier counts are performed for each

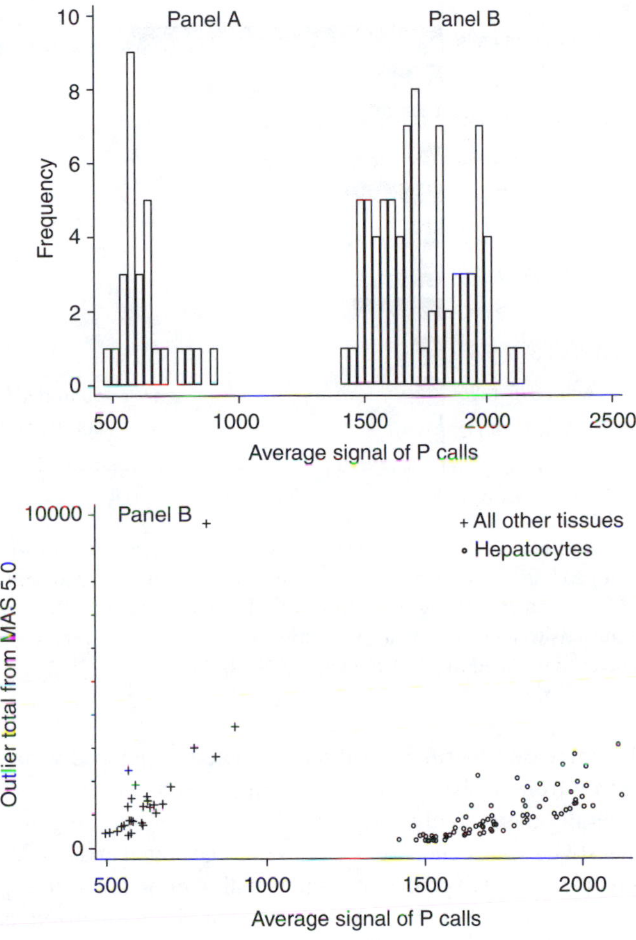

FIGURE 3.5 Quality issues due to biological effects. Panel A: the distribution of the average signal value for the GNF data set. Note that the distribution is bimodal. Panel B: the average signal for the same data set in relation to its outlier count in an ANOVA model using the "MAS" algorithm. Note that the separation of the data into two populations is primarily due to cell type.

quality metric, we can examine the fits by comparing *r*-squared values and use our understanding of each quality metric to diagnosis a cause.

The sensitivity of outlier detection methods are dependent on the signal algorithms and data analysis methods used [22]. Here we have performed this test on the same data set with two expression algorithms robust multi-array analysis (RMA) [65] and MAS 5.0 [66]. These two algorithms were chosen as RMA minimizes variance at the expense of introducing bias while MAS 5.0 takes the opposite approach [65]. Default normalization was used in both cases: scaling for MAS 5.0 and quantile normalization for RMA. MAS 5.0 values were then stabilized and log transformed by this formula log(signal + 50). The identical ANOVA model was applied to identify outliers. First we observed that RMA identifies a larger number of outliers than

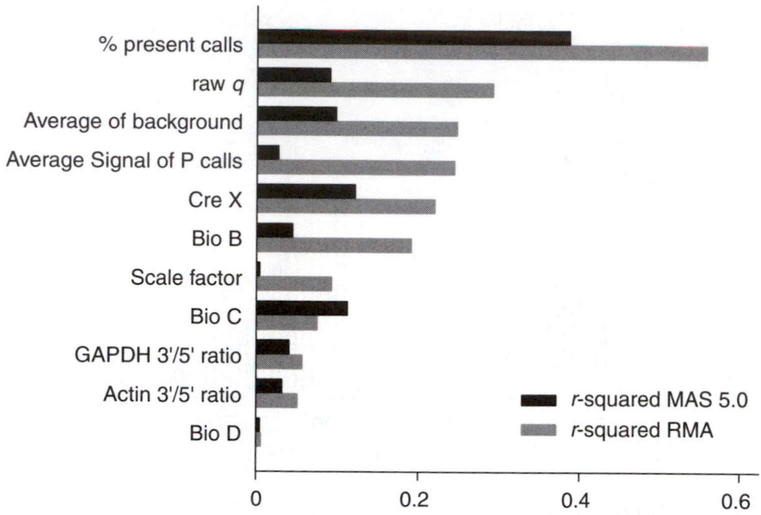

FIGURE 3.6 The effect of different algorithms on quality metrics. The x-axis is the r-squared value from a grouped logit regression where each quality metric was used to predict the number of outlier residuals per array. Using the identical CEL files we observe that RMA and MAS 5.0 have different sensitivities. Note that QC metrics like scale factor, average signal of P calls, and raw q are better correlated to RMA than to MAS 5.0.

MAS 5.0. This increased sensitivity is directly related to reduced variance of RMA. Next we observed from the bar chart of correlations to QC metrics that RMA outlier counts are generally more highly correlated than MAS counts (Figure 3.6). Finally, we observe that RMA is especially sensitive to average signal and that MAS 5.0 is not sensitive to this measure. Given that average signal is bimodal with respect to tissue type (Figure 3.5), we conclude that RMA, when quantile normalized, is sensitive to tissue type while MAS 5.0 is not. This example demonstrates the fact that the choice of an algorithm influences the degree of sensitivity of the data to quality and the whether a given source of variability will influence the data.

Data for another quality analysis example was provided by International Life Sciences Institute ILSI (ref dataset and contributors). In this experiment the same RNA samples were analyzed by different laboratories to test lab-to-lab variability. Unexpectedly, the analysis of quality metrics determined that the quality of the RNA influenced the results (Figure 3.7). The following strategy was used find the source of variation. Just as in the prior example quality metrics were used to predict outlier counts and their correlation of to regression plotted. In this case we can readily distinguish the cause of variability in some users. User 1 showed variability in Bio B, a prelabeled RNA control added prior to hybridization. The correlation indicates that variability in the hybridization or washing or scanner influenced signal. User 2 has a high raw q and average background and scale factor. Covariation of these metrics is associated with scanner settings. Later analysis found that the data for this study was collected from two scanners set at different levels. In this case the each scanner was operated by a different technician using different

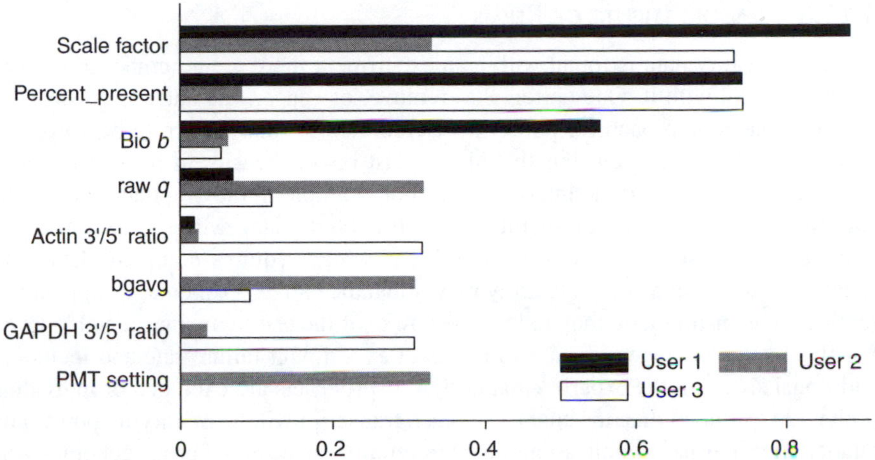

FIGURE 3.7 Mapping quality metrics to process. The x-axis is the r squared value from a grouped logit regression where each quality metric was used to predict the number of outlier residuals per array. Increased outliers per array should correlate with lower quality. If a given metric is relevant to signal quality it should have a high correlation. For User 1 was unusually highly correlated with Bio b. This spike-in control as well as percent present scale factor are associated with hybridization and washing steps. User 2 has a high raw q and average background, which is inconclusive until the correlation to pmt setting confirms that quality was tied to scanner. In this case the arrays were split between two scanners with divergent settings. User 3 has high correlations for the $3'/5'$ ratios indicating RNA quality is driving outlier detection in this case. Note that high correlations to percent present calls and scale factor are good generic indicators of quality, but do not address a specific problem.

protocols so that the variability could be attributed to several causes. User 3 also has a distinctly higher $3'5'$ actin and GAPDH ratio, which is predictive of RNA degradation.

3.11.2 Global or Probe Set Level Metrics

Nearly all of the metrics examined so far are global metrics. Global metrics refer to a single measure per array. These metrics are informative to evaluate any process that applies to all probesets equally, such as RNA sample and labeling processes. However, many processes do not affect all probesets equally, so probesets can usefully be stratified into sub-classes (Figure 3.1). In such cases outliers can be collected on a per probeset level to identify problematic probesets. In addition, the consistency over the array surface can be examined to detect hybridization faults, scratches, or debris. We recommend analysts do not rely solely on global metrics to make decisions about excluding or repairing data. By using probe set or probe level quality metrics to make decisions, we avoid unduly influencing an experiment by removing a complete, replicate.

3.11.3 Censor, Impute, or Repair

Once low quality data is found within an experiment there are a number of options on how to deal with it. One option is to remove, or censor, the data. Exclusion is a very reasonable approach for poorly measured data or data known to be corrupted by a physical cause. Exclusion should be a last resort, because it has implications on the data analysis and the interpretation. For example, removing data renders the analysis sensitive to the rule or filter used to exclude data. Missing data also has implications for some methods such as PCA, which requires complete dataset to function. Excluding a complete array may imbalance an experimental design and, if replication is insufficient, may reduce the power of the test to an unacceptable level. For this reason we recommend estimating an experiment failure rate and including additional arrays in the experimental design. If probesets are excluded or individual probes are excluded then the analysis of each transcript will be of varying power and quality. This may have implications on interpretation especially if broad genome-wide conclusions are to drawn.

An example of outlier removal is seen in Figure 3.8, where over a time course of hours, one replicate had a residual that was more than 2 standard deviations away from zero. Removing this single outlier sufficiently improved the *p*-value so that this probe set passed the Bonferroni threshold despite reduced degrees of freedom. Removing the array led to a loss of 1 degree of freedom, this caused the majority of the data to fall below the identity line. However many probesets now have substantially improved *p*-values. The changes in *p*-values reduced the *p*-values overall, but improved the interpretation.

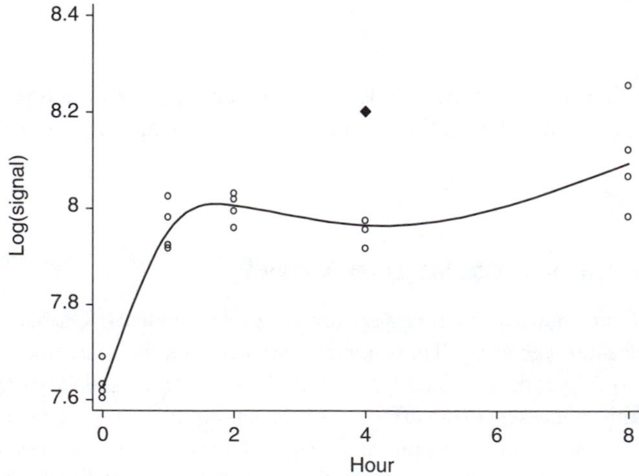

FIGURE 3.8 A probeset measured over an hourly time course. The single outlier at hour 4 (solid diamond) reduced the significance of the profile to below the Bonferroni threshold. After removal of this outlier the profile passed the threshold. Note that if the cause of the outlier behavior is unknown this gene profile should be regarded with more skepticism than profiles, which passed the threshold without data censoring.

Censored data can be replaced by imputed values for PCA or SVD [67] calculations. Keep in mind that imputed values in these parametric tests would tend to artificially inflate the degrees of freedom. The effect of imputation should always be tested by comparing imputation to zero values or to random numbers.

Data may be repaired based on knowledge of the cause of error or to conform to expected forms for computational ease. As a rule normalization methods and transformations (covered above) are intended to repair data by removing a systematic technical error of known origin. The risk to repair is in removing authentic biological effects (e.g., quantile normalization over signal) or generating false negatives (e.g., including saturated data). Again this is good time involve the biological experimenter in this process. The biologist can help assess whether the assumptions in data repair are valid and the analyst can explain the consequences of censoring on power. The removal of probesets and the uneven power in testing may influence downstream tests where the frequency of a given GO^{TM} annotation may be used to interpret the biology of the experiment.

3.11.4 Other Methods for Quality Analysis

One of the most valuable and sensitive methods of analysis for QC is PCA [68] or SVD [69]. The advantage of PCA is its ability to detect and rank all orthogonal factors that influence the linear fit of the data. A key benefit of PCA is that it will sensitively detect subtle patterns in the data. To be truly valuable the resultant PCA vectors must be correlated to quality metrics with defined causes. Here the annotations expression profiles that best fit these error patterns may be of use [34,51]. Another example is Figure 3.9. In this data set we have plotted the two highest vectors for a series of quality metrics (some of the metrics were log-transformed to better fit the normality requirement of PCA). Graphical interruption of the two top vectors clearly indicates two populations, which are defined by the raw q. Later investigation determined that the data was divided by two scanner settings. The second vector (y-axis) was best correlated to $3'/5'$ ratios indicating that RNA quality influenced this data.

Another valuable quality technique is the analysis of residuals for patterns. Just as the residuals were used to select outlier observations, the pattern of all residuals can be used to detect whether other variables are influencing the model. If residuals are not randomly distributed with respect to test variables either more complex normalization methods (SAS) [10] or more complex models may be applied. In general, the methods already developed for estimating goodness of fit for determining outliers by examining residuals apply [70–72].

Lastly, leave one out cross validation (LOOCV) on an array basis is a valid method for demonstrating stability of p-values and determining outlier arrays and is frequently applied in classification studies [73–76]. Of course, visualizing the influence of an outlier removal like in Figure 3.10 is impractical but a summary statistic that measures the influence of each array is useful. We recommend two summary measures of influence. First the correlation of negative log transformed p-values or f or t statistics to the negative log transformed p-values for the complete set as one summary statistic of influence of each array. Alternatively the sum of the difference between the negative

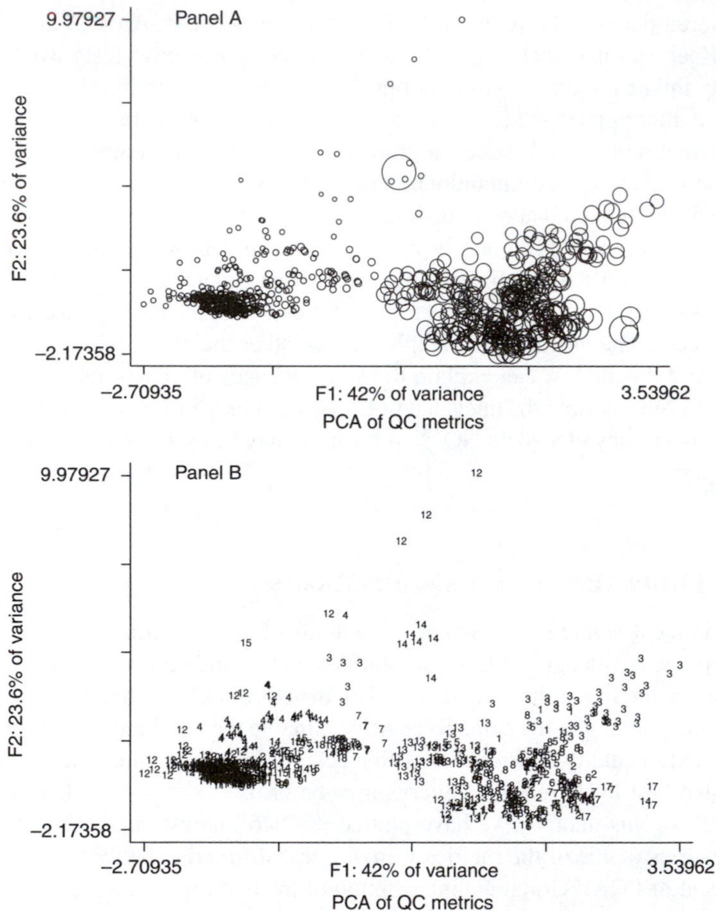

FIGURE 3.9 Quality differences between laboratories. Panel A: PCA analysis of the quality metrics of 700 arrays tested at multiple sites. The size of each circle is proportional to its noise level. The data separates into two populations are separated by noise level represented by raw q. Panel B: The same data as panel A, labeled by user. Comparing the data distributions between panel A and B show that data within a site tends to have the more consistent quality metrics than data across sites.

log transformed p-values (p scores) from all data and the p scores with each array left out can be computed for each array.

$$\sum_{p=1}^{n} [p \text{ score without a} - p \text{ score}]$$

p score $= -\log(p - \text{value}_p)$ for all arrays for probe set $= p$
p score without $a = -\log(p - \text{value}_p$ without $a)$ for all arrays except a and for probe set $= p$
where p-value is from the method selected for expression analysis.

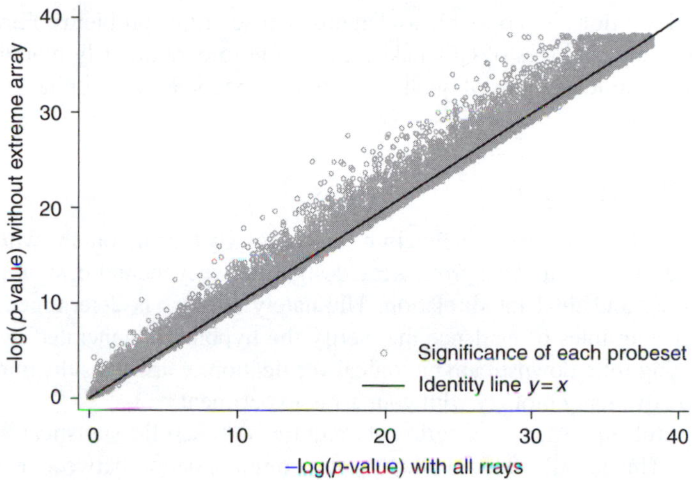

FIGURE 3.10 The effect of removing the array with the highest outlier count. The p-scores ($-\log(p$-values)) from an ANOVA model of the complete dataset (108 arrays) x-axis vs. the the p scores the dataset with the extreme value removed y-axis. Note that although many genes are below the identity line, due to the loss of one degree of freedom, the majority of strong changes are above the line. This indicates that overall removal of this array improved the significance of most probesets. The sum of the difference in p scores was 7,238.9 the mean difference was also positive (0.32). This sum is defined as:

$$p\sum[-\log(p\text{-value}_{\text{without extreme array for probeset } p})]-[-\log(p\text{-value}_{\text{with all arrays for probeset } p})].$$

Then the population of sums of score differences, one for each excluded array, can be assessed for outliers. As a rule a negative sum indicates that the array is not an outlier array as more genes are less significant due to loss of a degree of freedom than a removal of an outlier (Figure 3.10). However if the sum of the score differences is positive then the total significance score was improved by removing this array. This is a reasonable basis for assuming that the array is an outlier. Also note this summation of negative log transformed p-values reduces the influence of less significant changes relative to significant changes. Moreover unlike an outlier count based on standard deviation or mean absolute differences, a sum of differences in p-scores is not threshold sensitive.

Leave one out cross validation is one of the most direct ways to assess the influence of each array. As long as the entire process from raw data to analysis is paralleled, LOOCV is a very useful technique. While the effects may be subtle, when using an algorithm that incorporates replicate analysis such as D Chip or RMA, it is necessary to re-calculate signal in a LOOCV analysis.

3.11.5 Quality Control for Core Facilities

Quality metrics are also valuable for assessing the performance of core facilities. By tracking wash stations, date of hybridization, lot, scanner, operator, assay, chip design, and solutions it is possible to diagnose cause of QC problems. Furthermore, tracking moving averages of QC makes it possible to continuously monitor quality and to detect subtle erosions of quality before they become catastrophic.

3.12 CONCLUSIONS

Quality control is a process of maintaining data integrity throughout the whole investigative process. This includes from array design, to experimental design, principled data analysis, and final interpretation. Ultimately success is determined by proof from supporting lines of evidence that verify the hypothesis generated by microarrays. Planning for a downstream biological verification of array results improves the chances of advancing biology with your array experiment.

Successful high quality experiments require both a solid statistical and biological basis. This usually starts with authentic communication between a statistician and an experimenter. The first step is to define goals and make explicit biological and analytical assumptions. Next a pilot study is performed to refine the protocols and test both the experimental conditions and the analytical methods. Pilot studies also provide variance data useful for power estimation. Ideally an experimental plan should be flexible enough to incorporate changes in time points or doses and is replicated enough to tolerate outlier removal. Next, the full scale design is performed while materials are carefully tracked and the system is thoroughly monitored:

1. RNA QC will detect degraded RNAs before hybridization and extractions can be repeated, thereby saving arrays.
2. Spike-in controls are used to monitor the labeling and hybridization quality. Once the hybridization and scanning is complete and the arrays scanned, QC becomes part of data analysis.
3. Spike in controls monitor the labeling and hybridization quality directly without the dependencies common to other QC metrics.
4. Identifiable systematic measurement errors are corrected or minimized through normalization.
5. Transformations are applied to make analysis tractable. Measures of the degree of change these corrections imposed are recorded.
6. The quality of probe set performance is assessed through residual analysis or other means. These measures of performance quality are correlated to measures of process quality in order to diagnose sources of measurement error for future refinements.
7. Decisions to correct, censor or repair data must be made through consultation with the experimenter.
8. Next the interpretation of the data can begin, keeping in mind the implications of different levels of bioinformatic quality and annotation.
9. Ultimately a biological conclusion is drawn.
10. Conclusions that are validated first at the RNA level and eventually through other lines of biological inquiry.

This formula for a carefully monitored multistage investigative process is designed to enhance the chances of success. Success is defined as advancing biological understanding. Ideally, develop a full scale experimental plan that is flexible enough to incorporate changes in time points or doses and is replicated enough to tolerate outlier removal. Experience gained from pilot studies and designing based on past failure rates makes high quality microarray experiments practical. Finally, involving the biologist in the decision making process from design through quality assessment and interpretation.

REFERENCES

1. C.K. Lee, R.G. Klopp, R. Weindruch, and T.A. Prolla, Gene expression profile of aging and its retardation by caloric restriction. *Science* 285: 1390–1393, 1999.
2. T.R. Golub, D.K. Slonim, P. Tamayo, C. Huard, M. Gaasenbeek, J.P. Mesirov, H. Coller, M.L. Loh, J.R. Downing, M.A. Caligiuri, C.D. Bloomfield, and E.S. Lander, Molecular classification of cancer: class discovery and class prediction by gene expression monitoring. *Science* 286: 531–537, 1999.
3. A. Ridley, Molecular switches in metastasis. *Nature* 406: 466–467, 2000.
4. S.L. Pomeroy, P. Tamayo, M. Gaasenbeek, L.M. Sturla, M. Angelo, M.E. McLaughlin, J.Y. Kim, L.C. Goumnerova, P.M. Black, C. Lau, J.C. Allen, D. Zagzag, J.M. Olson, T. Curran, C. Wetmore, J.A. Biegel, T. Poggio, S. Mukherjee, R. Rifkin, A. Califano, G. Stolovitzky, D.N. Louis, J.P. Mesirov, E.S. Lander, and T.R. Golub, Prediction of central nervous system embryonal tumour outcome based on gene expression. *Nature* 415: 436–442, 2002.
5. J.K. Sax and W.S. El-Deiry, P53-induced gene expression analysis. *Meth. Mol. Biol.* 234: 65–71, 2003.
6. L. Zhang, M.F. Miles, and K.D. Aldape, A model of molecular interactions on short oligonucleotide microarrays. *Nat. Biotechnol.* 21: 818–821, 2003.
7. R. Mei, E. Hubbell, S. Bekiranov, M. Mittmann, F.C. Christians, M.M. Shen, G. Lu, J. Fang, W.M. Liu, T. Ryder, P. Kaplan, D. Kulp, and T.A. Webster, Probe selection for high-density oligonucleotide arrays. *Proc. Natl Acad. Sci. USA* 100: 11237–11242, 2003.
8. J.B. Hogenesch, K.A. Ching, S. Batalov, A.I. Su, J.R. Walker, Y. Zhou, S.A. Kay, P.G. Schultz, and M.P. Cooke, A comparison of the celera and ensembl predicted gene sets reveals little overlap in novel genes. *Cell* 106: 413–415, 2001.
9. M.J. Martinez, A.D. Aragon, A.L. Rodriguez, J.M. Weber, J.A. Timlin, M.B. Sinclair, D.M. Haaland, and M. Werner-Washburne, Identification and removal of contaminating fluorescence from commercial and in-house printed DNA microarrays. *Nucl. Acids Res.* 31: e18, 2003.
10. R.D. Wolfinger, G. Gibson, E.D. Wolfinger, L. Bennett, H. Hamadeh, P. Bushel, C. Afshari, and R.S. Paules, Assessing gene significance from CDNA microarray expression data via mixed models. *J. Comput. Biol.* 8: 625–637, 2001.
11. D. Finkelstein, R. Ewing, J. Gollub, F. Sterky, J.M. Cherry, and S. Somerville, Microarray data quality analysis: lessons from the AFGC project. *Arabidopsis* functional genomics consortium. *Plant. Mol. Biol.* 48: 119–131, 2002.
12. J. Gollub, C.A. Ball, G. Binkley, J. Demeter, D.B. Finkelstein, J.M. Hebert, T. Hernandez-Boussard, H. Jin, M. Kaloper, J.C. Matese, M. Schroeder, P.O. Brown,

D. Botstein, and G. Sherlock, The Stanford microarray database: data access and quality assessment tools. *Nucl. Acids Res.* 31: 94–96, 2003.
13. T.M. Chu, B. Weir, and R. Wolfinger, A systematic statistical linear modeling approach to oligonucleotide array experiments. *Math. Biosci.* 176: 35–51, 2002.
14. M.K. Kerr, Experimental design to make the most of microarray studies. *Meth. Mol. Biol.* 224: 137–147, 2003.
15. Y.H. Yang, S. Dudoit, P. Luu, D.M. Lin, V. Peng, J. Ngai, and T.P. Speed, Normalization for cDNA microarray data: a robust composite method addressing single and multiple slide systematic variation. *Nucl. Acids Res.* 30: e15, 2002.
16. J. Quackenbush, Microarray data normalization and transformation. *Nat. Genet.* 32 (Suppl): 496–501, 2002.
17. Y. Yang, J. Hoh, C. Broger, M. Neeb, J. Edington, K. Lindpaintner, and J. Ott, Statistical methods for analyzing microarray feature data with replications. *J. Comput. Biol.* 10: 157–169, 2003.
18. M.J. Hessner, X. Wang, S. Khan, L. Meyer, M. Schlicht, J. Tackes, M.W. Datta, H.J. Jacob, and S. Ghosh, Use of a three-color cDNA microarray platform to measure and control support-bound probe for improved data quality and reproducibility. *Nucl. Acids Res.* 31: e60, 2003.
19. T.L. Fare, E.M. Coffey, H. Dai, Y.D. He, D.A. Kessler, K.A. Kilian, J.E. Koch, E. LeProust, M.J. Marton, M.R. Meyer, R.B. Stoughton, G.Y. Tokiwa, and Y. Wang, Effects of atmospheric ozone on microarray data quality. *Anal. Chem.* 75: 4672–4675, 2003.
20. E. Taylor, D. Cogdell, K. Coombes, L. Hu, L. Ramdas, A. Tabor, S. Hamilton, and W. Zhang, Sequence verification as quality-control step for production of CDNA microarrays. *Biotechniques* 31: 62–65, 2001.
21. P. Pavlidis, Q. Li, and W.S. Noble, The effect of replication on gene expression microarray experiments. *Bioinformatics* 19: 1620–1627, 2003.
22. J. Quackenbush, Computational analysis of microarray data. *Nat. Rev. Genet.* 2: 418–427, 2001.
23. B. Durbin and D.M. Rocke, Estimation of transformation parameters for microarray data. *Bioinformatics* 19: 1360–1367, 2003.
24. D.M. Rocke and B. Durbin, Approximate variance-stabilizing transformations for gene-expression microarray data. *Bioinformatics* 19: 966–972, 2003.
25. D. Finkelstein, E. Hubbell, and J. Retief, Oligo arrays, global transcriptome analysis. In: J. Borlak (ed.), *Introduction to toxicogenomics*. Weinheim, John Wiley & Sons, New York (in review).
26. W. Pan, J. Lin, and C.T. Le, How many replicates of arrays are required to detect gene expression changes in microarray experiments? A mixture model approach. *Genome. Biol.* 3: 0022.1–0022.10, 2002.
27. X. Peng, C.L. Wood, E.M. Blalock, K.C. Chen, P.W. Landfield, and A.J. Stromberg, Statistical implications of pooling NA samples for microarray experiments. *BMC Bioinformatics* 4: 26, 2003.
28. C.M. Kendziorski, Y. Zhang, H. Lan, and A.D. Attie, The efficiency of pooling MRNA in microarray experiments. *Biostatistics* 4: 465–477, 2003.
29. K. Ichimura, T. Mizoguchi, R. Yoshida, T. Yuasa, and K. Shinozaki, Various abiotic stresses rapidly activate arabidopsis map kinases atmpk4 and atmpk6. *Plant J.* 24: 655–665, 2000.
30. A. Dash, I.P. Maine, S. Varambally, R. Shen, A.M. Chinnaiyan, and M.A. Rubin, Changes in differential gene expression because of warm ischemia time of radical prostatectomy specimens. *Am. J. Pathol.* 161: 1743–1748, 2002.

31. P.T. Spellman, G. Sherlock, M.Q. Zhang, V.R. Iyer, K. Anders, M.B. Eisen, P.O. Brown, D. Botstein, and B. Futcher, Comprehensive identification of cell cycle-regulated genes of the yeast *Saccharomyces cerevisiae* by microarray hybridization 7. *Mol. Biol. Cell* 9: 3273–3277, 1998.
32. S. Panda, M.P. Antoch, B.H. Miller, A.I. Su, A.B. Schook, M. Straume, P.G. Schultz, S.A. Kay, J.S. Takahashi, and J.B. Hogenesch, Coordinated transcription of key pathways in the mouse by the circadian clock. *Cell* 109: 307–320, 2002.
33. H.R. Ueda, W. Chen, A. Adachi, H. Wakamatsu, S. Hayashi, T. Takasugi, M. Nagano, K. Nakahama, Y. Suzuki, S. Sugano, M. Iino, Y. Shigeyoshi, and S. Hashimoto, A transcription factor response element for gene expression during circadian night. *Nature* 418: 534–539, 2002.
34. A.R. Whitney, M. Diehn, S.J. Popper, A.A. Alizadeh, J.C. Boldrick, D.A. Relman, and P.O. Brown, Individuality and variation in gene expression patterns in human blood. *Proc. Natl. Acad. Sci. USA* 100: 1896–1901, 2003.
35. J. Yang, G. Li, F. Zhang, Y. Liu, D. Zhang, W. Zhou, G. Xu, Y. Yang, and M. Luo, Identification of variations of gene expression of visceral adipose and renal tissue in type 2 diabetic rats using CDNA representational difference analysis. *Chin. Med. J. (Engl.)* 116: 529–533, 2003.
36. M. Wareing, C.J. Ferguson, M. Delannoy, A.G. Cox, R.F. McMahon, R. Green, D. Riccardi, and C.P. Smith, Altered dietary iron intake is a strong modulator of renal DMT1 expression. *Am. J. Physiol. Renal. Physiol.* 285: 1050–1059, 2003.
37. T. Mohri, N. Emoto, H. Nonaka, H. Fukuya, K. Yagita, H. Okamura, and M. Yokoyama, Alterations of circadian expressions of clock genes in dahl salt-sensitive rats fed a high-salt diet. *Hypertension* 42: 189–194, 2003.
38. W. Xu, P. Campbell, A. K. Vargheese, and J. Braam, The arabidopsis xet-related gene family: Environmental and hormonal regulation of expression. *Plant J.* 9: 879–889, 1996.
39. D. Seliktar, R.M. Nerem, and Z.S. Galis, Mechanical strain-stimulated remodeling of tissue-engineered blood vessel constructs. *Tissue Eng.* 9: 657–666, 2003.
40. S. Footitt, M. Ingouff, D. Clapham, and S. von Arnold, Expression of the viviparous 1 (pavp1) and p34cdc2 protein kinase (cdc2pa) genes during somatic embryogenesis in norway spruce (picea abies [l.] karst). *J. Exp. Bot.* 54: 1711–1719, 2003.
41. N. Jambunathan and T.W. McNellis, Regulation of arabidopsis copine 1 gene expression in response to pathogens and abiotic stimuli. *Plant Physiol.* 132: 1370–1381, 2003.
42. C. Li and W.H. Wong, Model-based analysis of oligonucleotide arrays: expression index computation and outlier detection. *Proc. Natl Acad. Sci. USA* 98: 31–36, 2001.
43. R.A. Irizarry, B.M. Bolstad, F. Collin, L.M. Cope, B. Hobbs, and T.P. Speed, Summaries of affymetrix genechip probe level data. *Nucl. Acids Res.* 31: e15, 2003.
44. D.A. Morrison and J.T. Ellis, The design and analysis of microarray experiments: applications in parasitology. *DNA Cell Biol.* 22: 357–394, 2003.
45. C.A. Glasbey and P. Ghazal, Combinatorial image analysis of DNA microarray features. *Bioinformatics* 19: 194–203, 2003.
46. M. Bakay, Y.W. Chen, R. Borup, P. Zhao, K. Nagaraju, and E.P. Hoffman, Sources of variability and effect of experimental approach on expression profiling data interpretation. *BMC Bioinformatics* 3: 4, 2002.
47. A. Brazma, H. Parkinson, U. Sarkans, M. Shojatalab, J. Vilo, N. Abeygunawardena, E. Holloway, M. Kapushesky, P. Kemmeren, G.G. Lara, A. Oezcimen, P. Rocca-Serra, and S.A. Sansone, Arrayexpress — a public repository for microarray gene expression data at the ebi. *Nucl. Acids Res.* 31: 68–71, 2003.

48. P.T. Spellman, M. Miller, J. Stewart, C. Troup, U. Sarkans, S. Chervitz, D. Bernhart, G. Sherlock, C. Ball, M. Lepage, M. Swiatek, W.L. Marks, J. Goncalves, S. Markel, D. Iordan, M. Shojatalab, A. Pizarro, J. White, R. Hubley, E. Deutsch, M. Senger, B.J. Aronow, A. Robinson, D. Bassett, C.J. Stoeckert, Jr., and A. Brazma, Design and implementation of microarray gene expression markup language (mage-ml). *Genome Biol.* 3: 0046, 2002.
49. A. Brazma, P. Hingamp, J. Quackenbush, G. Sherlock, P. Spellman, C. Stoeckert, J. Aach, W. Ansorge, C.A. Ball, H.C. Causton, T. Gaasterland, P. Glenisson, F.C. Holstege, I.F. Kim, V. Markowitz, J.C. Matese, H. Parkinson, A. Robinson, U. Sarkans, S. Schulze-Kremer, J. Stewart, R. Taylor, J. Vilo, and M. Vingron, Minimum information about a microarray experiment (miame)-toward standards for microarray data. *Nat. Genet.* 29: 365–371, 2001.
50. M.K. Kerr and G.A. Churchill, Experimental design for gene expression microarrays. *Biostatistics* 2: 183–201, 2001.
51. O. Alter, P.O. Brown, and D. Botstein, Singular value decomposition for genome-wide expression data processing and modeling. *Proc. Natl Acad. Sci. USA* 97: 10101–10106, 2000.
52. J. Landgrebe, W. Wurst, and G. Welzl, Permutation — validated principal components analysis of microarray data. *Genome Biol.* 3: 0019, 2002.
53. V.G. Tusher, R. Tibshirani, and G. Chu, Significance analysis of microarrays applied to the ionizing radiation response. *Proc. Natl Acad. Sci. USA* 98: 5116–5121, 2001.
54. A. Reiner, D. Yekutieli, and Y. Benjamini, Identifying differentially expressed genes using false discovery rate controlling procedures. *Bioinformatics* 19: 368–375, 2003.
55. P. Westfall, Simultaneous small-sample multivariate Bernoulli confidence intervals. *Biometrics* 41: 1001–1013, 1985.
56. Y. Benjamini, D. Drai, G. Elmer, N. Kafkafi, and I. Golani, Controlling the false discovery rate in behavior genetics research. *Behav. Brain Res.* 125: 279–284, 2001.
57. B. Efron and R. Tibshirani, Empirical Bayes methods and false discovery rates for microarrays. *Genet. Epidemiol.* 23: 70–86, 2002.
58. S. Draghici, P. Khatri, R.P. Martins, G.C. Ostermeier, and S.A. Krawetz, Global functional profiling of gene expression. *Genomics* 81: 98–104, 2003.
59. B.R. Zeeberg, W. Feng, G. Wang, M.D. Wang, A.T. Fojo, M. Sunshine, S. Narasimhan, D.W. Kane, W.C. Reinhold, S. Lababidi, K.J. Bussey, J. Riss, J.C. Barrett, and J.N. Weinstein, Gominer: a resource for biological interpretation of genomic and proteomic data. *Genome Biol.* 4: R28, 2003.
60. O.G. Troyanskaya, K. Dolinski, A.B. Owen, R.B. Altman, and D. Botstein, A Bayesian framework for combining heterogeneous data sources for gene function prediction (in *Saccharomyces cerevisiae*). *Proc. Natl Acad. Sci. USA* 100: 8348–8353, 2003.
61. T.R. Hvidsten, J. Komorowski, A.K. Sandvik, and A. Lægreid, Predicting gene function from gene expressions and ontologies. *Pac. Symp. Biocomput.* 6: 299–310, 2001.
62. R.F. Chuaqui, R.F. Bonner, C.J. Best, J.W. Gillespie, M.J. Flaig, S.M. Hewitt, J.L. Phillips, D.B. Krizman, M.A. Tangrea, M. Ahram, W.M. Linehan, V. Knezevic, and M.R. Emmert-Buck, Post-analysis, follow-up, and validation of microarray experiments. *Nat. Genet.* 32 (Suppl): 509–514, 2002.
63. S.H. Cheng, B. Moore, and J.R. Seemann, Effects of short- and long-term elevated CO_2 on the expression of ribulose-1,5-bisphosphate carboxylase/oxygenase genes and carbohydrate accumulation in leaves of *Arabidopsis thaliana* (l.) heynh. *Plant Physiol.* 116: 715–723, 1998.

64. S.H. Cheng and J.R. Seemann, Extraction and purification of RNA from plant tissue enriched in polysaccharides. *Meth. Mol. Biol.* 86: 27–32, 1998.
65. R.A. Irizarry, B. Hobbs, F. Collin, Y.D. Beazer-Barclay, K.J. Antonellis, U. Scherf, and T.P. Speed, Exploration, normalization, and summaries of high density oligonucleotide array probe level data. *Biostatistics* 4: 249–264, 2003.
66. E. Hubbell, W.M. Liu, and R. Mei, Robust estimators for expression analysis. *Bioinformatics* 18: 1585–1592, 2002.
67. O. Troyanskaya, M. Cantor, G. Sherlock, P. Brown, T. Hastie, R. Tibshirani, D. Botstein, and R.B. Altman, Missing value estimation methods for DNA microarrays. *Bioinformatics* 17: 520–525, 2001.
68. F. Model, T. Konig, C. Piepenbrock, and P. Adorjan, Statistical process control for large scale microarray experiments. *Bioinformatics* 18 (Suppl 1): S155–S163, 2002.
69. O. Alter, P.O. Brown, and D. Botstein, Generalized singular value decomposition for comparative analysis of genome-scale expression data sets of two different organisms. *Proc. Natl Acad. Sci. USA* 100: 3351–3356, 2003.
70. G.E.P. Box, W.G. Hunter, and J.S. Hunter, *Statistics for Experimenters: An Introduction to Design, Data Analysis, and Model Building*. John Wiley & Sons, New York, 1978.
71. H. Abu-Libdeh, B.W. Turnbull, and L.C. Clark Analysis of multi-type recurrent events in longitudinal studies; application to a skin cancer prevention trial. *Biometrics* 46: 1017–1034, 1990.
72. E. Krusinska, U.L. Mathiesen, L. Franzen, G. Bodemar, and O. Wigertz, Influence of "outliers" on the association between laboratory data and histopathological findings in liver biopsy. *Meth. Inf. Med.* 32: 388–395, 1993.
73. F. Azuaje, Genomic data sampling and its effect on classification performance assessment. *BMC Bioinformatics* 4: 5, 2003.
74. D. Hwang, I. Alevizos, W.A. Schmitt, J. Misra, H. Ohyama, R. Todd, M. Mahadevappa, J.A. Warrington, G. Stephanopoulos, and D.T. Wong, Genomic dissection for characterization of cancerous oral epithelium tissues using transcription profiling. *Oral. Oncol.* 39: 259–268, 2003.
75. C.L. Nutt, D.R. Mani, R.A. Betensky, P. Tamayo, J.G. Cairncross, C. Ladd, U. Pohl, C. Hartmann, M.E. McLaughlin, T.T. Batchelor, P.M. Black, A. von Deimling, S.L. Pomeroy, T.R. Golub, and D.N. Louis, Gene expression-based classification of malignant gliomas correlates better with survival than histological classification. *Cancer Res.* 63: 1602–1607, 2003.
76. H. Zhang, C.Y. Yu, and B. Singer, Cell and tumor classification using gene expression data: construction of forests. *Proc. Natl. Acad. Sci. USA* 100: 4168–4172, 2003.

4 Epistemological Foundations of Statistical Methods for High-Dimensional Biology*

Stanislav O. Zakharkin, Tapan Mehta, Murat Tanik, and David B. Allison

CONTENTS

4.1 The Challenge We Face ... 58
4.2 Our Vantage Point: From Samples to Populations ... 59
4.3 What Is Validity? ... 59
  4.3.1 Inference ... 60
  4.3.2 Estimation ... 61
  4.3.3 Prediction ... 61
  4.3.4 Classification ... 61
  4.3.5 Evaluating Validity ... 62
4.4 Comparison of Different Methods ... 63
4.5 Data Sets of Unknown Nature: Circular Reasoning ... 64
4.6 The Search for Proof: Deduction ... 64
  4.6.1 What Is Proof? ... 64
  4.6.2 Mathematical Description Is Not Mathematical Proof ... 65
  4.6.3 No Such Thing as a Free Lunch: Assumptions, Assumptions, Assumptions ... 65
  4.6.4 Proof by Reference ... 66
4.7 The Proof of the Pudding Is in the Eating: Induction ... 67
4.8 Combined Modes ... 69
4.9 Where to from Here? ... 69
Acknowledgments ... 71
References ... 71

---

*An abbreviated version of this chapter was published as: T. Mehta, M. Tanik, and D.B. Allison. Toward sound epistemological foundations of statistical methods for high-dimensional biology. *Nature Genetics*, 36: 943–947, 2004.

4.1 THE CHALLENGE WE FACE

Science, from the Latin *scientia*, means having knowledge. Epistemology is the study of how we come to and what constitutes knowledge. The biological sciences use empiricism and induction [1]. We make inferences, predictions, and estimations from data, commonly via statistical techniques. Confidence in our biological knowledge stems, partly, from confidence in our statistical methodology's validity. This begs the question: "How do we derive knowledge about the validity of statistical methods such that they also enjoy a solid epistemological foundation?" Although in more traditional fields there seems to be an implicitly shared understanding among purveyors of statistical methodology, writings on these second-order epistemological questions in statistical analysis are remarkably scarce as is a shared understanding in the field of high-dimensional biology (HDB).

High-dimensional biology encompasses the many "omic" technologies [2] and can involve thousands of genetic polymorphisms, sequences, expression levels, protein measurements, or combination thereof. How do we derive knowledge about the validity of statistical methods for HDB? Although our comments are applicable to HDB overall, we emphasize microarrays, where the need is acute. "The field of expression data analysis is particularly active with novel analysis strategies and tools being published weekly" ([3]; Figure 4.1), and the value of many of these methods is questionable [4]. Some results produced by using these methods are so anomalous

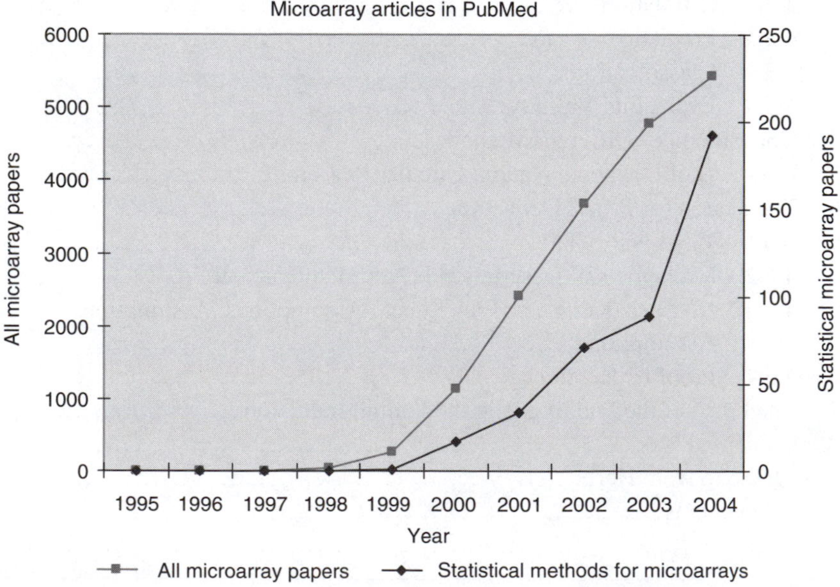

FIGURE 4.1 Growth of microarray and microarray methodology literatures 1995 to 2004. The category "All microarray papers" includes those found by searching PubMed for "microarray" or "gene expression profiling." The category "statistical microarray papers" includes those found by searching PubMed for "statistical method" or "statistical technique" or "statistical approach" and "microarray" or "gene expression profiling."

that a breed of "forensic" statisticians [5–7], who doggedly detect and correct other HDB investigators' prominent mistakes, has been created.

Here we offer a "meta-methodology" and framework in which to evaluate epistemological foundations of proposed statistical methods. We address (1) What it means to be a valid statistical procedure, (2) Methods to evaluate validity, (3) A frequently used but logically invalid method, and (4) Frequently unstated (implicit) assumptions. On the basis of this framework, we opine that many statistical methods offered to the HDB community do not have an adequate epistemological foundation. We hope the framework will help methodologists to develop robust methods and help applied investigators to evaluate the extent to which statistical methods are valid.

4.2 OUR VANTAGE POINT: FROM SAMPLES TO POPULATIONS

We study samples and data to understand populations and nature. People often say that they want to understand their data. Truthfully, nobody cares one wit about their data — about, for example, how much of protein Y was in 10 particular mouse brains after being fed X. We really want to know the effect of feeding X on the expression of protein Y in the overall mouse population and ultimately the truth about nature. We observe imperfect measurements of variables from specific objects drawn from larger populations, and these observations form our base data. Using these data, we wish to make inferences from these imperfect measurements to the real variables they represent and then from the sample cases to the population they were sampled from.

Borrowing the language of Plato's Allegory of the Cave, we can view the data at level II as manifestations of a random sampling process and "shadows" of the population-level reality we are trying to model. Similarly, we can view the data at level III as manifestations of our inevitably error-prone measurement processes and "second-order shadows" of the actual sample values. Although once stated this may seem obvious, it is apparently not universally appreciated. This has lead some authors to resample across genes when trying to assess the stability of certain microarray results [8] whereas appreciation of this samples-to-population perspective exposes that the sampling units, and therefore that which should be resampled, are cases (e.g., mice) not genes. Similarly, methods have been proposed whereby inferences about gene expression differences between populations are made by comparing observed sample differences to an estimated null distribution of differences based on technical rather than biological replicates [9]. This conflates the standard error of measurement with the standard error of the sample statistic; it takes observations from Level I (Table 4.1), makes an inference to Level II and conflates this inference with the desired inference to Level III. This is one example of a common class of mistakes that can be avoided by considering the sample-to-population perspective.

4.3 WHAT IS VALIDITY?

There are multiple philosophical orientations to derive knowledge with statistical methods, but regardless of the orientation, the standards for evaluation should be

TABLE 4.1
Iconic Representation of Levels of Observation and Inference

	Level I	Level II	Level III
Aspects of variables studied	Measurements of specific variables on cases	True values of specific variables on cases	Population distribution of variables
Units of observation	Cases/samples	Cases/samples	Population/universe
Subject of study	Data	Data	Nature
Conceptual description	Second-order shadows	Shadows	Reality

explicit [10]. Before describing methods for evaluating the validity of proposed statistical methods, it is necessary to expose what we mean by validity. This requires explicit statements of what a method is supposed to do or what properties it is supposed to have.

Much statistical methodology can be grouped into five, admittedly-overlapping, categories: measurement, inference, estimation, prediction, and classification. For brevity, we focus only on the last four. We briefly describe how one might choose to define validity of methods for each, adhering to the samples-to-populations perspective.

4.3.1 Inference

By "inference" statisticians refer to the process of making decisions or drawing conclusions about the truth or falsity of hypotheses [11]. Commonly this is done in the frequentist paradigm where p-values are calculated and a null hypothesis (e.g., mean expression level of gene X is equal in conditions A and B) is rejected if the p-value falls below some threshold, α. However, other increasingly popular inferential paradigms exist [11–13]. By law of excluded middle, hypotheses are either true or not true. We use inferential procedures to conclude or choose to behave as though a hypothesis is either true or not true and can define validity in terms of error rates. Traditionally valid tests are ones that reject true null hypothesis $<\alpha^*100\%$ of the time and reject false null hypothesis $>\alpha^*100\%$ of the time. Among such tests, those that reject false null hypotheses more frequently (i.e., more powerful; lower type 2 error rates) than others might be labeled "more valid" — validity can come in degrees. Of course, these are other properties or criteria one could focus on. For example, some HDB methodologists prefer to develop procedures that hold the false discovery rate (FDR) or the proportion of false positives (PFP) to some level across all hypotheses tested [14–17]. The choice among these and other potential definitions of validity is inherently subjective. One cannot "prove" that one criterion is better than another. However, having stated exactly what criteria one has chosen, determining whether these criteria are met can be evaluated objectively [10].

Small sample sizes typical for HDB offer little power, especially if corrected for multiple testing. Greater power and flexibility might be obtained by borrowing information across genes [18], but doing so often makes assumptions about exchangeability across or independence of genes that may not be explicit or valid [19].

4.3.2 Estimation

Estimation, closely related to but distinct from inference, entails computing sample statistics (e.g., "fold-change" of gene expression in one group vs. another) on observed data as estimates of corresponding unobserved population parameters (e.g., the true fold-change difference in the population) termed estimands. Estimation procedures are valid to the extent that estimates they provide correspond to the estimands.

The HDB can present special challenges. For example, estimators that are ordinarily unbiased can be markedly biased when used only to estimate significant effects in genome-wide contexts [20]. Statisticians have elucidated properties of estimators such as bias, mean square error, efficiency, sufficiency, and consistency [21]. There is subjectivity in choosing which properties one is interested in, but once these properties are specified, objective evaluation is possible. HDB offers especially rich ground for modern "information-borrowing" techniques that can radically improve estimation when many estimates are simultaneously made on small samples [13,18].

4.3.3 Prediction

Prediction is a common activity in transcriptomics and proteomics, typically prediction of status of some categorical variable. With categorical outcomes, prediction is often labeled (supervised) classification, but for clarity we reserve "classification" for other activities described further. With categorical prediction, defining validity is straightforward and typically entails minimizing and accurately estimating expected prediction error probabilities. Of course, subjectivity is also present here. For example, one could differentially weigh different types of errors. Importantly, the predictive validity of a prediction rule developed in a single dataset should not be confused with the validity of a general method for developing prediction rules on datasets.

The seemingly simple problem of estimating predictive accuracy on a single data set is not so simple in HDB [22]. As Ambroise and McLachlan [5] clarified, some methodologists have markedly overestimated predictive accuracy by using invalid methods for estimating predictive accuracy of derived predictive rules. Had such investigators checked their methods' validity with, for example, simulations as discussed below, they would have likely detected their errors before publication.

4.3.4 Classification

We use "classification" to denote the construction of classification schemes and assignment of objects to classes within schemes (so-called unsupervised classification; typically a form of cluster analysis). In microarrays, classification is often applied to the genes (variables) as opposed to the cases and we focus on that. What do we know about these methods' validity?

> availability of computer calculation methods has led many psychiatrists to using the new data techniques uncritically. Spurious findings may result. The available clustering techniques should be validated as by applying them to sets of data of known structure. Everitt [23]
>
> Often it is useful to identify groups of samples with similar expression patterns and genes that are similar across samples. Statisticians are often skeptical about clustering: Whatever method is used, clusters are invariably found, and it seems difficult to assess the strength of evidence for these. Expositions of clustering methods tend to focus on the algorithms, apparently due to a lack of a good theoretical framework to assess and compare different methods. Edwards [24]

Everitt's [23] comments about benchmarking methods against datasets of known structure presage our comments below on validation modes and Edwards' [24] clarify that the HDB field has yet to meet this challenge. What criteria can serve as validity evidence for classification methods? This is conceptually challenging. Classifications do not exist — we create them. Thus, there is no null hypothesis to test,* no independent reality to compare to a derived classification, in short, no "right answer." Although classifications are not incorrect or correct, they can be useless or useful and, if useful, have varying degrees of utility. Therefore, a valid method ultimately produces useful classifications. This is difficult to address, because it raises the question "Useful for what?" If we could answer this, it might be easier to assess a method's validity, but most authors do not describe in testable terms exactly what the goal of the classification is and, in fairness, doing so may be premature. Classifications that are replicable may turn out to have important uses that we cannot anticipate. However, even absent the ability to check ultimate validity, there may be a path forward.

The extent to which they yield classifications are replicable across multiple samplings from a population [8]. Although a replicable classification is not necessarily useful, a useful classification that characterizes some aspect of the population must be replicable [26]. By replicable we mean reproducible across multiple samplings from a population. Therefore, we can quantify the validity of clustering methods by the degree to which they yield classifications that are replicable beyond chance levels ([27,28]; see Reference 8 for overview of stability metrics). Some authors have, in our opinion, mistakenly resampled across genes when trying to assess stability of cluster solutions [8], which makes little sense from the samples-to-population perspective shown in Table 4.1.

4.3.5 Evaluating Validity

Assessing validity necessitates explicit standards for evaluating methods. This requires an explanation of what a method is supposed to do or what properties it is

---

* Some [25] disagree on this point and state that "the null hypothesis that is being tested here is that of no structure in the data." Exactly what "no structure" means is not clear, nor can it be taken to be equivalent to "no classification."

TEXT BOX 4.1 Commonly Used Statistical Terms and Quantities

	H_a true	H_0 true
H_0 rejected	A	B
H_0 not rejected	C	D

True positive rate, TPR $= \frac{A}{A+B}$
True negative rate, TNR $= \frac{D}{C+D}$
False positive rate, FPR $= \frac{B}{A+B}$
False negative rate, FNR $= \frac{C}{C+D}$
Expected discovery rate (EDR) $= \frac{D}{B+D}$; the expected proportion of accepted null hypotheses among all true null hypotheses [29]
False discovery rate (FDR) $= E\left(\frac{B}{A+B}\right)$; the expected proportion of rejected true null hypotheses among all rejections
power — the probability that a statistical significance test will reject the null hypothesis for a specified value of an alternative hypothesis.

supposed to have. A full description of various qualities that a statistical procedure should have is beyond our scope; however, we mention some of the most commonly used briefly. One of the common tasks in microarray analysis is to determine what genes are differentially expressed between two or more different conditions and a plethora of tests has been developed for that purpose. Methods can be evaluated based on their sensitivity, specificity, ability to detect false positives, and other criteria (Text Box 4.1).

It is desirable that a statistical test is relatively insensitive to small deviations in data or a model and to effects that are not considered in a design, that is, robust. The basic idea is that a method should work not only in ideal conditions, but also when there are minor departures from assumptions [30–32].

Validity can be relative and situation-specific. This is noteworthy in considering the merit of a newly proposed procedure when one or more procedures already exist for similar purposes. In such cases, it may be important to ask not only whether the new method is valid in an absolute sense, but whether and under what circumstances it confers any relative advantage with respect to the chosen properties. There is inherent subjectivity in choosing which properties are of interest or desired, but once criteria are chosen, methods can and should be evaluated objectively.

4.4 COMPARISON OF DIFFERENT METHODS

The enormous variety of modern quantitative approaches leaves researchers with the nontrivial task of matching method to the research question of interest. It is important to compare new methodologies to existing and established ones. Recently, a number of papers comparing different methods applied to microarray research have been presented [33–36]. Although complex and state-of-the-art methods are sometimes necessary to address research questions effectively, simpler classical approaches often can provide elegant and sufficient answers to important questions. Fisher [37] pointed out that

> ...Experimenters should remember that they and their colleagues usually know more about the kind of material they are dealing with than do the authors of text-books written without such personal experience, and that a more complex, or less intelligible, test is not likely to serve their purpose better, in any sense, than those of proved value in their own subject...

In agreement with that, de Lichtenberg et al. [35] found most new and advanced methods actually had a worse performance than older methods.

It is unlikely that different methods will produce exactly identical results. Our goal is to determine how much the new method is different from old and if we can replace an old method or use two interchangeably. Often, some methods seem to perform better than others in some situations but there is no clear winner overall. If this is the case, it is important to understand strengths and weaknesses of each approach and to explain disagreement between approaches paying particular attention to their underlying assumptions. Performance of the methods should be compared using well formulated criteria.

4.5 DATA SETS OF UNKNOWN NATURE: CIRCULAR REASONING

Authors often purport to demonstrate a new method's validity in HDB by applying it to one real data set of unknown nature. A new method is applied to a data set, and a new interesting finding is reported; for example, a gene previously not known to be involved in disease *X* is found to be related to the disease, and the authors believe that the finding shows their method's value. The catch is this: if the gene was previously not known to be involved in disease *X*, how do the authors know that they got the right answer? If they do not know that the answer is right, how do they know that this validates their method? If they do not know that their method is valid, how do they know that they got the right answer? We are in a loop (circular argument). Illustration of a method's use is not demonstration of its value. Illustration with single data sets of unknown nature, though interesting, is not a sound epistemological foundation for method development.

4.6 THE SEARCH FOR PROOF: DEDUCTION

4.6.1 What Is Proof?

A proof is a logical argument that proceeds from axioms to eventual conclusion via an ordered deductive process [38]. A proof's certainty stems from the deductive nature by which each step follows from a prior step. In reality, as things proven and methods of their proof have become more complex, certainty is not always easy to achieve and what is obvious to one person may not be to another [39]. The key structure that reader's should seek as evidence that there is proof that a method has some property should be precise statements of axioms formulated, the method's purported property, and logical steps "obviously" connecting the two.

4.6.2 Mathematical Description Is Not Mathematical Proof

Mathematical description of some process is not equivalent to mathematical proof that the process' result has any particular properties. Methodological papers in HDB often present new algorithms with exquisite mathematical precision, Greek letters, and arcane symbols. Less mathematically comfortable readers may mistake this for proof. However, writing an equation may define something but does not *ipso facto* prove anything. To paraphrase Abraham Lincoln, "How many legs does a calf have if you call the tail a leg? Four. Calling a tail a leg doesn't make it a leg" [40]. Finally, just as mathematically elegant description of an algorithm does not *ipso facto* constitute proof, a proof "in practice" need not contain mathematical symbols and can sometimes be constructed simply by elegant use of language or even pictures [41].

4.6.3 No Such Thing as a Free Lunch: Assumptions, Assumptions, Assumptions

A proof begins with axioms or postulates, that is, assumptions, and is valid only when the assumptions hold. The proof's practical conclusions may hold across broader circumstances, but additional evidence is required to support this. For example, Student's t-test [42] is a valid frequentist test of the null hypothesis that two means are equal and more powerful than any other possible valid frequentist test. However, this is only definitively true when the test's assumptions hold [43]. Therefore, it is critical to state and appreciate the assumptions underlying any method's validity. Making assumptions explicit allows assessment of whether those assumptions are plausible and, if not, what the effect of violations might be.

Almost all methods developed assume that the distribution of the random variables under study have finite mean and variance [44]. Many methods assume that residuals from some fitted model are normally distributed. This assumption deserves greater scrutiny. It is unclear that microarray or proteomic data are normally distributed even after the near-ubiquitous log transformation. For least squares-based procedures, the central limit theorem [21] guarantees robustness with large samples. However, in HDB, samples are typically very small. Some analyses utilize the enormous numbers of measurements to compensate for the few cases [45], but the extent to which such procedures can compensate with respect to robustness to departures from distributional assumptions is unclear.

An equally important "Gauss–Markov" assumption [43], homoscedasticity (homogeneity of variance) is critical to most least squares-based tests. Although normality is more frequently discussed, homoscedasticity or lack thereof may have greater effect on power and Type 1 error levels. In this regard, it is important to highlight a common misconception about nonparametric statistics. Nonparametric statistics, including permutation and certain forms of bootstrap testing [46] are distribution free. Their validity is not dependent upon any particular data distribution. However, "distribution free" is not "assumption free." Many HDB methodologists seem to utilize nonparametric, particularly permutation or bootstrap, testing as though it eliminates all assumptions and is valid in all circumstances [47]. This is not the case [48,49]. For example, conventional permutation tests of differences in means assume homogeneity of variance and can be invalidated by the presence of outliers [47]. Moreover, as Cohen's [50] apt phrase "nonparametric nonpanacea" reminds us, conducting inference for one's method via permutation, even if this yields correct type 1 error rates, may not be optimal for all purposes. For example, in some microarray testing contexts, permutation tests may yield more stable results than parametric alternatives yet may perform less well than parametric tests for ranking genes by their magnitude of effect [36].

Another assumption of many statistical techniques is that certain elements of the data are independent [51] and violations can markedly invalidate tests. This includes permutation and bootstrap tests unless the dependency is built into the bootstrap or permutation process as some have done [52]. Thus, we should ask whether such dependency is accommodated in methods developed. A popular approach in microarray data is to calculate a test statistic for each gene and then permute the data multiple times, each time recalculating and recording the test statistics, and thereby creating a pseudo-null distribution against which the observed test statistics can be compared for statistical significance. If one only uses the distribution of test statistics within each gene, then, given the typically small samples available, there are insufficient possible permutations and the distribution is coarse and minimally useful [18,53]. Some investigators (e.g., [54]) pool the permutation-based test statistics across all genes to create a pseudo-null distribution with far lesser coarseness. However, this treats all genes as though they are independent, which is not the case [55]. Therefore, p-values derived from such permutations may not be strictly valid.

4.6.4 Proof by Reference

Not every statement about a proposed approach requires *de novo* proof because proof may already be published. For example, Wolfinger et al. [56] propose a particular mixed model approach. They do not need to prove that (under certain conditions), this model is asymptotically valid for frequentist testing, because this has already been shown and Wolfinger need only cite those references. However, it is important to recognize the limits of what has been previously shown and mixed models offer a good example of an acute concern in HDB. Certain mixed model tests are known to be asymptotically valid, but can be highly invalid under some circumstances with a sample as small as 20 per group [57], which is larger than typically used in HDB. Thus, it is important to validate methods with small samples when their validity is based on asymptotic approximations. Methods should be validated across different α levels since some methods have an inflated type 1 error rate at lower level while performing well at higher levels.

4.7 THE PROOF OF THE PUDDING IS IN THE EATING: INDUCTION

In induction, there is no proof that a method has certain properties. Instead we rely on extra-logical information [58–60]. If a method performs in a particular manner across many instances, we assume it will probably do so in the future. We therefore seek to implement methods in situations that can provide feedback about their performance [60]. Simulation and plasmode studies (below) are two such methods.

Many methodologists use simulation to examine methods for HDB (e.g., [53,60]). Because the data are simulated, one knows the right answers and can unequivocally evaluate the correspondence between the underlying "truth" and estimates, conclusions or predictions derived with the method. Moreover, once a simulation is programmed, one can generate and analyze many data sets and, thereby, observe expected performance across many studies. Furthermore, one can manipulate many factors in the experiment (e.g., sample size, measurement reliability, effect magnitude) and observe performance as a function. There are two key challenges to HDB simulation: computational demand and representativeness.

Regarding computational demand, consider that we need to analyze many variables (e.g., genes) and may use permutation tests that necessitate repeating analyses many times per data set. This demand is compounded when we assess method performance across many conditions and wish to work at α levels around 10^{-4} or less, necessitating on the order of 10^6 simulations per condition to accurately estimate (i.e., with 95% confidence to be within 20% of the expected value) type 1 error rates. Simulating at such low α levels is important, because a method based on asymptotic approximations may perform well at higher levels but have inflated type 1 error rates at lower levels. In such situations, even a quick analysis for an individual variable becomes a computational behemoth at the level of the simulation study. Good programming, ever-increasing computational power and advances in simulation methodology (e.g., importance sampling) [61] are, therefore, essential.

The second challenge entails simulating data that reasonably represent actual HDB data, despite limited knowledge about the distribution of individual mRNA or protein levels and the transcriptome- or proteome-wide covariance structure. Consequently, some investigators believe that HDB simulation studies are not worthwhile. This extreme and dismissive skepticism is ill-founded.

First, although we have limited knowledge of the key variables' distributions, this is not unique to HDB [62], and we can learn about such distributions by observing real data. We rarely know unequivocally the distribution of biological variables, yet we are able to develop and evaluate statistical tests for these. One can simulate data from an extraordinarily broad variety of distributions [63]. If tests perform well across this variety, we can be relatively confident of their validity. Moreover, if we identify specific distributions for which our statistical procedures perform poorly, subsequently when using those procedures in practice, we can attempt to ascertain whether the data to which they are applied have such "pathological" distributions.

Regarding correlation among genes, it is easy to simulate a few, even non-normal, correlated variables [64]. In HDB, the challenge is simulating many correlated variables. Using block diagonal correlation matrices [53] oversimplifies the situation.

"Random" correlation matrices [65] are unlikely to reflect reality. Alternatively, one can use real data to identify a correlation structure from which to simulate. This can be done by using the observed expression values and simulating other values (e.g., group assignments, quantitative outcomes) in hypothetical experiments or by generating simulated expression values from a correlation matrix that is based in some way on the observed matrix [66] using factoring procedures. Exactly how to do this remains to be elucidated, but the challenge seems to be surmountable. Investigators are addressing this challenge [67,68], and several microarray data simulators exist ([69–71]; see also gene expression data simulator at http://bioinformatics.upmc.edu/GE2/index.html).

Another challenge in simulation is to make the covariance structure "gridable." We refer to a situation as "gridable" when the theoretically possible space of a parameter set can be divided into a reasonably small set of mutually exclusive and exhaustive adjacent regions. Typically, simulation is used when we are unable to derive a method's properties analytically. Therefore, it is usually desirable to evaluate performance across the plausible range of a key factor. If that factor is the correlation between two variables, one can easily simulate along the possible range $(-1, 1)$ at suitably small adjacent intervals (a grid). With multiple variables under study, the infinite number of possible correlation matrices is not obviously represented by a simple continuum, and it is not obvious how to establish a reasonably sized grid. But if one could extract the important information from a matrix in a few summary metrics, such as some function of eigenvalues, it might be possible to reduce the dimensionality of the problem and make it "gridable." This is an important topic for future research.

A plasmode is a real data set whose true structure is known [26]. As in simulations, the right answer is known *a priori*, allowing the inductive process to proceed. Plasmodes may represent actual experimental data sets better than simulations do. In transcriptomics, the most common type of plasmode is the "spike-in" study. For example, real cases from one population are randomly assigned to two groups and then known quantities of mRNA for specific genes (different known quantities for each group) are added to the mRNA samples. In this situation, the null hypothesis of no differential expression is known to be true for all genes except those that were spiked, and the null hypothesis is known to be false for all those that are spiked. One can then evaluate a method's ability to recover the truth. Spike-in datasets have been used for objective comparison of microarray image analysis systems [72,73].

Plasmode studies have great merit and are being used more widely (e.g., [74,75]). Certain datasets are commonly used and become *de facto* standards for methods evaluation, for example, yeast cell cycle data [76,77]. Although true structure of the data may not be completely known, additional independently obtained information can be used to determine genes known to play important role in yeast cell cycle and thus construct a benchmark for objective methods comparison and verification [35]. In another example, Carpentier et al. [34] proposed to use a bacterial expression data as an objective benchmark that reflects biological reality. In bacteria, a number of genes are organized in operons, or groups of contiguous genes, which are transcribed from the same promoter and thus exhibit the same expression patterns.

However, there are certain limitations. First, to be maximally informative, there is a need for greater plurality. Because the field of statistics deals with random variables, we cannot be certain that the performance we see in one dataset is the same thing we will see in the next. Usually we must be content to make statements about expected or average performance and estimating such expected or average performance well requires multiple realizations. Analysis of a single plasmode is minimally compelling. Because the creation of plasmodes can be expensive and laborious, for any investigator to create many will be difficult. Second, although plasmodes may offer better representations of experimental datasets, there is no guarantee. For example in the typical spike in the study described earlier, it is unclear how many genes should be spiked or what the distribution of spike-created effects should be to reflect some reality.

4.8 COMBINED MODES

One can also combine the approaches above [18]. When two or more modes yield consistent conclusions, confidence is strengthened. One could also creatively combine deduction and induction. For example, suppose there were two alternative inferential tests, *A* and *B*, which could be proven deductively to have the correct type 1 error rate under the circumstances of interest. If one applied the tests to multiple real data sets and consistently found that test *A* rejected more null hypotheses than did test *B*, one could reasonably conclude that test A was more powerful than test *B*. This makes sense only if both tests have correct type 1 error rates.

4.9 WHERE TO FROM HERE?

We offer four suggestions for progress:

(i) *Vigorous solicitation* of rigorous substantiation. Guidelines have been offered or requested for genome scan inference [78], transcriptomic data storage [79], specimen preparation and data collection [80], and result confirmation [81]. We agree that these should remain guidelines and not rules [82]. Such guidelines help evaluate evidential strength of claims. But there are no guidelines for presentation and evaluation of methodological developments [60]. Thus, we offer the guidelines in Text Box 4.2 to be used in evaluating proffered methods.

(ii) *Meta-methods.* For methodologists to strive for high standards of rigor, they must have the tools to do so. An important area for new research is HDB "meta-methodology," methodological research about how to do methodological research. Such second-order methodological research could address how to simulate realistic data and how to meet computational demands. Public plasmode database archives would also be valuable.

(iii) *Qualified claims?* A risk in requesting more rigorous evidential support for new HDB statistical techniques is that if such requests became inflexible demands, progress might be slowed. "Omic" sciences move fast, and investigators need new methodology. We believe it is in the interests of scientific progress and intellectual freedom that compelling methods, though merely conjectured to be useful, can be published. But how might we present such methods to the community of scientific consumers? We might take a lesson from the legislative world. Recently, the U.S. Court of Appeals in *Pearson vs. Shalala* declared certain U.S. Food and Drug Administration (FDA) regulations regarding the conditions under which purveyors of dietary supplements could make certain health claims to be in violation of the First Amendment (free speech) of the U.S. Constitution [83]. The court reasoned that FDAs regulation was inappropriate because it forbid marketers from making health claims unless they had evidence of "significant scientific agreement" (a very high standard of evidence) in support of that claim regardless of whether the marketer provided any truthful qualifying statements to accompany that claim. The court opined that if marketers provided truthful disclosers about the incomplete nature of evidence, they should be permitted to make qualified claims. Similarly, the mathematician Jacob Bernoulli wrote

> In our judgments we must beware lest we attribute to things more than is fitting to attribute, lest we consider something which is more probable than other things to be absolutely certain, and lest we foist this more probable thing upon other people as something absolutely certain. For one must see to it that the credibility which we attribute to things is proportional to the degree of certainty which every single thing possesses and which is less than absolute certainty to the same degree that the probability of the thing is less than 1; as we put it in our vernacular: Man muß ein jedes in seinem Wert und Unwert beruhen lassen [One must let each thing lie in its worth and worthlessness.] [84].

Thus, following *Pearson vs. Shalala*, we see it as reasonable to publish methods where there may be incomplete evidence regarding their properties provided that we also follow Bernoulli and make clear to readers *exactly* what claims we are making for our

---

TEXT BOX 4.2 Suggested Guidelines for Promoting a Sound Epistemological Foundation for New Statistical Methodology in HDB

- State exactly what the method is intended to do or what properties it is intended to have in objectively testable terms
- State the assumptions under which these properties or expected outcomes should occur
- Provide evidence that the method has the claimed performance or properties from simulation studies, analytic proofs and multiple plasmode data analyses
- In the absence of compelling evidence from point 3 above, state clearly that the claimed properties are conjectured and await substantiation
- Where an alternative method already exists, compare the properties of the new method with those of the existing method, or, at minimum, note that an alternative exists, conjecture why the new method may be superior in some situations and suggest future testing of the conjecture

---

proffered methods and the extent to which such claims are supported by simulations, analytic proofs, plasmode analyses, or merely conjecture.

(iv) *Caveat emptor.* Ultimately, we offer the ancient wisdom, "caveat emptor." Statistical methods are, by definition, probabilistic, and in using them, we will err at times. But we should have the opportunity to proceed knowing how error-prone we will be, and we appeal to methodologists to provide that knowledge.

ACKNOWLEDGMENTS

We thank J. Dickson for editorial assistance and S. Vollmer, S. Barnes, and E.K. Allison for discussions. This work was supported by grants T32HL072757 and 1U54CA100949 from the U.S. National Institutes of Health and grants 217651 and 0090286 from the U.S. National Science Foundation.

REFERENCES

1. H.A. Nielson. *Methods of Natural Science*. Englewood Cliffs, NJ: Prentice Hall, Inc., 1967.
2. G.A. Evans. Designer science and the "omic" revolution. *Nat. Biotechnol.* 18: 127, 2000.
3. A.Y. Gracey and A.R. Cossins. Application of microarray technology in environmental and comparative physiology. *Annu. Rev. Physiol.* 65: 231–259, 2003.
4. C.Tilstone. DNA microarrays: vital statistics. *Nature* 424: 610–612, 2003.
5. C. Ambroise and G.J. McLachlan. Selection bias in gene extraction on the basis of microarray gene-expression data. *Proc. Natl Acad. Sci. USA* 99: 6562–6566, 2002.
6. K.A. Baggerly, J.S. Morris, J. Wang, D. Gold, L.C. Xiao and K.R. Coombes. A comprehensive approach to the analysis of matrix-assisted laser desorption/ionization-time of flight proteomics spectra from serum samples. *Proteomics* 3: 1667–1672, 2003.
7. K. Shedden and S. Cooper. Analysis of cell-cycle-specific gene expression in human cells as determined by microarrays and double-thymidine block synchronization. *Proc. Natl Acad. Sci., USA* 99: 4379–4384, 2002.
8. A.F. Famili, G. Liu and Z. Liu. Evaluation and optimization of clustering in gene expression data analysis. *Bioinformatics* 20: 1535–1545, 2004.
9. K. Toda, S. Ishida, K. Nakata, R. Matsuda, Y. Shigemoto-Mogami, K. Fujishita, S. Ozawa, J. Sawada, K. Inoue, K. Shudo and Y. Hayashi. Test of significant differences with *a priori* probability in microarray experiments. *Analyt. Sci.* 19: 1529–1535, 2003.
10. K.E. Muller. Richness for the one-way ANOVA layout. The future of statistical software. In *Proceedings of a forum*. Commission on Physical Sciences, Mathematics, and Applications. Washington, DC: The National Academies Press, 1991, pp. 3–14.
11. M. Oakes. *Statistical Inference*. Chestnut Hill, MA: Epidemiology Resources Inc., 1990.
12. A.W. Edwards. Statistical methods in scientific inference. *Nature* 222: 1233–1237, 1969.
13. D. Yang, S.O. Zakharkin, G.P. Page, J.P. Brand, J.W. Edwards, A.A. Bartolucci and D.B. Allison. Applications of Bayesian statistical methods in microarray data analysis. *Am. J. Pharmacogenomics* 4: 53–62, 2004.

14. G.L. Gadbury, G.P. Page, J. Edwards, T. Kayo, T.A. Prolla, R. Weindruch, P.A. Permana, J. Mountz and D.B. Allison. Power and sample size estimation in high dimensional biology. *Stat. Meth. Med. Res.* 13: 325–338, 2004.
15. R.L. Fernando, D. Nettleton, B.R. Southey, J.C. Dekkers, M.F. Rothschild and M. Soller. Controlling the proportion of false positives in multiple dependent tests. *Genetics* 166: 611–619, 2004.
16. C. Sabatti, S. Service and N. Freimer. False discovery rate in linkage and association genome screens for complex disorders. *Genetics* 164: 829–833, 2003.
17. E.J. van den Oord and P.F. Sullivan. False discoveries and models for gene discovery. *Trends Genet.* 19: 537–542, 2003.
18. M.A. Newton, A. Noueiry, D. Sarkar and P. Ahlquist. Detecting differential gene expression with a semiparametric hierarchical mixture method. *Biostatistics* 5: 155–176, 2004.
19. J. Kowalski, C. Drake, R.H. Schwartz and J. Powell. Non-parametric, hypothesis-based analysis of microarrays for comparison of several phenotypes. *Bioinformatics* 20: 364–373, 2004.
20. D.B. Allison, T.M. Beasley, J. Fernandez, M. Heo, S. Zhu, C. Etzel and C.I. Amos. Bias in estimates of quantitative-trait-locus effect in genome scans: demonstration of the phenomenon and a method-of-moments procedure for reducing bias. *Am. J. Hum. Genet.* 70: 575–585, 2002.
21. R.V. Hogg and A.T. Craig. *Introduction to Mathematical Statistics*. New York: Macmillan Publishing, 1978.
22. U.M. Braga-Neto and E.R. Dougherty. Is cross-validation valid for small-sample microarray classification? *Bioinformatics* 20: 374–380, 2004.
23. B.S. Everitt. Cluster analysis: a brief discussion of some of the problems. *Brit. J. Psychiat.* 120: 143–145, 1972.
24. D. Edwards. Statistical analysis of gene expression microarray data. *Biometrics* 60: 287–289, 2004.
25. M. Smolkin and D. Ghosh. Cluster stability scores for microarray data in cancer studies. *BMC Bioinformatics* 4: 36, 2003.
26. R.K. Blashfield and M.S. Aldenderfer. The methods and problems of cluster analysis. In: J.R. Nesselroade and R.B. Cattell, eds. *Handbook of Multivariate Experimental Psychology*, 2nd ed. New York: Plenum Press, 1988, pp. 447–473.
27. S. Datta and S. Datta. Comparisons and validation of statistical clustering techniques for microarray gene expression data. *Bioinformatics* 19: 459–466, 2003.
28. B. Gorman and K, Zhang. Cluster stability. In: D.B. Allison, T.M. Beasley, J. Edwards, and G.P. Page, eds. *DNA Microarrays and Statistical Genomic Techniques: Design, Analysis, and Interpretation of Experiments*. New York: Marcel Dekker (in press).
29. J.D. Storey and R. Tibshirani. Statistical significance for genome wide studies. *Proc. Natl. Acad. Sci. USA* 100: 9440–9445, 2003.
30. P.J. Huber. *Robust Statistics*. New York: John Wiley & Sons, 1981.
31. R.G. Staudte, S.J. Sheather. *Robust Estimation and Testing*. New York: John Wiley & Sons, 1990.
32. F.R. Hampel, E.M. Ronchetti, P.J. Rousseeuw and W.A. Stahel. *Robust Statistics: The Approach Based on Influence Functions*. New York: John Wiley & Sons, 1986.
33. W. Pan. A comparative review of statistical methods for discovering differentially expressed genes in replicated microarray experiments. *Bioinformatics* 18: 546–554, 2002.

34. A.S. Carpentier, A. Riva, P. Tisseur, G. Didier and A. Henaut. The operons, a criterion to compare the reliability of transcriptome analysis tools: ICA is more reliable than ANOVA, PLS and PCA. *Comput. Biol. Chem.* 28: 3–10, 2004.
35. U. de Lichtenberg, L.J. Jensen, A. Fausboll, T.S. Jensen, P. Bork, and S. Brunak. Comparison of computational methods for the identification of cell cycle regulated genes. *Bioinformatics*, 1: 1164–1117, 2004.
36. R.H. Xu and X.C. Li. A comparison of parametric versus permutation methods with applications to general and temporal microarray gene expression data. *Bioinformatics* 19: 1284–1289, 2003.
37. R.A. Fisher. *The Design of Experiments*. Edinburgh, Scotland: Oliver & Boyd, 1935, p. 49.
38. D. Solow. *How to Read and Do Proofs: An Introduction to Mathematical Thought Processes*. New York: Wiley Text Books, 2001.
39. I. Lakatos. Proofs and refutations: I. *Br. J. Philos. Sci.* 14: 1–25, 1963.
40. M. Stein. *Abe Lincoln's Jokes*. Chicago, IL: Max Stein, 1943.
41. R.B. Nelson. *Proofs without Words: Exercises in Visual Thinking* Washington, DC: The Mathematical Association of America, 1997.
42. Student. The probable error of a mean. *Biometrika* 6: 1–25, 1908.
43. W.D. Berry. A formal presentation of the regression assumptions. In: M.S. Lewis-Beck, ed. *Understanding Regression Assumptions*. Thousand Oaks, CA: Sage University Publications, 1993, pp. 3–11.
44. T.M. Beasley, G.P. Page, J.P.L Brand, G.L. Gadbury, J.D. Mountz, and D.B. Allison. Chebyshev's inequality for non-parametric testing with small N and α in microarray research. *J. R. Stat. Soc. Ser. C Appl. Stat.* 53: 95–108, 2004.
45. P. Baldi and A.D. Long. A Bayesian framework for the analysis of microarray expression data: regularized t-test and statistical inferences of gene changes. *Bioinformatics* 17: 509–519, 2001.
46. F.J. Manly. *Randomization, Bootstrap and Monte Carlo Methods in Biology*. Boca Raton, FL: Chapman & Hall/CRC, 1997.
47. T. Roy. The effect of heteroscedasticity and outliers on the permutation t-test. *J. Stat. Comput. Simul.* 72: 23–26, 2002.
48. P. Hall and S.R. Wilson. Two guidelines for bootstrap hypothesis testing. *Biometrics* 47: 757–762, 1991.
49. J.F. Troendle, E.L. Korn, and L.M. McShane. An example of slow convergence of the bootstrap in high dimensions. *Am. Statistic.* 58: 25–29, 2004.
50. J. Cohen. Some statistical issues in psychological research. In: B.B. Wolman, ed. *Handbook of Clinical Psychology*. New York: McGraw-Hill, 1965.
51. W. Kruskal. Miracles and statistics: the casual assumption of independence. *J. Am. Stat. Assoc.* 83: 929–940, 1988.
52. A. Reiner, D. Yekutieli, and Y. Benjamini. Identifying differentially expressed genes using false discovery rate controlling procedures. *Bioinformatics* 19: 368–375, 2003.
53. G.L. Gadbury, G.P. Page, M. Heo, J.D. Mountz, and D.B. Allison. Randomization tests for small samples: an application for genetic expression data. *J. R. Stat. Soc. Ser. C Appl. Stat.* 52: 365–376, 2003.
54. V.G. Tusher, R. Tibshirani, and G. Chu. Significance analysis of microarrays applied to the ionizing radiation response. *Proc. Natl Acad. Sci. USA* 98: 5116–5121, 2001.
55. M. Kotlyar, S. Fuhrman, A. Ableson, and R. Somogyi. Spearman correlation identifies statistically significant gene expression clusters in spinal cord development and injury. *Neurochem. Res.* 27: 1133–1140, 2002.

56. R.D. Wolfinger, G. Gibson, E.D. Wolfinger, L. Bennett, H. Hamadeh, P. Bushel, C. Afshari, and R.S. Paules. Assessing gene significance from cDNA microarray expression data via mixed models. *J. Comput. Biol.* 8: 625–637, 2001.
57. D.J. Catellier and K.E. Muller. Tests for Gaussian repeated measures with missing data in small samples. *Stat Med* 19: 1101–1114, 2000.
58. *The Basic Writings of Bertrand Russell, 1903–1959*. London: George Allen and Unwin; New York: Simon and Schuster, 1961.
59. A. Ertas, T. Maxwell, V. Rainey, and M.M. Tanik. Transformation of higher education: the transdisciplinary approach in engineering. *IEEE Trans. Education* 46: 289–295, 2003.
60. M.A. Spence, D.A. Greenberg, S.E. Hodge, and V.J. Vieland. The emperor's new methods. *Am. J. Hum. Genet.* 72: 1084–1087, 2003.
61. J.D. Malley, D.Q. Naiman, and J.E. Bailey-Wilson. A comprehensive method for genome scans. *Hum. Hered.* 54: 174–185, 2002.
62. T. Miccerri. The unicorn, the normal curve, and other improbable creatures. *Psychol. Bull.* 105: 156–166, 1989.
63. Z.A. Karian and E.J. Dudewicz. *Fitting Statistical Distributions: The Generalized Lambda Distribution and Generalized Bootstrap Methods*. New York: CRC Press, 2000, pp. 1–38.
64. T.C. Headrick and S.S. Sawilowsky. Simulating correlated multivariate non-normal distributions — extending the Fleishman power method. *Psychometrika* 64: 25–35, 1999.
65. P.I. Davies and N.J. Higham. Numerically stable generation of correlation matrices and their factors. *BIT Num. Math.* 40: 640–651, 2000.
66. V. Cherepinsky, J. Feng, M. Rejali, and B. Mishra. Shrinkage-based similarity metric for cluster analysis of microarray data. *Proc. Natl Acad. Sci., USA* 100: 9668–9673, 2003.
67. Y. Balagurunathan, E.R. Dougherty, Y. Chen, M.L. Bittner, and J.M. Trent. Simulation of cDNA microarrays via a parameterized random signal model. *J. Biomed. Opt.* 7: 507–523, 2002.
68. M. Perez-Enciso, M.A. Toro, M. Tenenhaus, and D. Gianola. Combining gene expression and molecular marker information for mapping complex trait genes: a simulation study. *Genetics* 164: 1597–1606, 2003.
69. P. Mendes, W. Sha, and K. Ye. Artificial gene networks for objective comparison of analysis algorithms. *Bioinformatics* 19: II122–II129, 2003.
70. D.J. Michaud, A.G. Marsh, and P.S. Dhurjati. eXPatGen: generating dynamic expression patterns for the systematic evaluation of analytical methods. *Bioinformatics* 19: 1140–1146, 2003.
71. S. Singhal, C.G. Kyvernitis, S.W. Johnson, L.R. Kaiser, M.N. Liebman, and S.M. Albelda. Microarray data simulator for improved selection of differentially expressed genes. *Cancer Biol. Ther.* 2: 383–391, 2003.
72. E.L. Korn, J.K. Habermann, M.B. Upender, T. Ried, and L.M. McShane. Objective method of comparing DNA microarray image analysis systems. *Biotechniques* 36: 960–967, 2004.
73. D. Rajagopalan. A comparison of statistical methods for analysis of high density oligonucleotide array data. *Bioinformatics* 19: 1469–1476, 2003.
74. J.B. Aimone and F.H. Gage. Unbiased characterization of high-density oligonucleotide microarrays using probe-level statistics. *J. Neurosci. Meth.* 135: 27–33, 2004.

75. R.A. Irizarry, B.M. Bolstad, F. Collin, L.M. Cope, B. Hobbs and T.P. Speed. Summaries of Affymetrix GeneChip probe level data. *Nucl. Acids. Res.* 31: e15, 2003.
76. R.J. Cho, M.J. Campbell, E.A. Winzeler, L. Steinmetz, A Conway, L. Wodicka, T.G. Wolfsberg, A.E. Gabrielian, D. Landsman, D.J. Lockhart, and R.W. Davis. A genome-wide transcriptional analysis of the mitotic cell cycle. *Mol. Cell.* 2: 65–73, 1998.
77. P.T. Spellman, G. Sherlock, M.Q. Zhang, V.R. Iyer, K. Anders, M.B. Eisen, P.O. Brown, D Botstein, and B. Futcher. Comprehensive identification of cell cycle-regulated genes of the yeast *Saccharomyces cerevisiae* by microarray hybridization. *Mol. Biol. Cell.* 9: 3273–3297, 1998.
78. E. Lander and L. Kruglyak. Genetic dissection of complex traits — guidelines for interpreting and reporting linkage results. *Nat. Genet.* 11: 241–247, 1995.
79. A. Brazma, P. Hingamp, J. Quackenbush, G. Sherlock, P. Spellman, C. Stoeckert, J. Aach, W. Ansorge, C.A. Ball, H.C. Causton, T. Gaasterland, P. Glenisson, F.C. Holstege, I.F. Kim, V. Markowitz, J.C. Matese, H. Parkinson, A. Robinson, U Sarkans, S. Schulze-Kremer, J. Stewart, R. Taylor, J. Vilo, and M. Vingron. Minimum information about a microarray experiment (MIAME) — toward standards for microarray data. *Nat. Genet.* 29: 365–371, 2001.
80. V. Benes, M. Muckenthaler. Standardization of protocols in cDNA microarray analysis. *Trends. Biochem. Sci.* 28: 244–249, 2003.
81. J.C. Rockett, G.M. Hellmann. Confirming microarray data — is it really necessary? *Genomics* 83: 541–549, 2004.
82. J.S. Witte, R.C. Elston, and N.J. Schork. Genetic dissection of complex traits. *Nat. Genet.* 12: 355–356, 1996.
83. J.W. Emord. Pearson versus Shalala: the beginning of the end for FDA speech suppression. *J. Public Policy Mark* 19: 139–143, 2000.
84. J. Bernoulli. Ars Conjectandi. 1713 [from the translation by Bing Sung, Department of Statistics, Harvard University: http://cerebro.xu.edu/math/Sources/JakobBernoulli/ars_sung/ars_sung.html].

5 The Role of Sample Size on Measures of Uncertainty and Power

Gary L. Gadbury, Qinfang Xiang, Jode W. Edwards, Grier P. Page, and David B. Allison

CONTENTS

5.1 Introduction 77
5.2 TP, TN, and EDR in Microarray Experiments 79
5.3 Sample Size and Sources of Uncertainty in Microarray Studies 80
5.4 On the Distribution of p-Values 84
5.5 A Mixture Model for the Distribution of p-Values 85
5.6 Planning Future Experiments: The Role of Sample Size on TP, TN, and EDR 88
5.7 Sample Size and Threshold Selection: Illustrating the Procedure 90
5.8 Discussion 90
Acknowledgments 92
References 92

5.1 INTRODUCTION

Since an initial focus on "fold change" (cf., Reference 1), researchers have recognized the need to quantify statistical significance of estimated differences in genetic expression between two or more treatment groups using microarrays [2]. In light of this, replication has been recognized as a critical element of the design of microarray studies [3]. Replication may imply spotting a single gene multiple times on one array [4] or multiple tissue samples that each have their own array (e.g., [5,6]). We consider the latter and use the term "sample size" to refer to number of arrays in a study. For another discussion of levels of replication see Simon and Dobbin [7].

Replicate arrays in an experiment provide for statistical tests of differential expression at the level of a specific gene [8–10]. Regardless of the test used, the result is often a measure of "certainty" (e.g., a p-value) in rejecting a null hypothesis of no differential expression. However, in microarray studies there are questions that remain to be answered and new ones that emerge. Examples are as follows:

1. At what threshold is a p-value statistically significant?
2. Of those genes declared differentially expressed, what proportion are truly differentially expressed?
3. Of those not so declared, what proportion are *not* differentially expressed?
4. Of the genes that are truly differentially expressed, what proportion do we expect to detect in a particular study?
5. What role does sample size play in all of these questions?

Obtaining answers to these questions can be a challenging task due to small sample sizes that are typical in many microarray experiments. If testing for a difference in mean genetic expression at a specific gene using a parametric test, the validity of a p-value may be questioned when sample sizes are small. Nonparametric tests also have difficulties since randomization distributions or bootstrapped distributions can be course with small samples, and p-values cannot attain small enough values to be "statistically significant."

This chapter presents techniques that facilitate answers to questions 1 to 5. We consider quantities of interest in a microarray study shown in Table 5.1. Two of these quantities are the expected number of genes that are (1) differentially expressed and will be detected as "significant" at a particular threshold and (2) not differentially expressed and will not be detected as such, denoted D and A, respectively. The other two quantities are the expected number of genes that are differentially expressed but are not so declared (B) and are not differentially expressed but are so declared (C). Proportions based on these quantities, defined in Gadbury et al. [11] are:

$$\mathrm{TP} = \frac{D}{C+D}, \quad \mathrm{TN} = \frac{A}{A+B}, \quad \mathrm{EDR} = \frac{D}{B+D} \tag{5.1}$$

where each is defined as zero if its denominator is zero. TP is true positive; TN, true negative; and EDR, the expected discovery rate, which is the expected proportion of genes that will be declared significant at a particular threshold among all genes that

TABLE 5.1
Quantities of Interest in Microarray Experiments

	Genes for which there is no real effect	**Genes for which there is a real effect**
Genes not declared significant at designated threshold	A	B
Genes declared significant at designated threshold	C	D

Note: $A + B + C + D =$ the number of genes analyzed in a microarray experiment.

are truly differentially expressed. EDR sounds like but is not identical to the notion of power. It is an "expected proportion" and there may be no gene — specific test with a power identical to the EDR. Moreover, TP and TN are expected proportions and there may be no specific gene that has a "true positive probability" (or true negative) equal to TP (or TN). The focus of this chapter is to discuss techniques to estimate TP, TN, and EDR from microarrray experiments, and to present methods that evaluate the role of sample size in bringing these proportions to desired levels. Traditional power calculations for planning future sample sizes to detect gene specific effects are notably problematic with microarray data since information about variances and meaningful effect sizes are typically absent [12].

5.2 TP, TN, AND EDR IN MICROARRAY EXPERIMENTS

In this chapter it is the interplay between threshold, sample size, and the proportions in Equation 5.1 that is of interest. An example is shown in Figure 5.1. The figure is based on an "experimental situation" (details discussed later) with two treatment groups of equal size, $n =$ integers 2 to 10, 20, 40 and a threshold at which a gene is declared differentially expressed equal to 0.1, 0.05, 0.01, 0.001, 0.0001, and 0.00001 shown on a logarithm (base 10) scale. The figure shows that as sample size increases, so does EDR. However, EDR is smaller for smaller thresholds since the criteria for declaring a gene differentially expressed are more strict (this assumes that there are indeed genes that are truly differentially expressed and it is our ability to detect them

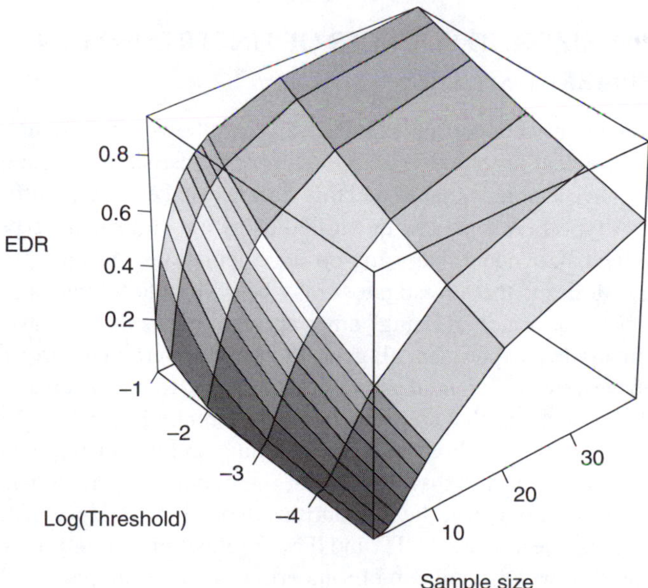

FIGURE 5.1 Three-dimensional plot showing an expected discovery rate (EDR) for varying sample sizes and threshold (on logarithm base 10 scale).

as such that is in question). Using larger thresholds, (and thus increasing EDR) comes at the cost of reducing TP (not shown in Figure 5.1).

The five questions in the introduction and the quantities in Equation 5.1 have interested others and have led to a body of literature on this topic. Much of the literature in this area names three quantities of interest that are related to those in Equation 5.1. First is the false discovery rate (FDR) that is analogous to $1 - \text{TP}$, a false negative rate (i.e., $1 - \text{TN}$), and what some have called "power," analogous to EDR (e.g., [13]), though some have called "power" to be one minus the probability of a false negative, interpreting "false negative" as a type II error [14]. Motivating the need for these results can be illustrated by considering Question 1 from the introduction. The answer may seem straightforward enough were it not for the many thousands of simultaneous tests that are conducted in a microarray experiment. Some of the traditional corrections for multiple comparisons such as the Bonferroni technique were not developed for this context and are far too conservative, particularly when results from an initial study might be used to plan follow-up studies, that is, too small a threshold may "miss" many interesting genes worthy of further attention. Less conservative methods have been developed that, rather than controlling for an experiment-wise error, controls instead the expected proportion of falsely rejected null hypotheses [15–17].

The approach to estimating the above described quantities from microarray data has generally fallen into two (at least) areas: permutation based methods, and model based methods involving Bayesian posterior probabilities. All have generally recognized the importance of sample size in bringing these quantities to desired levels.

5.3 SAMPLE SIZE AND SOURCES OF UNCERTAINTY IN MICROARRAY STUDIES

It is well known that when testing a single null hypothesis, small sample sizes have lower power to detect an effect vs. larger samples. Test statistics computed using small samples have a larger variance and this variance makes it more difficult to "see" a true effect. A type I error may be quantified using a p-value and a type II error, at a particular effect size, quantified using power calculations. An emerging paradigm in microarray studies is that investigators may be willing to tolerate a proportion of type I errors in favor of not "missing" any important genes that do have differential expression. In an ideal experiment all genes declared differentially expressed would be ones that have a true differential expression due to the treatment condition, $\text{TP} = 1$ or $\text{FDR} = 0$. Those genes that are ruled out as differentially expressed are also correctly determined, $\text{TN} = 1$. Finally, when planning and conducting an experiment, one would hope to "expect to discover" all of the important differentially expressed genes (high EDR). In practice, these quantities depend on variance of test statistics and some have used estimates of TP and TN to reflect uncertainty in a microarray experiment (e.g., low TP implies more uncertainty). The sample size has a direct effect on measures of TP, TN, and EDR in a microarray experiment. Figure 5.2 shows a diagram depicting a hypothetical two-dye microarray experiment. Only two arrays are shown (due to space restrictions) in each of two treatment groups. Issues related to

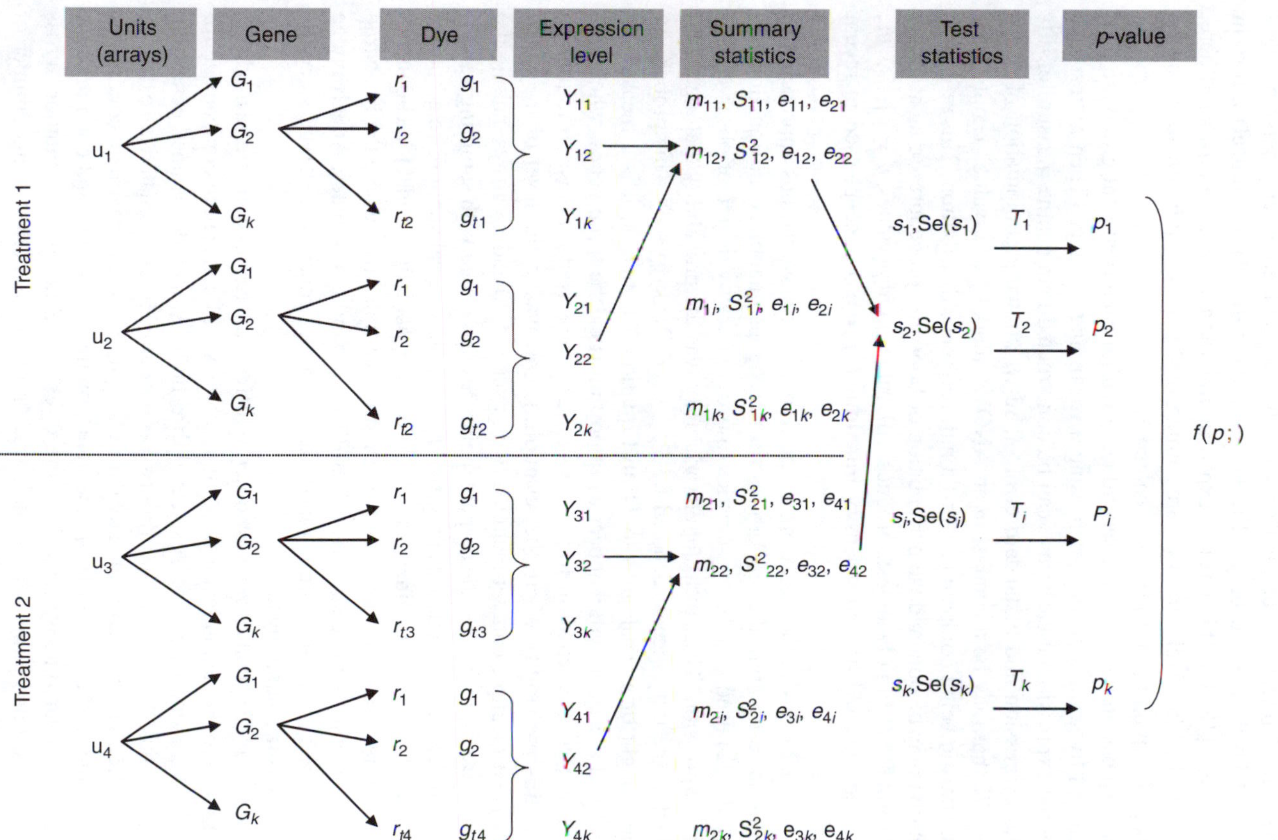

FIGURE 5.2 Diagram of a hypothetical 2-dye microarray experiment with four arrays divided into two treatment groups of two arrays each.

background correction and normalization are not considered. On each array, k genes are spotted. Figure 5.2 expands on the second gene, G_2, on each array. The r_i and g_i denote pixel intensities at each spot on the red and green channels, respectively. Pixel level data can be used to evaluate measurement (technical) errors in an experiment [18] but this level of experimental variance is not considered here. We assume that the pixel intensities are summarized into an expression level for the ith experimental unit (array) for G_2, denoted Y_{i2}.

The expression levels for the second gene within each treatment group are then summarized by some statistic, often using a mean and variance. Then a "test" for differential expression at each gene can be constructed using some measure of differential expression and a standard error, $\delta, \mathrm{Se}_{(\delta)}$, allowing computation of a test statistic, T, that may be a contrast in an ANOVA model. A "p-value" results from the test statistic being compared against a null reference distribution. This reference distribution may be obtained via permutation or bootstrap procedures or an assumed model form for T. Finally all tests produce a distribution of p-values, $p = p_1, \ldots, p_k$, that can be modeled by some density function, $f(p;\theta)$, a concept used in recent papers [19–21].

Figure 5.2 helps in the following discussion that reviews various approaches to computing or estimating the proportions (or analogous quantities) in Equation 5.1. Pepe et al. [22] used rank based statistics to order genes by degree of observed differential expression. Their measure of variability was the probability that a gene, g, is ranked in the top c genes, denoted by $P_g(c) = P[\mathrm{Rank}(g) \leq c]$. This probability may be thought of as analogous to TP in Equation 5.1, that is, it is a measure of "certainty" that a gene will reappear as important (i.e., the top c genes, where c is selected by the researcher) in follow-up studies. They used a bootstrap routine to estimate this probability where the resampling unit was at the level of the tissue (array). Their example data set included two groups of 30 and 23 arrays making the bootstrap a useful technique to compute quantities associated with sampling variability. A simulation technique was proposed for sample size calculations (e.g., power computations) where the original data set served as a population model for a bootstrap routine. The technique has limitations with small sample sizes where the bootstrap distribution would be too coarse to be useful for probability estimates. An advantage of the approach is that the correlation structure among genes is preserved since the entire array is the sampling unit.

Tusher et al. [23] ranked genes by order of differential expression using a modified t-statistic where the denominator was inflated by a small constant to compensate for genes with very small variance. A selected threshold, Δ, then determined genes that were differentially expressed. Uncertainty was measured by an estimated false discovery rate (FDR). This was computed by permuting arrays across treatment conditions, computing the "t-statistic," ordering the genes by these statistics, and counting how many genes appeared above (or below) the threshold. Since the permutation across treatment conditions mimics the situation of "no treatment effect," this number of genes was an estimate of a number falsely declared. This number was recorded for all permutations and then averaged across permutations. This average divided by the number originally declared differentially expressed is an estimated FDR, analogous to $1 - \mathrm{TP}$ in Equation 5.1.

Kerr et al. [24] used an ANOVA model that may include array, dye, treatment, and gene effects to model the gene expression values, for example, Y_{ij} in Figure 5.2. They used a residual bootstrap technique [25] to simulate a reference distribution for an F-statistic and to obtain standard errors of contrasts, but they did not directly deal with the proportions in Equation 5.1. Wolfinger et al. [14] employed mixed models and some parametric assumptions to model gene expression measurements, and they did suggest a method for power analysis. The method involved specification of an exemplary data set, variance components, fitting the model to the exemplary data holding variance components at their specified values, computing standard errors of specified contrasts, and computing power using a noncentral t-distribution and a specified false positive rate.

Efron and Tibshirani [26] modeled the test statistics arising from a Wilcoxon test. The model was of the form, $f_T(t) = p_0 f_0(t) + p_1 f_1(t)$ where $f_i(t)$ is the distribution of the test statistics under the null hypothesis ($i = 0$ meaning no differential expression), or under the alternative ($i = 1$ meaning there is differential expression). The $p_i, i = 0, 1$, are prior probabilities of no differential expression or differential expression, respectively. A use of Bayes theorem resulted in posterior probabilities such as the probability that a gene is differentially expressed given the test statistic, analogous to TP in Equation 5.1. They fitted a model using empirical Bayes methods. They also provided comparisons between their method and that of Benjamini and Hochberg's false discovery rate [15].

Lee and Whitmore [13] considered a table like Table 5.1, and investigated sample size requirements on types I and II error probabilities. "Power" was equal to $1 - P$ (type II error), which is analogous to our quantity EDR in Equation 5.1. They defined FDR as an expected proportion of falsely rejected null hypothesis, analogous to $1 -$ TP. Their defined type I error was $C/(A + C)$, where A and C are cell entries in Table 5.1. They also presented a Bayesian perspective on power and sample size using mixture models fitted to summary statistics of differential expression, where the prior probability, p_1, was an "anticipated" proportion of truly differentially expressed genes. Their focus, however, was on evaluating required sample size and power for linear summaries of differential expression involving computation of a null variance, effect size, and specification of an expected number of false positives. They noted, in particular, that specification of a null variance is problematic since it requires knowledge of the inherent variability of the data in the planned study. They also extended their results to situations where there may be more than two treatment groups where interest is in determining differential expression among several treatment groups.

Pan et al. [6] used a t-type statistic to quantify differential expression. However, the threshold to declare significance was obtained by creating a reference distribution that was a mixture of normal distributions. This model, when fitted to a "pilot" data set, could then be used to assess the number of replicates required to achieve desired power at a given significance level. The fitted model was considered fixed, a type I error was specified, and power computed for any specified effect size, for example, standardized difference in mean expression levels between two groups.

Zein et al. [5] considered sample size effects on pairwise comparisons of different groups and discussed the role of both technical and biological variability. Actual data sets were used to develop parameter specifications for simulated data sets. They used the term sensitivity as analogous to EDR in Equation 5.1, and specificity that is analogous to TP. They evaluated the effect of varying sample size on these two quantities for various simulated data sets and using different types of statistical tests for differential expression, for example, t-tests and a rank-based test.

Many other contributions have been made where quantities related to those in Equation 5.1 were computed to reflect uncertainty in conclusions, or to assess power and the role of sample size in ensuring power rises to acceptable levels. A slightly different approach was taken by Pavlidis et al. [27] who used several real data sets to assess the effect of replication on microarray experiments. Mukherjee et al. [28] estimated sample size requirements for a classification methodology. Van der Laan and Bryan [29] developed a technique that incorporates sampling variability into a cluster type analysis and provide results for sensitivity, proportion of false positives, and a sample size formula.

Not discussed thus far regarding Figure 5.2 are some results based on the distribution of p-values resulting from statistical tests on all genes. Results based on this distribution are valid assuming that a valid test was used to produce a p-value. A distribution of p-values can be modeled and this distribution can provide estimates of the proportions in Equation 5.1 as well as shed light on the answers to questions 1 to 5, posed in the introduction. We now expand on this idea and indicate that further details are available in Allison et al. [19], Gadbury et al. [20], and Gadbury et al. [11]. First, however, we highlight some history regarding the use of p-values as random variables.

5.4 ON THE DISTRIBUTION OF p-VALUES

An often overlooked characteristic of a p-value is that, since it is computed from the sample, it too is a random variable [30]. The earliest work on the stochastic properties of a p-value may have been by Dempster and Shatzoff [31]. Other work has subsequently appeared in Schervish [32] and Donahue [33]. A key result related to our work here is the well-known probability integral transform that states that a cumulative distribution function evaluated at a random variable is a uniform random variable. Applied to p-values, this states that a test statistic, under a null hypothesis of no differential expression will produce a uniformly distributed p-value on the interval (0,1) as long as the distribution of the test statistic is known. We will refer to this latter condition as using a test that produces a "valid p-value."

Schweder and Spjøtvoll [34] may have been the first to consider this in the context of multiple testing. They produced the "p-value plot" as a means to (visually) quantify the number (proportion) of false null hypotheses. This used the idea that if several null hypotheses were not true, then there should be a larger number of "small" p-values than would have been expected if all null hypotheses were true. Hung et al. [35] derived the exact distribution of p-values under the alternative hypothesis and under various distributional assumptions for the data. They showed that, for their specific cases, these distributions depended on the effect size (under the alternative) and sample size. Parker and Rothenberg [36] suggested modeling a distribution of

p-values using a mixture of a uniform distribution and one or more beta distributions, the beta distribution being chosen for its flexibility in modeling shapes on the unit interval.

Allison et al. [19] adopted this idea from Parker and Rothenberg [36] and developed a method for modeling the distribution of p-values from microarray experiments. An idea later echoed in Pounds and Morris [21], they used the model to estimate proportions in Equation 5.1. Yang et al. [37] also noted how a distribution (histogram) of p-values will have a "peak" near zero when many null hypotheses are not true, but they did not directly model this distribution. Next, we review the method from Allison et al. [19] and discuss a procedure by which sample size effects can be assessed based on the method. Details of the latter are in Gadbury et al. [11].

5.5 A MIXTURE MODEL FOR THE DISTRIBUTION OF *p*-VALUES

The following example is used to illustrate the mixture model method of Allison et al. [19]. Human rheumatoid arthritis synovial fibroblast cell line samples were stimulated with tumor necrosis factor-α where one group ($n = 3$) had the Nf-κB pathway taken out by a dominant negative transiently transfected vector and the other group ($n = 3$) had a control vector added.

Figure 5.3 shows a histogram of p-values obtained from two sample t-tests of a null hypothesis $H_0 : \mu_{1j} = \mu_{2j}$ vs. a two tailed alternative where μ_{ij} is a population mean expression for gene j in treatment group i and where $j = 1, \ldots, k = 12{,}625$ genes. A p-value from any valid statistical test can be used such as a test of a contrast from an ANOVA model.

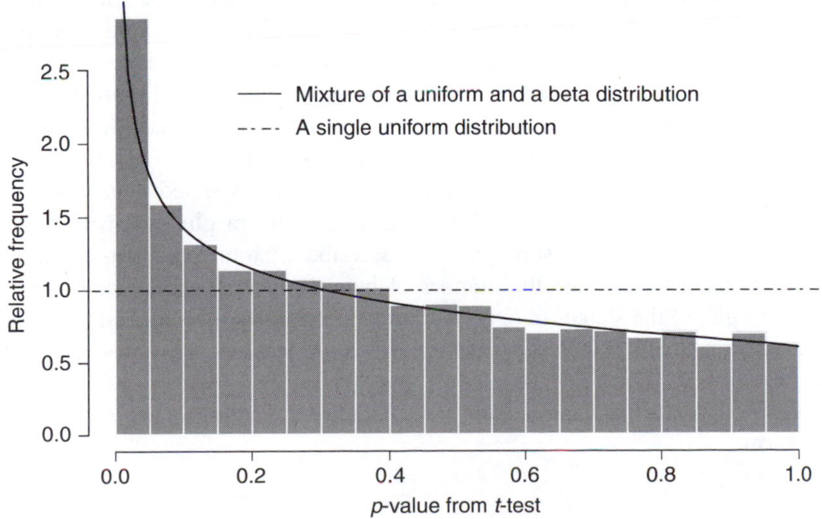

FIGURE 5.3 Distribution of continuous p-values obtained from two tailed t-tests on 12,625 genes from the rheumatoid arthritis data set, and the fitted mixture model is shown by the curve.

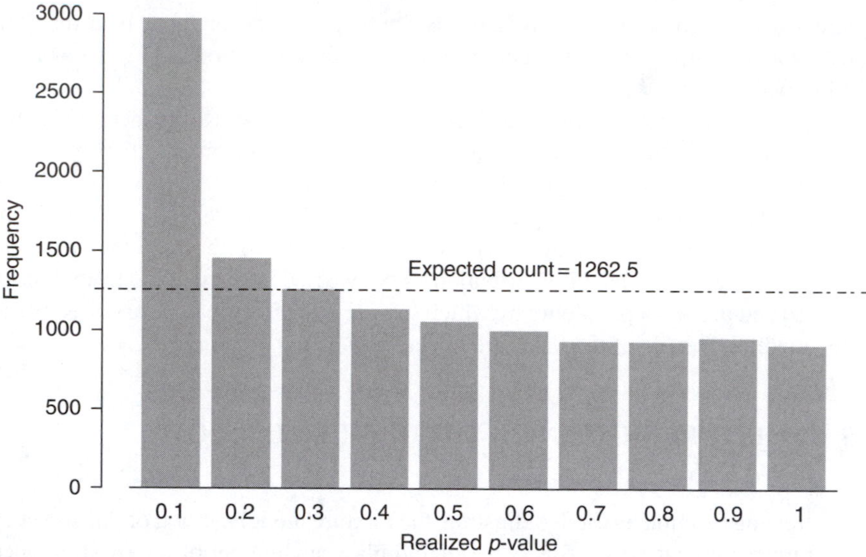

FIGURE 5.4 Distribution of discrete p-values obtained from two tailed randomization tests on 12,625 genes from the rheumatoid arthritis data set.

Gadbury et al. [20] described a procedure based on randomization tests for a difference in expression at each gene. The result is a discrete distribution of p-values with only 10 distinct values. The distribution shown in Figure 5.4 shows a similar shape as the continuous distribution if Figure 5.3. The number of p-values equal to 0.1 was 2975 vs. an expected 1262.5 if there were no differences in genetic expression for any genes. As a comparison, the number of genes with p-values obtained from the t-tests that were less than or equal to 0.1 was 2797, in close agreement to the randomization test results. Thus the randomization test can served as a "sensitivity check" for other testing procedures. The discrete distribution in Figure 5.4 can be modeled using a multinomial distribution but the information that can be gleaned from it is less rich than that available in a continuous model.

Returning to Figure 5.3, many more p-values than expected under the global null hypothesis cluster near zero. A mixture of a uniform plus one beta distribution captures this shape. Allison et al. [19] describe a parametric bootstrap method that can be used to estimate the number of beta distribution components that are required to model the shape. In many of the data examples they studied, they found that a uniform and one beta distribution was sufficient, as was the case for this example.

The fitted mixture model in Figure 5.3 is the solid line, represented by an equation of the form

$$f^*(p) = \prod_{i=1}^{k} [\lambda_0 + \lambda_1 \beta(p_i; r, s)], \quad p_i \in (0, 1), \ \ i = 1, \ldots, k = 12{,}625 \qquad (5.2)$$

where $\beta(p; r, s)$ is a beta distribution with shape parameters r and s, and $\lambda_1 = 1 - \lambda_0$. Parameters of the model were estimated using maximum likelihood. For these data, λ_1 is estimated as 0.395, suggesting that 39.5% of the genes are differentially expressed — an unusually strong signal and not representative of the many microarray data sets we have analyzed. The estimates for r and s are 0.539 and 1.844, respectively. The p-values were obtained from t-tests using pooled degrees of freedom. Using a Welsh correction on the t-test, for comparison, the estimated λ_1, r, s are 0.384, 0.686, and 1.944, respectively. The value of the maximum log-likelihood was 1484 vs. zero, which would be the value for a strictly uniform distribution (i.e., a distribution with no signal). The log-likelihood thus serves as a measure of relative model fit.

Calling the expression in Equation 5.2 a likelihood presupposes independence of p-values, but Allison et al. [19] also used simulations to examine the extent to which the log-likelihood would be affected by moderate dependence among genes. Gadbury et al. [20] did the same for the resulting distribution of discrete p-values obtained from randomization tests. It is challenging to model dependence among gene expression levels due to the small sample sizes and very large number of genes. Allison et al. [19] and Gadbury et al. [20] used a multivariate normal distribution with a mean and variance structure similar to the data and the correlation structure implemented through a block diagonal equicorrelation matrix with a parameter ρ occupying the off-diagonal entries of the blocks. The expressions levels were assumed independent across different blocks. The correlation ρ varied from 0.0 to 0.8. Moderate correlation was considered to be values of ρ around 0.4 with stronger dependence at 0.8. Negative correlation was not feasible with this approach since the correlation matrix was not positive definite for large negative values of ρ. In simulations where no genes were differentially expressed, the variance of the sampling distribution of the maximum log-likelihood increased with ρ. This suggests that for data fitted with the mixture model where the value of the log-likelihood is not large, some caution must be exercised since the value could be attributed to some genes being differentially expressed, or no genes differentially expressed but correlated instead. The effects of correlated expression levels on results from statistical methods for microarray data have been given limited attention in the literature and is a subject of continuing investigation.

Immediately available from the fitted mixture model, Equation 5.2, are maximum likelihood estimates of TP, TN, and EDR. Suppose a threshold τ is selected that determines genes that are declared differentially expressed (p-value $\leq \tau$) or not (p-value $> \tau$). Then at this threshold,

$$\widehat{\text{TP}} = \frac{\hat{\lambda}_1 B(\tau; \hat{r}, \hat{s})}{\hat{\lambda}_0 \tau + \hat{\lambda}_1 B(\tau; \hat{r}, \hat{s})}, \quad \widehat{\text{TN}} = \frac{\hat{\lambda}_0 (1 - \tau)}{\hat{\lambda}_0 (1 - \tau) + \hat{\lambda}_1 [1 - B(\tau; \hat{r}, \hat{s})]}, \quad \widehat{\text{EDR}} = B(\tau; \hat{r}, \hat{s}) \tag{5.3}$$

where $B(\tau; \hat{r}, \hat{s})$ is the cumulative distribution function of a beta distribution with estimated shape parameters, evaluated at τ. Choosing $\tau = 0.05$ resulted in $\widehat{\text{TP}} = 0.790$ and $\widehat{\text{TN}} = 0.671$, and $\widehat{\text{EDR}} = 0.287$. The high estimate of TP results from the strong signal present in this cell line dataset. The low estimate of EDR is attributed to small sample size (3 per group). The bootstrap can be used to estimate standard errors and confidence intervals for TP, TN, and EDR. Sampling variability, in this context, is variability associated with a realization of k p-values from a model of the form given by Equation 5.2. The p-values from the t-tests were resampled 1000 times, each time fitting a mixture model to the bootstrap sample. An approximate 95% confidence interval for TP is (0.768, 0.812), for TN it is (0.608, 0.734), and for EDR it is (0.254, 0.320). As a comparison, the estimate of TP using the method of Tusher et al. [23] on these data was 0.77 using a threshold of $\Delta = 1.2$. In the next section we review a method from Gadbury et al. [11] that evaluates, for a given model fitted to actual data, the effect of threshold selection and sample size on estimated TP, TN, and EDR.

5.6 PLANNING FUTURE EXPERIMENTS: THE ROLE OF SAMPLE SIZE ON TP, TN, AND EDR

Gadbury et al. [11] used a computational procedure to consider the effect of threshold and sample size on TP, TN, and EDR. The procedure assumes that an experiment has been conducted with $N = 2n$ units divided into two groups of equal size and a mixture model $f^*(p)$ (Equation 5.2) fitted to the distribution of p-values obtained from a t-test of differential expression on each gene. A p-value from any valid test could be used as long as it can be back-transformed to the test statistic that produced it. Equal sample sizes in each group is convenient but is not required. The model is fitted using maximum likelihood and the estimated parameters are now considered fixed and equal to the true values, conceptually similar to Pan et al. [6].

A random sample $p^* = p_1^*, \ldots, p_k^*$ is generated from the mixture model $f^*(p)$, with the parameters estimated from the preliminary sample. The outcome of a Bernoulli trial first determines whether a p_i^* is generated from the uniform component with probability λ_0, or the beta distribution component with probability $\lambda_1 = 1 - \lambda_0$. From this sample of p-values, a set of adjusted p-values, $p^{**} = p_1^{**}, \ldots, p_k^{**}$, is created by transforming the p_i^* that were generated from the beta distribution to the corresponding t-statistic t_i^* and computing a new p-value, p_i^{**}, using a new sample size, n^*. The p_i^* generated from the uniform distribution are left unchanged.

From the new p^{**}, estimates of TP, TN, and EDR can be computed. To illustrate this, let $Z = \{1, 2, \ldots, k\}$ be a set of indices corresponding to the genes in the study, and let T be a subset of Z representing the set of genes that have a true differential expression across two experimental groups, that is, $T \subseteq Z$. Let

$$I_{\{T\}}(i) = \begin{cases} 1, & i \in T \\ 0, & i \notin T \end{cases} \quad \text{for } i = 1, \ldots, k$$

then $\sum_{i=1}^{k} I_{\{T\}}(i)$ represents the number of genes under study that are truly differentially expressed, unknown in practice but known and calculable in computer simulations.

A gene is declared to be differentially expressed if the p-value (calculated on observed data) from a statistical test falls below a predetermined threshold (τ). The resulting decision function, when equal to 1, declares a gene differentially expressed:

$$\psi_i(x_i) = \begin{cases} 1, & p_i \leq \tau \\ 0, & p_i > \tau \end{cases}$$

where x_i is a vector of length N representing the data for the ith gene, $i = 1, \ldots, k$, hereafter abbreviated as ψ_i.

Estimates for the values in Table 5.1 that can be calculated in computer simulation experiments are given by

$$\begin{aligned} \hat{A} &= \sum_{i=1}^{k}(1-\psi_i)[1 - I_{\{T\}}(i)], & \hat{B} &= \sum_{i=1}^{k}(1-\psi_i)I_{\{T\}}(i) \\ \hat{C} &= \sum_{i=1}^{k}\psi_i[1 - I_{\{T\}}(i)], & \hat{D} &= \sum_{i=1}^{k}\psi_i I_{\{T\}}(i) \end{aligned} \tag{5.4}$$

The values A, B, C, and D in Table 5.1 are defined using the expectations of the estimates in Equations 5.4, which are taken with respect to the fitted mixture model. These are,

$$\begin{aligned} &E(\hat{A}) = A = k\lambda_0(1-\tau), \quad E(\hat{B}) = B = k\lambda_1(1 - \mathrm{B}(\tau; r, s)) \\ &E(\hat{C}) = C = k\lambda_0\tau, E(\hat{D}) = D = k\lambda_1 B(\tau; r, s) \end{aligned}$$

It can be seen that

$$\mathrm{TP} = \frac{D}{C+D}, \quad TN = \frac{A}{A+B}, \quad \mathrm{EDR} = \frac{D}{B+D}$$

defined in Equation 5.1, have the same form as $\widehat{\mathrm{TP}}$, $\widehat{\mathrm{TN}}$, and $\widehat{\mathrm{EDR}}$ in Equation 5.3, if estimated model parameters are taken as fixed. In simulations, Gadbury et al. [11] proposed estimating TP, TN, and EDR using

$$\widehat{\mathrm{TP}} = \frac{\hat{D}}{\hat{C}+\hat{D}}, \quad \widehat{\mathrm{TN}} = \frac{\hat{A}}{\hat{A}+\hat{B}}, \quad \widehat{\mathrm{EDR}} = \frac{\hat{D}}{\hat{B}+\hat{D}} \tag{5.5}$$

where the quantities in Equation 5.4 are readily available in the simulations.

The process is repeated M times thus obtaining M values of $\widehat{\mathrm{TP}}$, $\widehat{\mathrm{TN}}$, and $\widehat{\mathrm{EDR}}$ given in Equation 5.5. The value of M is chosen sufficiently large so that Monte Carlo estimates of $E[\widehat{\mathrm{TP}}]$, $E[\widehat{\mathrm{TN}}]$, and $E[\widehat{\mathrm{EDR}}]$ can be accurately estimated using the average over the M values of $\widehat{\mathrm{TP}}$, $\widehat{\mathrm{TN}}$, and $\widehat{\mathrm{EDR}}$. This expectation is with respect to the simulation process. Since the model has been fixed, sampling variability in the estimates in Equation 5.5 is due to simulation uncertainty rather than model uncertainty. This is analogous to traditional power calculations where a desired effect size is fixed, a sample size is fixed, and then power computed at that effect size and sample size using some statistical model or distribution. Thus, standard errors in the averages of the M values of $\widehat{\mathrm{TP}}$, $\widehat{\mathrm{TN}}$, and $\widehat{\mathrm{EDR}}$ are generally small. The above described process is repeated for different values of n^* and τ.

5.7 SAMPLE SIZE AND THRESHOLD SELECTION: ILLUSTRATING THE PROCEDURE

Effects of sample size and threshold selection are evaluated using the above procedure on the rheumatoid arthritis cell line data set described earlier. Results are shown in Figure 5.5. The graph (a) in Figure 5.5 shows the minimum and maximum number (from $M = 100$ simulations) of 12,625 genes that were determined to be differentially expressed at three chosen thresholds for different sample sizes. The graph labeled (b) plots the average of 100 $\widehat{\mathrm{TP}}$ values for the three thresholds at each sample size. Graphs (c) and (d) show the average of the 100 $\widehat{\mathrm{TN}}$ and 100 $\widehat{\mathrm{EDR}}$ values, respectively.

The number declared significant (graph a) was plotted since it reveals key information about TP. At very small sample sizes and thresholds, very few (and sometimes zero) genes are declared significant. This quantity estimates $C + D$ in Table 5.1, the denominator of TP. TP is defined to be zero when $C + D$ is zero; estimates, $\widehat{\mathrm{TP}}$, are not expected to be very accurate when $\widehat{\mathrm{C}} + \widehat{\mathrm{D}}$ is a small positive number. This effect is seen in the plots for TP at small sample sizes. The TP plot also shows that the lines representing different thresholds cross over each other. Values of TP will be higher at lower thresholds as long as the sample size is large enough to detect differentially expressed genes.

Estimates of the quantities $A+B$ and $B+D$ (i.e., the denominators of TN and EDR, respectively) are more accurate at small sample sizes and small thresholds because A and B are usually large. However, estimates of EDR are small at these n and τ because D is small. So lines do not cross over in plots for EDR because a smaller threshold makes it more difficult to detect differentially expressed genes regardless of sample size. In the actual data set, $n = 3$, and one can see that the estimated EDR is quite small. One can also see from the procedure and resulting graph that EDR values rise to more acceptable levels as sample sizes approach around 20 arrays per treatment group.

5.8 DISCUSSION

The introduction posed five questions that are of interest in high-dimensional studies such as using microarrays. This chapter reviewed several approaches taken by others and provided some details of a technique based on mixture models and the use of simulated experiments from a given model. Answers to questions 2 to 4 are available using this technique. As far as questions 1 and 5 regarding threshold and sample size, the answer may very well be "it depends." We saw that there are trade-offs for choice of threshold. A very small threshold may miss many important genes (low EDR) but of the genes that are declared differentially expressed, one can be fairly certain that they are real (high TP). The described procedure relies on an initial study where a mixture model was fit to data and this model then becomes a standard for evaluating sample size effects on hypothetical future studies.

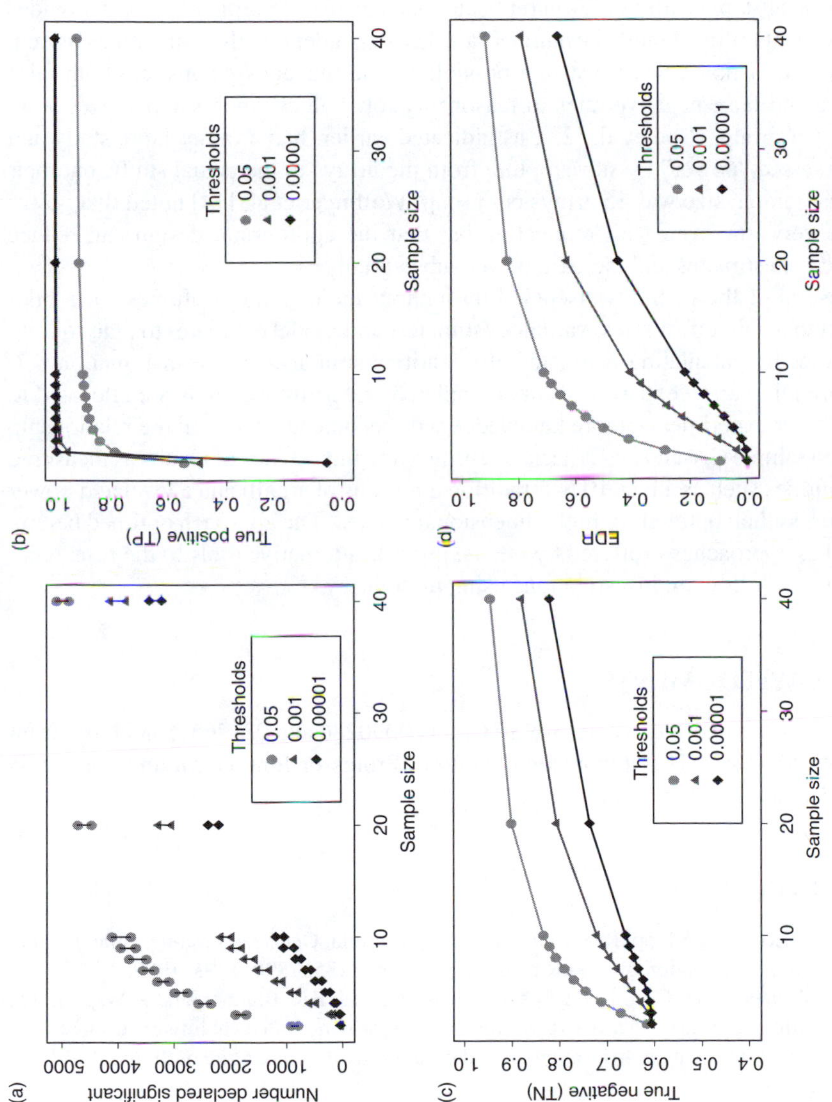

FIGURE 5.5 Effect of sample size (n = number of arrays per treatment group) on number declared significant (a), true positive (b), true negative (c), and EDR (d) for three selected thresholds, $\tau = 0.05, 0.001$, and 0.00001.

An interesting twist to this idea, mentioned earlier, was given by Pavlidis et al. [27]. They identified several publicly available microarray data sets with varying sample sizes (6 arrays per group to 50 per group). Using the larger studies, they could sample subsets of arrays thus simulating a smaller experiment and evaluating the stability of results using small samples. They concluded that results become somewhat unstable with 5 or fewer replicates but that 10 to 15 replicates often provides reasonable stability, though the numbers are data dependent. In the simulations of Zein et al. [5], they noted that it was not possible to simultaneously constrain both false positive and false negative rates to reasonably low values when sample sizes were only 8 per group. Pepe et al. [22], as indicated earlier, had a rather large study and they assessed "power" by subsampling from the arrays in the actual study, but their smallest sample size was 15 arrays per group. Wolfinger et al. [14] noted that power can be very low even with replication but that the appropriate design can reduce variance of estimates and increase power substantially.

Several of the methods discussed throughout indicate the usefulness of a pilot data set to obtain effect sizes, variance estimates, and model estimates to plan follow-up studies by calculating estimates of quantities similar to those in Equation 5.1. As more microarray experiments are completed and as future funds are allocated to allow larger experiments, more knowledge will become available on the relationship between sample size and key criteria attesting to the importance of results as measured by quantities such as TP and TN. Traditional notions of significance level and power may be less than optimal for high dimensional studies. The approach outlined herein, as well as approaches proffered by others, provide alternative tools to the researcher to help plan follow-on investigations using microarrays.

ACKNOWLEDGMENTS

This research supported in part by NSF Grants 0090286 and 0217651, and NIH Grant U54CA100949-01. The authors are grateful to Professor John D. Mountz for access to the data used in this chapter.

REFERENCES

1. C. Lee, R.G. Klopp, R. Weindruch, and T.A. Prolla. Gene expression profile of aging and its retardation by caloric restriction. *Science* 285: 1390–1393, 1999.
2. M.A. Newton, C.M. Kendziorski, C.S. Richmond, F.R. Blattner, and K.W. Tsui. On differential variability of expression ratios: improving statistical inference about gene expression changes from microarray data. *Journal of Computational Biology* 8: 37–52, 2001.
3. M.L.T. Lee, F.C. Kuo, G.A. Whitmore, and J. Sklar. Importance of replication in microarray gene expression studies: statistical methods and evidence from repetitive cDNA hybridizations. *Proceedings of the National Academy of Sciences, USA* 97: 9834–9839, 2000.
4. M.A. Black, and R.W. Doerge. Calculation of the minimum number of replicate spots required for detection of significant gene expression fold change in microarray experiments. *Bioinformatics* 18: 1609–1616, 2002.

5. A. Zien, J. Fluck, R. Zimmer, and T. Lengauer. Microarrays: how many do you need? *Journal of Computational Biology* 10: 653–667, 2003.
6. W. Pan, J. Lin, and C.T. Le. How many replicates of arrays are required to detect gene expression changes in microarray experiments? A mixture model approach. *Genome Biology* 3: 1–10, 2002.
7. R.M. Simon, and K. Dobbin. Experimental design of DNA microarray experiments. *Bio Techniques* 34: 16–21, 2003.
8. M.K. Kerr, C.A. Afshari, L. Bennett, P. Bushel, J. Martinez, N.J. Walker, and G.A. Churchill. Statistical analysis of a gene expression microarray experiment with replication. *Statistica Sinica* 12: 203–217, 2002.
9. T. Ideker, V. Thorsson, A.F. Siegel, and L.E. Hood. Testing for differentially-expressed genes by maximum-likelihood analysis of microarray data. *Journal of Computational Biology* 7: 805–817, 2000.
10. S. Dudoit, Y.H. Yang, M.J. Callow, and T.P. Speed. Statistical methods for identifying differentially expressed genes in replicated cDNA microarray experiments. *Statistica Sinica* 12: 111–139, 2002.
11. G.L. Gadbury, G.P. Page, J. Edwards, T. Kayo, T.A. Prolla, R. Weindruch, P.A. Permana, J. Mountz, and D.B. Allison. Power and sample size estimation in high dimensional biology. *Statistical Methods in Medical Research* 13: 325–338, 2004.
12. Y.H. Yang, and T. Speed. Design issues for cDNA microarray experiments. *Nature Reviews in Genetics* 3: 579–588, 2002.
13. M.L.T. Lee, and G.A. Whitmore. Power and sample size for DNA microarray studies. *Statistics in Medicine* 21: 3543–3570, 2002.
14. R.D. Wolfinger, G. Gibson, E.D. Wolfinger, L. Bennett, H. Hamadeh, P. Bushel, C. Afshari, and R.S. Paules. Assessing gene significance from cDNA microarray expression data via mixed models. *Journal of Computational Biology* 8: 625–637, 2001.
15. Y. Benjamini, and Y. Hochberg. Controlling the false discovery rate: a practical and powerful approach to multiple testing. *Journal of Royal Statistical Society B* 57: 289–300, 1995.
16. H.J. Keselman, R. Cribble, and B. Holland. Controlling the rate of type I error over a large set of statistical tests. *British Journal of Mathematical and Statistical Psychology* 55: 27–39, 2002.
17. J.D. Storey. A direct approach to false discovery rates. *Journal of Royal Statistical Society B* 64: 479–498, 2002.
18. J.P. Brody, B.A. Williams, B.J. Wold, and S.R. Quake. Significance and statistical errors in the analysis of DNA microarray data. *Proceedings of the National Academy of Sciences, USA* 99: 12975–12978, 2002.
19. D.B. Allison, G.L. Gadbury, M. Heo, J.R. Fernanfez, C. Lee, T.A. Prolla, and R. Weindruch. A mixture model approach for the analysis of microarray gene expression data. *Computational Statistics and Data Analysis* 39: 1–20, 2002.
20. G.L. Gadbury, G.P. Page, M. Heo, J.D. Mountz, and D.B. Allison. Randomization tests for small samples: an application for genetic expression data. *Journal of Royal Statistical Society C* 52: 365–376, 2003.
21. S. Pounds, and S.W. Morris. Estimating the occurrence of false positive and false negative in microarray studies by approximating and partitioning the empirical distribution of *p*-values. *Bioinformatics* 19: 1236–1242, 2003.
22. M.S. Pepe, G. Longton, G.L. Anderson, and M. Schummer. Selecting differentially expressed genes from microarray experiments. *Biometrics* 59: 133–142, 2003.

23. V.G. Tusher, R. Tibshirani, and G. Chu. Significance analysis of microarrays applied to the ionizing radiation response. *Proceedings of the National Academy of Sciences, USA* 98: 5116–5121, 2001.
24. M.K. Kerr, M. Martin, and G.A. Churchill. Analysis of variance for gene expression microarray data. *Journal of Computational Biology* 7: 819–837, 2000.
25. B. Efron, and R.J. Tibshirani. *An Introduction to the Bootstrap*. New York: Chapman & Hall, 1993.
26. B. Efron, and R.J. Tibshirani. Empirical Bayes methods and false discovery rates for microarrays. *Genetic Epidemiology* 23: 70–86, 2002.
27. P. Pavlidis, Q. Li, and W.S. Noble. The effect of replication on gene expression microarray experiments. *Bioinformatics* 19: 1620–1627, 2003.
28. S. Mukherjee, P. Tamayo, S. Rogers, R. Rifkin, A. Engle, C. Campbell, T.R. Golub, and J.P. Mesirov. Estimating dataset size requirements for classifying DNA microarray data. *Journal of Computational Biology* 10: 119–142, 2003.
29. M.J. Van der Laan, and J. Bryan. Gene expression analysis with the parametric bootstrap. *Biostatistics* 2: 445–461, 2001.
30. H. Sackrowitz, and E. Samuel-Cahn. *P* values as random variables — expected *p* values. *The American Statistician* 53: 326–331, 1999.
31. A.P. Dempster, and M. Schatzoff. Expected significance level as a sensibility index for test statistics. *Journal of American Statistical Association* 60: 420–436, 1965.
32. M.J. Schervish. *P* values: what they are and what they are not. *The American Statistician* 50: 203–206, 1996.
33. R.M.J. Donahue. A note on information seldom reported via *p* values. *The American Statistician* 53: 303–306, 1999.
34. T. Schweder and E. Spjøtvoll. Plots of *p*-values to evaluate many tests simultaneously. *Biometrika* 69: 493–502, 1982.
35. H.M. Hung, R.T. O'Neill, P. Bauser, and K. Köhne. The behavior of the *p*-value when the alternative hypothesis is true. *Biometrics* 53: 11–22, 1997.
36. R.A. Paker, and R.B. Rothenberg. Identifying important results from multiple statistical tests. *Statistics in Medicine* 17: 1031–1043, 1988.
37. Y. Yang, J. Hoh, C. Broger, M. Neeb, J. Edington, K. Lindpaintner, and J. Ott. Statistical methods for analyzing microarray feature data with replications. *Journal of Computational Biology* 10: 157–169, 2003.

6 Pooling Biological Samples in Microarray Experiments

Christina M. Kendziorski

CONTENTS

6.1 Introduction 95
6.2 Derivation of the Equivalence Formula 96
6.3 Assumptions Used to Derive the Formula 99
6.4 Utility of Pooling 105
6.5 Conclusion 108
References 108

6.1 INTRODUCTION

The number of scientific studies utilizing microarray technologies continues to increase. To optimize the information from such studies, good experimental design is critical. Unfortunately, design questions have received *relatively* little attention in the literature (commentary by Nguyen et al., 2002); and, as a result, a number of important questions remain open. One is whether or not to pool messenger RNA (mRNA) samples across subjects.

The general question of pooling biological samples is certainly not new. Related work began as early as the 1940s when WWII inductees were required to undergo blood tests for syphilitic antigens. Dorfman (1943) showed that the number of total tests could be reduced if blood samples were first pooled and tested, followed by individual retesting of all composite samples in positive pools only. The idea has been extended to a number of areas (for a review, see Gastwirth, 2000) including the detection of mutant alleles in a population (Amos et al., 2000), the estimation of disease prevalence (Gastwirth and Hammick, 1989), and the estimation of joint allele frequencies and linkage disequilibrium (Pfeiffer et al., 2002). In each of these cases, the general goal is to detect the presence of a characteristic, perhaps followed by estimation of population prevalence or allele frequencies. Pooling designs for microarray experiments are different. Most often, relatively few subjects are pooled, but pools are replicated many times.

For microarray studies, there are convincing reasons both for and against pooling, and thus both pooled and nonpooled designs are used in practice. For example, in studies of tissues such as the hypothalamus, or in studies of small animals such as

Drosophila, it can be difficult if not impossible to obtain a sufficient amount of mRNA from a single subject. In such cases, pooling can be useful (Jin et al., 2001; Saban et al., 2001). Even when not limited by the amount of mRNA available, pooling is often done to reduce the effect of biological variability (Brown et al., 2001; Chabas et al., 2001; Waring et al., 2001; Agrawal et al., 2002; Ernard et al., 2002; discussion in Churchill and Oliver, 2001). The idea is that differences due to subject-to-subject variation will be minimized making substantive differences easier to find. Peng et al. (2003) used simulation studies to demonstrate the increase in power that can be obtained. A third reason for pooling is to produce a reference mRNA population (for an overview, see Weil et al., 2002 or Yang et al., 2002). (This last reason is inherently different from the first two since reference pools are *not* used for comparison between two or among multiple conditions; rather, the reference pool is used as a calibration tool for two or more populations that are being compared. This type of pooling will not be discussed in detail here.)

While there are potential advantages to pooled designs, there are situations in which one would certainly want to analyze only individual samples. An obvious reason not to pool is specific interest in individual profiles. This is often the case when samples from humans are obtained for classification and diagnosis. In this setting or a similar one, an individual's level of expression for a set of genes is of interest and pooling is not beneficial. Even if individual profiles are not of direct interest, pooling may be discouraged due to concerns regarding the effects of contaminated samples (see Section 6.3).

A statistical consideration of this debate appeared in Kendziorski et al. (2003a). In that work, a formula was derived giving the total numbers of subjects and arrays required in a pooled experiment to obtain an estimate of expression comparable to that obtained in the no pooling case. It was shown that under certain assumptions, equivalent estimates can be obtained in pooled designs using fewer arrays, provided an appropriate increase in the total number of subjects is made. In this chapter, a simpler equivalence criterion is considered to avoid any assumptions regarding distributions on errors. Similar results are shown to hold. In the following section, the derivation of the analogous formula is given. The assumptions used to derive the formula are considered in detail in Section 6.3. Section 6.4 considers the utility of pooling in identifying differentially expressed genes.

6.2 DERIVATION OF THE EQUIVALENCE FORMULA

Consider the problem of estimating the nominal expression levels for m genes, denoted by the m-vector θ. An experiment to estimate θ consists of extraction and labeling of mRNA from n_s subjects, hybridization to n_a arrays, followed by scanning and image processing. The technology that one uses dictates in large part the details of each of these steps. Our concern is not with a specific technology or image processing method, but with the actual measurements of gene expression, however obtained.

We assume here that sufficient data preprocessing has been done to remove artifacts within the array and across a set of arrays. In this case, the gene expression measurements for gene g denoted by $x_{g,1}, x_{g,2}, \ldots, x_{g,n_a}$ are considered independent and identically distributed samples from a distribution parameterized by mean θ_g. The processed measurements are assumed to be affected primarily by two sources of variation: subject-to-subject and array-to-array variability (hereafter referred to as biological variability and technical variability, respectively). In spite of fluctuations, the average $\bar{x}_g = (1/n_a)\sum_{k=1}^{n_a} x_{g,k}$ estimates θ_g.

The problem of estimating gene expression could be addressed by implementing any one of many potential experimental designs. Here, two designs are considered. The first design (design I) requires that an mRNA sample from each subject is probed with one array. Due to variability among subjects and measurement error inherent to the array, any observed $x_{g,i}$ is defined as:

$$x_{g,i} = \theta_g + \epsilon_{g,i} + \xi_{g,i} \tag{6.1}$$

$$= T_{g,i} + \xi_{g,i} \tag{6.2}$$

$i = 1, 2, \ldots, n$ and $n = n_s = n_a$. Here, $\epsilon_{g,i}$ represents gene specific subject (biological) variability and $\xi_{g,i}$ represents measurement error (technical variation), which might also depend on gene g. Design I considers a single subject per pool and thus, $\epsilon_{g,i}$ can also be thought of as pool-to-pool variability. We assume that $\epsilon_{g,i}$ and $\xi_{g,i}$ have zero mean with variances $\sigma^2_{\epsilon,g}$ and $\sigma^2_{\xi,g}$, respectively.

For the second design, design II, mRNA samples from r_s different subjects are first pooled; then, r_a replicate samples are drawn from the mRNA pool and hybridized onto a set of arrays. This is repeated a number of times denoted by n_p to represent the number of distinct mRNA pools. It is possible that r_s and r_a are pool dependent. The extension is not difficult, but to simplify notation, it is not considered here. For this design,

$$x_{g,i,j} = \theta_g + \epsilon'_{g,i} + \xi_{g,i,j} \tag{6.3}$$

$$= T'_{g,i} + \xi_{g,i,j} \tag{6.4}$$

$$= \frac{1}{r_s}\sum_{k=1}^{r_s} T_{g,i_k} + \xi_{g,i,j} \tag{6.5}$$

where $i = 1, 2, \ldots, n_p$ and $j = 1, 2, \ldots, r_a$; T_{g,i_k} is the kth subject's contribution to the ith pool for gene g. Assuming that the mRNAs average out across the pool, we would expect the variability of $\epsilon'_{g,i}$ to be reduced to $\sigma^2_{\epsilon,g}/r_s$. Note that design I is a special case of design II (with $r_s = r_a = 1$); it is useful to distinguish between the two.

Designs I and II are first evaluated by considering the bias and variance of $\bar{x}_{g\cdot\cdot}$. For both designs, $E[\bar{x}_{g\cdot\cdot}] = \theta_g$:

$$\sigma^2_{\bar{x},(1),g} = \frac{1}{n_{p1}}\left(\sigma^2_{\epsilon,g} + \sigma^2_{\xi,g}\right) \quad \text{and} \quad \sigma^2_{\bar{x},(2),g} = \frac{1}{n_{p2}}\left(\frac{\sigma^2_{\epsilon,g}}{r_{s2}} + \frac{\sigma^2_{\xi,g}}{r_{a2}}\right) \tag{6.6}$$

where $\sigma^2_{\bar{x},(1),g}$ and $\sigma^2_{\bar{x},(2),g}$ denote the variance of $\bar{x}_{g\cdot\cdot}$ in designs I and II, respectively. In each case, the estimator is unbiased for θ_g and the gene specific variance decreases as the number of arrays increases.

The variance components $\sigma^2_{\xi,g}$ and $\sigma^2_{\epsilon,g}$ are never known in practice and so

$$R = \frac{\sum_{g=1}^{m} E\left[\sigma^2_{\bar{x},(1),g}\right]}{\sum_{g=1}^{m} E\left[\sigma^2_{\bar{x},(2),g}\right]}$$

is used to compare the designs. Expectations are considered since the variance components must be estimated; averages are considered since the variances are gene dependent. Using the unbiased estimator W for the variance of $\bar{x}_{g\cdot\cdot}$ where

$$W = \left(\frac{\hat{\sigma}^2_{\epsilon,g}}{n_s} + \frac{\hat{\sigma}^2_{\xi,g}}{n_a}\right) = \left(\frac{\hat{\sigma}^2_{\epsilon,g}}{r_s n_p} + \frac{\hat{\sigma}^2_{\xi,g}}{r_a n_p}\right)$$

one can show that $R = 1$ when

$$n_{s2} = n_{s1}\left[\frac{\lambda}{K(\lambda + 1) - (n_{a1}/n_{a2})}\right]. \tag{6.7}$$

where n_{s1} (n_{s2}) and n_{a1} (n_{a2}) denote the total number of subjects and arrays in design I (II); $\lambda = \sum_{g=1}^{m} \sigma^2_{\epsilon,g} / \sum_{g=1}^{m} \sigma^2_{\xi,g}$ and $K = 1$. Note that in some cases, discussed below, K will differ from 1.

Equation 6.7 gives the total number of subjects required in a pooled design (design II) to obtain expression estimates that on average are as precise as those obtained in the no pooling case. The equation shows that by increasing the number of subjects in design II, the number of arrays can be decreased, without changing the average precision.

As an example, consider an experiment with the mRNA from 16 individuals probed using 16 arrays (design I with $n_{s1} = n_{a1} = n_{p1} = 16$) and suppose that the average biological variability is twice as large as the average technical variability (averages are taken across genes: $\bar{\sigma}^2_{\epsilon,\cdot} = 2$ and $\bar{\sigma}^2_{\xi,\cdot} = 1$ which gives $\lambda = 2$). The number of subjects and arrays required in a pooled experiment to obtain comparable precision is given by Equation 6.7. Since n_{s2} as defined by Equation 6.7 might not be integer valued, the values given are considered lower bounds on the total number of subjects. If the total number of arrays in the pooled experiment is reduced to 12, Equation 6.7 indicates that the mRNA from at least 20 subjects is required to obtain an interval comparable to that obtained without pooling. If the number of arrays is reduced to 8, mRNA samples from at least 32 subjects are required.

As these calculations report the total number of subjects and arrays in a given design, they give no information about the exact way in which to allocate the totals to pools and arrays. Multiple arrays obtained from a given pool allow one to estimate technical variability. As discussed earlier, estimates of the sum of biological and technical variability are required to make inference about any given expression level; estimates of technical variability itself is often not of interest. When this is the case, it is not advantageous to measure a pool using multiple arrays, as doing so provides no independent information to estimate biological variability. As a result, for a fixed total number of subjects and arrays, the number of arrays used to probe a pool, r_a, should be 1. This holds by definition for design I and we impose this constraint throughout for design II. This provides the information necessary to allocate totals to pools.

Notice that no specific parametric assumptions are made regarding the distribution on the pool or array variability. For the biological component, it is only required that $E[T_{g,i_k}] = \theta_g$ and $\text{var}[T_{g,i_k}] = \sigma^2_{\epsilon,g}$. For the technical variability, the above calculations hold when $E[\bar{x}_{g,i,\cdot}|T'_{g,i}] = T'_{g,i}$ and $\text{var}[\bar{x}_{g,i,\cdot}|T'_{g,i}] = \sigma^2_{\xi,g}/r_a$. Note that $x_{g,i,j}$ can be the result of a transformation of the original intensity values. A discussion of log transformed measurements is given in Section 6.3.

It is clear that whatever specific assumptions are made, the determination of the number of subjects and arrays required for pooled designs will depend in some way on technical and biological variability. There are certainly many systems for which good estimates of variability are not yet available and without them, the determination of equivalent designs is not possible using this type of approach. However, a closely related question can be answered without estimates of variance components.

Consider again the experiment with 16 subjects on 16 arrays. Suppose that instead of trying to determine the increase in the number of subjects required to offset a decrease in the number of arrays, one would simply like to know whether advantage can be gained by pooling 32 subjects onto 16 arrays. If obtaining an individual subject sample costs \$40.00, this increase only requires an additional \$640.00, which could be less than the cost of a single array depending on the technology used.

For this much simpler question, if there is no interest in an individual profile and the assumptions discussed above hold, it clearly makes sense to perform the latter experiment as the variability for each estimate should be reduced. The main assumptions to be considered when addressing either question are:

1. Biological averaging: mRNAs average out when pooled
2. Pool-to-pool variability is smaller than subject-to-subject variability
3. There is no contamination
4. The average across intensity measurements (mathematical averaging) provides an unbiased estimate of expression

These assumptions are considered in detail in the next section.

6.3 ASSUMPTIONS USED TO DERIVE THE FORMULA

The calculations in Section 6.2 assume that biological averaging does occur. In other words, it is assumed that individual mRNAs average out across pools. Certain artifacts could cause this assumption to appear to be violated. As one example, suppose the scanner being used in the experiment is calibrated so that the dynamic range is limited at 55,000 units. Suppose furthermore that the mRNA from six animals is obtained — three have mRNA levels of 40,000 units and three have mRNA levels of 60,000 units. The quantifications are expected to be near 40,000 units for the first three (with deviations due to processing and measurement error), but near 55,000 units for the last three since mRNA levels above 55,000 units are outside the dynamic range. Averaging the quantifications across the six chips gives 47,500 units ($\frac{1}{6}(40{,}000 \times 3 + 55{,}000 \times 3)$). If pooling was done, the average in the pool (if in fact averaging does occur) would be 50,000 units, which are detectable. Thus, readings from the pool would be expected to be near 50,000 units. If we compared the results from such a study, we would see that the average calculated from individual arrays was less than the average calculated from the pools.

Such artifacts are not expected to affect a significant proportion of genes considering the dynamic range of most technologies and the relationships that are often cited between gene expression quantifications and qRT-PCR. In addition, there are three studies that we are aware of to date, which comment on a small comparison between individual and pooled samples. In both studies, the individual samples showed expression profiles similar to that of the pools. Waring et al. (2001), in a study of rat liver tissue, report:

> . . . the individual animal expression profiles all clustered with the pooled profile. . . the decision to pool RNA samples did give a representative analysis of gene expression changes and suggest that pooling RNA samples from the same treatment group may be a viable method for detecting key changes caused by a compound.

Gieseg et al. (2002) report similar results in a study of human normal and tumor kidney tissues.

The results in Agrawal et al. (2002) concur, and provide additional information. In that manuscript, the authors measured gene expression (using Affymetrix Human HuFL 6800 and HuU95Av2 chips) from five individual human colon tumor samples at the same stage. They considered one pool of the five tumors (P1), a second pool of five independent tumors (P2), and a "calculated" pool (CP) constructed by mathematically averaging the values obtained from the five individuals contributing RNA to P1. They also considered expression levels from matched normal mucosa samples. The results showed that the individual tumor samples correlated extremely well with both P1 and CP; there was reduced, but still strong, correlation between the individuals and P2. In addition, they observed that tumor samples and pools showed a higher correlation to one another than they did to the normal mucosa. Although encouraging, the data sets considered in these studies are small. Unpublished reports in other labs (pers. commun.) using cDNA array data have not yet demonstrated conclusive evidence.

If there is in fact a biological averaging of mRNA levels, one would expect a linear decrease in variance. However, due to the high degree of processing during hybridization, scanning, imaging, probe level analysis, and possible transformations, nonlinearities could result. If the decrease is not linear (suppose $\epsilon_{ig'} \sim \left(0, \sigma^2_{\epsilon,g}/r_s{}^{\alpha}\right)$), Equation 6.7 still holds but now depends on α, which would have to be estimated. If $\alpha > 1$, the advantage is greater than expected as the variance reduction is greater than assumed; if $\alpha < 1$, there is less advantage.

Contamination is more likely the larger concern regarding pooled designs. For the following discussion, we assume that the total numbers of subjects and arrays in each design have been chosen so that the designs are comparable in terms of equivalent average precisions. Two types of contamination are considered.

- *Type I (nonbiological)*: Gene expression levels for some proportion of subjects are altered due to some technical error. All genes $(\theta_1, \theta_2, \ldots, \theta_m)$ are affected.
- *Type II (biological)*: Gene expression for some proportion of subjects differs from the majority of the population being sampled. The difference is a feature of the population and occurs for only a subset of the genes.

Type I contamination could arise, for example, if a significantly larger (or smaller) amount of labeled product than expected is produced for some proportion of subjects, p_s. In this case, the abundance levels for those subjects would be altered by some level η.

$$x_{g,i,j} = \frac{1}{r_s} \sum_{k=1}^{r_s} [T_{g,i_k} + \delta_{g,i_k}] + \xi_{g,i,j}$$

where

$$\delta_{g,i_k} = \begin{cases} \eta_g, & \text{with probability } p_s \\ 0, & \text{with probability } 1 - p_s \end{cases}$$

Here, δ_{g,i_k} represents the contamination effect of the kth subject's contribution to the ith pool for gene g. In type I contamination, it is assumed that the effect, η_g, applies to all genes. For both pooled and nonpooled designs, the estimators are biased and the variance is increased from the noncontamination case: $E[\bar{x}_{g..}] = \theta_g + \eta_g p_s$ and

$$\sigma^2{}_{\bar{x_g}} = \frac{1}{n_p} \left(\frac{\sigma^2_{\epsilon,g} + \eta_g{}^2 p_s(1 - p_s)}{r_s} + \frac{\sigma^2_{\xi,g}}{r_a} \right) \tag{6.8}$$

In other words, this type of contamination is problematic for either design. As shown above, at the population level, there is equivalent bias in the estimators and inflated variance. The variance for the pooled design is somewhat smaller (since $r_s > 1$), but beware that in this case one is considering the variance of a biased estimator.

In practice, if the deviation for a particular array (using individual or pooled samples) is noticeable relative to other arrays, an array effect would be identified in the normalization phase of data analysis and the values would either be adjusted or removed. A distinction between the designs lies in the ability to identify such effects (see Figure 6.1). It might be easier to identify these effects in design I, since the effect of a contaminant would not be attenuated by averaging. On the other hand, suppose a contaminant does contribute mRNA to a pool, but the attenuation due to pooling reduces the effect down to a negligible level that is indistinguishable from array variation. The importance of identifying the contaminant in this situation is questionable.

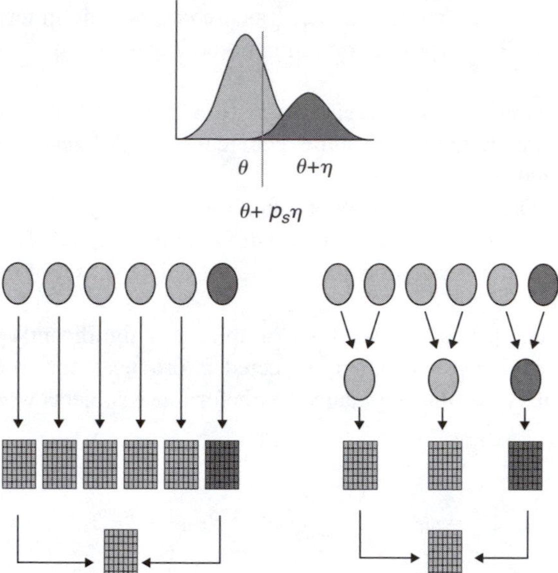

FIGURE 6.1 A schematic of an experiment with mRNA samples from six individuals (five from a population with mean θ and a "contaminant" member sampled from a population with mean $\theta + \eta$). The average mRNA level for both designs is $\theta + p_s\eta$. However, consider the arrays containing mRNA from the contaminant. When probed individually (left), the array is more obviously an outlier.

Type II contamination concerns a population where some proportion of subjects have altered levels of expression for some collection of genes. This type of contamination could be due to genetic heterogeneity. It is of much bigger concern, since it is not likely to be identified in data preprocessing. Consider only the unknown subset of genes that are affected. The expression level for a gene in this collection is distributed as a mixture: the major component of the mixture has mean θ_g and the other has mean $\theta_g + \eta_g$.

For a given gene g in this subset, measurements are written as before (see Equation 6.8). As in type I, the estimators are biased and the variance is increased from the non-contamination case. The only difference here is that the genes that are affected are not known and as a result, the effects would most likely not be identified as array effects in data preprocessing. The comparison between pooled and nonpooled designs given for type I contamination apply here for the unknown subset of genes.

In short, type I contamination poses similar problems for either design (biased estimator with increased variance); but in practice, problematic arrays can often be identified and either adjusted or removed. There is a slight advantage to the nonpooled design in that it may be easier to identify outliers. Furthermore, when removing a particular array due to contamination in design I, only a single mRNA sample is lost. Type II contamination is similar to type I in terms of comparing bias and variance between designs. However, in type II, the genes that are in fact affected by contamination cannot be easily identified.

It might be worthwhile to consider more complicated contamination scenarios. The most obvious extension is to consider gene dependent p. The simple scenarios considered here illustrate that contamination is a potential problem for both pooled and nonpooled designs. This is due to the assumption that averaging occurs at some level: in design I, there is mathematical averaging of intensity measurements; in design II, the mathematical averaging follows averaging of mRNAs across the pool.This latter type of averaging is referred to as biological averaging. Each type is considered below.

The question of whether or not biological averaging occurs is one that requires experimental data and currently there is insufficient data to answer this question. The three studies cited above indicate that this assumption is reasonable. The related question of whether one should mathematically average intensity measurements to obtain an estimate of expression must also be addressed. The answer will depend on the distribution of expression levels within individual genes across subjects. There is some evidence indicating that this distribution is at least approximately Gaussian. By this, we do not mean that the distribution of gene intensity measurements across an array is approximately Gaussian; it is not.

Figure 6.2 gives normal quantile–quantile (qq) plots for intensity measurements obtained from rat mammary tissue. The details of the experiment are given in Kendziorski et al. (2003b). In short, mammary tissue was obtained from four genetically identical rats. Each was probed using Affymetrix Rat Genome U34 chips. The left panel gives qq plots of MAS 5.0 Signal intensities across the array. Clearly,

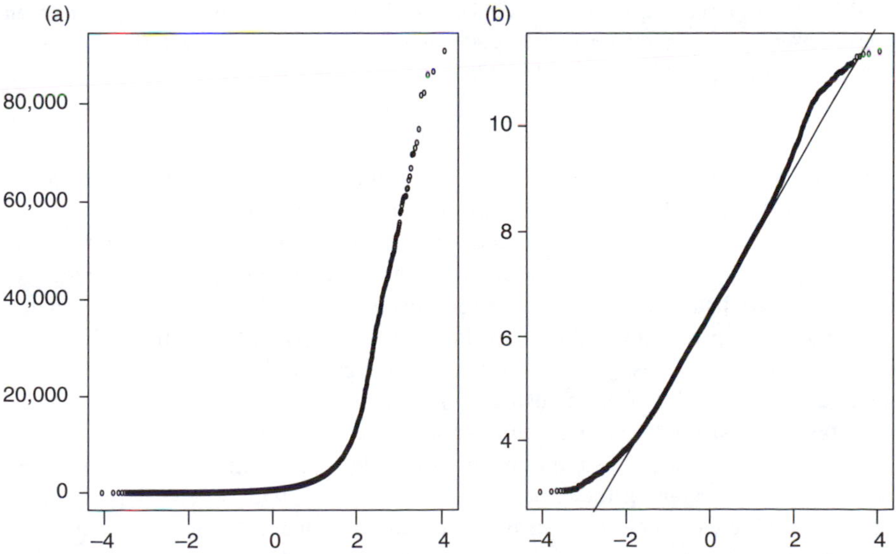

FIGURE 6.2 Normal qq plots of MAS 5.0 (a) signal intensities and (b) log transformed intensities.

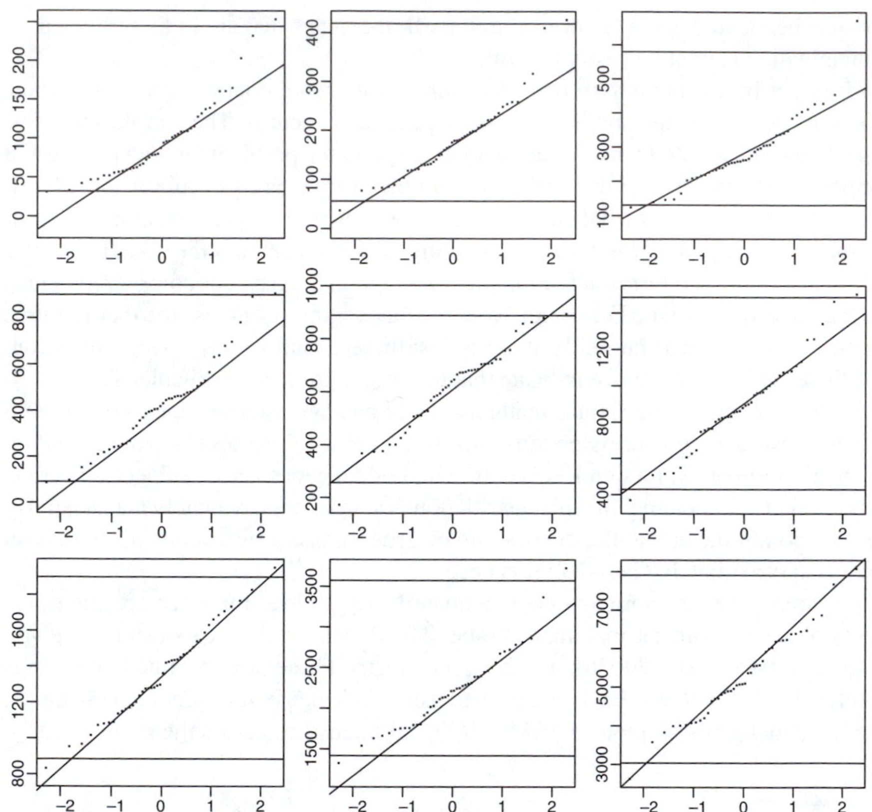

FIGURE 6.3 Normal qq plots of MAS 5.0 signal intensities grouped by common mean (means taken across the four replicates); 100 genes are shown in each group. The horizontal lines cut off the upper and lower 2.5 percentiles.

this is not well approximated by a Gaussian distribution. The log-transformed data is better approximated, but there is still considerable deviation in the tails.

Here, we are not primarily concerned with the distribution of measurements across the entire array, but rather with the distribution of measurements within a gene. As there is limited data for any given gene (here, just four values), data with similar mean intensities were grouped together to evaluate the underlying distribution. Figure 6.3 shows qq plots for intensity measurements in nine groups. Each group contains 100 genes that have similar mean intensities, with means taken across the four replicates. As shown, the assumption of normality *within* gene is not grossly violated.

When the Gaussian assumption applies at least approximately within gene, mathematically averaging the expression measurements within gene is reasonable. Provided biological averaging does in fact occur, the calculations given above apply. Furthermore, if errors are at least approximately Gaussian, it might be useful to consider other equivalence criterion, such as the squared lengths of the confidence intervals for θ, within this framework. Denoting these lengths by l_1^2 and l_2^2 for designs I and II, respectively, it can be shown that $E[l_2^2]/E[l_1^2] = 1$ when Equation 6.7 is satisfied for $K = t_1^2/t_2^2$ where t_1 and t_2 denote critical t values for designs I and II, respectively. This criterion explicitly accounts for the differences in degrees of freedom resulting from the different number of arrays in each design. It provides a slightly more conservative estimate than simply using the precisions and guarantees appropriate coverage of confidence intervals.

Whatever criterion is used, most likely the determination of equivalence will rely in some way on the biological and technical variability. For our development, Equation 6.7 depends on $\lambda = \sum_{g=1}^{m} \sigma_{\epsilon,g}^2 / \sum_{g=1}^{m} \sigma_{\xi,g}^2$. If $\lambda < 1$, the increase in biological subjects required to offset a reduction in arrays is probably not justified. However, the utility of pooling is increased with increasing λ and so for systems with large λ, pooling could be worthwhile. To see this, consider once again the example with $\lambda = 2$; 32 subjects were required to reduce the total number of arrays by one-half. If $\lambda = 4$, the number of subjects required decreases to 22. Cheung et al. (2003) estimate biological and technical variability for 813 genes measured in lymphoblastoid cells in 35 individuals from the CEPH Utah pedigrees; λ values ranged from 0.4 to 64 with a median value of 2.5. Pritchard et al. (2001) report similar values for testis tissue in mice; kidney and liver tissue had reduced values.

It should be noted that in practice, transformations are often done prior to mathematical averaging. This can affect the calculations shown here. For example, suppose a log transformation is applied to intensity measurements obtained from an array. The expression estimate for each gene would be defined as the average of the log measurements taken over subjects within a given condition. In design II, the measurements are from pools and if there is biological averaging, the outcome is a log of averages. Jensen's inequality indicates that the values from design I (average of logs) will be smaller than the values from design II (average of logs of averages). Figure 6.4 shows this numerically. A single pool was simulated (using the array data described above) by mathematically averaging values from arrays 1 and 2; this was repeated using arrays 3 and 4 to form a second simulated pool. The x-axis gives averages of the log intensities taken across the four replicates; the y-axis shows averages of the log intensities taken across the two simulated pools. As shown, the values from the pools are larger, but the increase is not substantial.

In summary, a few studies using small data sets indicate that assumptions 1 and 3 are reasonable. Larger studies are required to verify these results, to check assumption 2, and to evaluate the impact of various forms of contamination on each design. Most likely, it will be shown that contamination is problematic for both pooled and nonpooled designs.

6.4 UTILITY OF POOLING

The utility of pooling in practice will be measured by the ability to find differentially expressed genes. Until now, we have only considered the effects of pooling within a single biological condition. However, if the assumptions considered so far do in fact hold, then precision of the gene expression estimator can be maintained with a

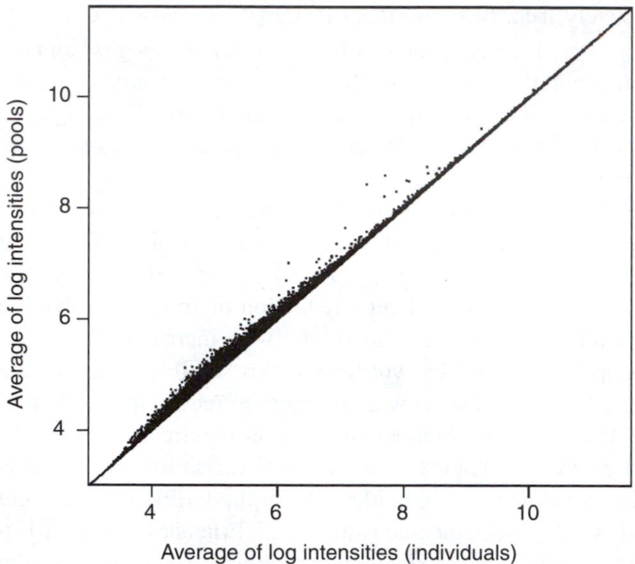

FIGURE 6.4 The average log intensities taken across 4 replicates is plotted against the average of log intensities taken across two simulated pools of 2.

reduced number of arrays (or increased by increasing the number of subjects while holding the number of arrays constant). This would imply that pooled designs should find comparable numbers (or increased numbers if the precision is increased) of differentially expressed genes in practice.

The Agrawal et al. (2002) study considered this problem. Recall that they probed five human colon tumor samples and one pool containing the RNA from the five individuals. Matched data was obtained from paired normal mucosa samples. (Note that the second pool containing RNA from five independent tumors was not used for addressing the question of identifying DE genes.) Using a fold change criterion (fold change between tumor and normal bigger than 2 in at least 3 individual samples), they identified 282 genes as differentially expressed in the individuals. Of these, 230 were identified using the data from pools.

This difference is somewhat expected. By Equation 6.7, these two designs (5 individuals on 5 arrays and 5 individuals pooled onto 1 array) are not equivalent. As such, one would not expect to find the same proportion of differentially expressed genes. A small simulation study shows this more clearly.

Consider simulated log intensities for 10,000 genes in two biological conditions, with 16 replicates in each condition. The model within condition is given by Equation 6.3. For an equivalently expressed gene, a common mean, θ_g, was considered across conditions ($\theta = \theta_{g1} = \theta_{g2}$); θ_g is sampled from a Normal distribution with mean $\mu = 2.3$ and variance $\sigma^2 = 0.8$. For a differentially expressed gene g, θ_{g1} and θ_{g2} are independent samples from that same Normal distribution ($N(\mu, \sigma^2)$). Parameter estimates were chosen based on values observed in previous microarray

TABLE 6.1
The Identification of Differently Expressed (DE) Genes

Method	Design	Sensitivity	FDR
EBarrays	I	0.527 (0.002)	0.033 (<0.001)
	II	0.525 (0.002)	0.033 (<0.001)
	III	0.504 (0.002)	0.038 (0.001)
Fold change	I	0.559 (0.002)	0.127 (0.002)
	II	0.558 (0.002)	0.121 (0.002)
	III	0.553 (0.002)	0.146 (0.002)
EBarrays	Ib	0.324 (0.002)	0.056 (0.002)
	IIb	0.329 (0.002)	0.058 (0.002)
	IIIb	0.258 (0.003)	0.064 (0.002)
Fold change	Ib	0.580 (0.002)	0.455 (0.002)
	IIb	0.572 (0.002)	0.418 (0.001)
	IIIb	0.582 (0.003)	0.509 (0.002)

The genes are identified using *EBarrays* and a fold-change criterion applied to 12 groups of simulations (each group contains 20 simulated data sets). The first group (I) represents 16 mRNA samples on 16 arrays; group II has 32 samples pooled onto 8 arrays; group III has 24 samples pooled onto 8 arrays. Averages across each data set in the group are given; standard errors are shown in parentheses. Designs Ib, IIb, and IIIb correspond to 6 samples on 6 arrays; 12 samples on 3 arrays; and 6 samples on 3 arrays. Using the criterion given in Equation 6.7, designs I and II are equivalent and designs Ib and IIb are equivalent.

studies (see Kendziorski et al., 2003b). Biological and technical variability are defined to give $\lambda = 2$; the proportion of DE genes was set to 0.25.

Recall that a design with 16 subjects on 16 arrays was considered in Section 1. As shown there, when reducing the number of arrays to 8 when $\lambda = 2$, the number of subjects must be increased to 32 for the pooled and nonpooled designs to be equivalent. If the appropriate increase is not made, the designs are not considered equivalent and are therefore not expected to have similar properties. To illustrate this, 60 data sets were simulated as described above. Twenty of the sets were simulated for 16 subjects on 16 arrays; 20 sets contained data from a design with 32 subjects on 8 arrays (equivalent design); the remaining 20 simulations considered 24 subjects on 8 arrays (nonequivalent design).

To evaluate the comparability of these designs, two methods for identifying differentially expressed genes were applied to the simulated data: an empirical Bayes approach, *EBarrays* (Kendziorski et al., 2003b), and a simple fold change criterion.

For the latter, a gene was considered differentially expressed if the fold change between the conditions was bigger than two. The fold change criterion is known to be problematic when identifying differentially expressed genes. It is used here because it was the criterion used in Agrawal et al. (2001); furthermore, its simplicity yields a comparison among designs that is quite easy to interpret. The results, shown in Table 6.1, indicate that pooled and nonpooled designs yield comparable numbers of identified genes with comparable false discovery rates (FDRs) when the total numbers of subjects and arrays are defined so that the designs are equivalent. When fewer subjects are used, the results show significantly reduced sensitivity and increased FDR. As shown in the lower part of Table 6.1, the differences are more pronounced in smaller designs.

6.5 CONCLUSION

Investigators are often faced with a number of difficult questions when designing a microarray experiment. One concerns deciding on an appropriate number of subjects and arrays to use. Due to the large number of genes on an array, gene dependent variability, and gene dependent effects, the sample size problem is not trivial. This problem is not considered here (see Müller et al. [2003] for microarray sample size calculations).

Instead, we have considered the problem of defining fairly general conditions under which it is advantageous to pool. We have also derived a formula which establishes equivalency between pooled and nonpooled designs. In particular, the formula presented in Equation 6.7 defines the total number of subjects and arrays required in a pooled experiment to obtain gene expression estimates that are, on average, as precise as in the no-pooling case. Experiments are underway to check assumptions made in these calculations, estimate variance components, and investigate the utility of pooled and nonpooled designs to identify differentially expressed genes.

REFERENCES

Agrawal, D., T. Chen, R. Irby, J. Quackenbush, A.F. Chambers, M. Szabo, A. Cantor, D. Coppola, and T.J. Yeatman (2002) Osteopontin identified as lead marker of colon cancer progression, using pooled sample expression profiling. *Journal of the National Cancer Institute* 94: 513–521.

Amos, C.I., M.L. Frazier, and W. Wang (2000). DNA pooling in mutation detect ion with reference to sequence analysis. *American Journal of Human Genetics* 66: 1689–1692.

Brown, K.M., T. MacDonald, P.H. Cogen, Y.W. Chen, K. Peterson et al. (2001) Identification of expression changes of prognostic and therapeutic value in metastasizing medulloblastoma. Abstract # 39 at Oncogenics Conference: Dissecting Cancer Through Genome Research, sponsored by Nature Genetics and the American Association for Cancer Research.

Chabas, D., S.E. Baranzini, D. Mitchell, C.C.A. Bernard, S.R. Rittling et al. (2001) The influence of the proinflammatory cytokine, osteopontin, on autoimmune demyelinating disease. *Science* 294: 1731–1735.

Cheung, V., L.K. Conlin, T.M. Weber, M. Arcaro, K-Y. Jen, M. Morley, and R. Spielman (2003) Natural variation in human gene expression assessed in lymphoblastoid cells. *Nature Genetics* 33: 422–425.

Churchill, G.A. and B. Oliver (2001) Sex, flies and microarrays. *Nature Genetics* 29: 355–356.

Dorfman, R. (1943) The detection of defective members of large populations. *Annals of Mathematical Statistics* 14: 436–440.

Ernard, W., P. Khaitovich, J. Klose, S. Zollner, F. Heissig et al. (2002) Intra- and interspecific variation in primate gene expression patterns. *Science* 296: 340–343.

Gastwirth, J.L. and P.A. Hammick (1989) Estimation of the prevalence of a rare disease, preserving the anonymity of the subjects by group testing: application to estimating the prevalence of aids antibodies in blood donors. *Journal of Statistical Planning and Inference* 22: 15–27.

Gastwirth, J.L. (2000) The efficiency of pooling in the detection of rare mutations. *American Journal of Human Genetics* 67: 1036–1039.

Gieseg, M.A., T. Cody, M.Z. Man, S.J. Madore, M.A. Rubin, and E.P. Kaldjian (2002) Expression profiling of human renal carcinomas with functional taxonomic analysis. *BMC Bioinformatics*, 3: 26, http://www.biomedcentral.com/1471-2105/3/26.

Jin, W., R.M Riley, R.D. Wolfinger, K.P. White, G. Passador-Gurgel, and G.Gibson (2001) The contributions of sex, genotype and age to transcriptional variance in *Drosophila melanogaster*. *Nature Genetics* 29: 389–395.

Kendziorski, C.M., Y. Zhang, H. Lan, and A. Attie (2003a) The efficiency of mRNA pooling in microarray experiments. *Biostatistics* 4: 465–477.

Kendziorski, C.M., M.A. Newton, H. Lan, and M.N. Gould (2003b) On parametric empirical Bayes methods for comparing multiple groups using replicated gene expression profiles. *Statistics in Medicine* 22: 3899–3914.

Müller, P., G. Parmigiani, C. Robert, and J. Rousseau (2003) Optimal sample size for multiple testing: the case of gene expression microarrays. Johns Hopkins University, Department of Biostatistics, Working Paper, p. 31.

Nguyen, D.V., A.B. Arpat, N. Wang, and R.J. Carroll (2002) DNA microarray experiments: biological and technological aspects. *Biometrics* 58: 701–717.

Parmigiani, G., E.S. Garrett, R. Irizarry, and S.L. Zeger (2003) *The Analysis of Gene Expression Data: Methods and Software*. Springer-Verlag, Berlin.

Peng, X., C.L. Wood, E.M. Blalock, K.C. Chen, P.W. Landfield, and A.J. Stromberg (2003) Statistical implications of pooling RNA samples for microarray experiments. *BMC Bioinformatics* 4: 26.

Pfeiffer, R.M., J.L Rutter, M.H. Gail, J. Struewing, and J.L. Gastwirth (2002) Efficiency of DNA pooling to estimate joint allele frequencies and measure linkage disequilibrium. *Genetic Epidemiology* 22: 94–102.

Pritchard, C., L. Hsu, J. Delrow, and P.S. Nelson (2001) Project normal: defining normal variance in mouse gene expression. *Proceedings of the National Academy of Sciences, USA*, 98: 13266–13271.

Saban, M.R., H. Hellmich, N. Nguyen, J. Winston, T.G. Hammond, and R. Saban (2001) Time course of LPS-induced gene expression in a mouse model of genitourinary inflammation. *Physiological Genomics* 5: 147–160.

Waring, J.F., R.A. Jolly, R. Ciurlionis, P.Y. Lum, J.T. Praestgaard et al. (2001) Clustering of hepatotoxins based on mechanism of toxicity using gene expression profiles. *Toxicology and Applied Pharmacology* 175: 28–42.

Weil, M.R., T. Macatee, and H.R. Garner (2002) Toward a universal standard: comparing two methods for standardizing spotted microarray data. *Biotechniques* 32: 1310–1314.

Yang, I.V., E. Chen, J. P. Hasseman, W. Liang, B.C. Frank, S. Wang, V. Sharov, A.I. Saeed, J. White, J. Li, N.H. Lee, T. J. Yeatman, and J. Quackenbush (2002) Within the fold: assessing differential expression measures and reproducibility in microarray assays. *Genome Biology* 3: 1–13.

7 Designing Microarrays for the Analysis of Gene Expressions

Jane Y. Chang and Jason C. Hsu

CONTENTS

7.1 Two Approaches to Gene Expressions Analysis 111
7.2 Designing 2-Channel Microarrays 112
7.3 Modeling 2-Channel Microarray Gene Expression Data 113
  7.3.1 Model in Matrix Form 114
  7.3.2 Reduced Normal Equation and Least Squares Estimates 114
  7.3.3 Orthogonality Property of Balanced $2 \times a$ Generalized Latin Square Designs 115
  7.3.4 Estimators of Parameters of Interest 120
    7.3.4.1 Using the Unweighted Average of Housekeeping Gene Differentials as Control 120
    7.3.4.2 Using the Weighted Average of Housekeeping Gene Differentials as Control 122
    7.3.4.3 Using the Unweighted Average of Differentials of All the Gene as Control 123
7.4 Estimation When the Microarray Design Is Not Orthogonal 124
7.5 Summary 124
References 129

7.1 TWO APPROACHES TO GENE EXPRESSIONS ANALYSIS

One approach to gene expression analysis is testing for equalities of gene expression levels without explicit modeling [1–4]. Statistical methods taking this nonmodeling approach typically use either conservative multiplicity adjustments, or use permutation testing or resampling to capture the structure of the joint distribution of the test statistics.

Another approach is to explicitly model the data [6–9]. Modeling is used to remove variability due to extraneous factors. It can also provide more precise multiplicity adjustment than nonmodeling-based multiple testing. In contrast to the nonmodeling approach, modeling makes it possible to obtain confidence intervals for differential expression, allowing for inference not only on the existence of differential expressions, but also on the magnitude of differential expressions.

The design of microarray experiments we recommend in this chapter facilitates gene expressions analysis that uses the linear modeling approach. With this design, normalization and unbiased estimation of differential expressions can be computed using explicit formulas. This allows for more efficient computations (both in terms of time and memory useage) in the analysis of gene expressions data, compared with matrix inversion algorithms.

7.2 DESIGNING 2-CHANNEL MICROARRAYS

We believe that microarray experiments should be designed in accordance with the established statistical principles of *blocking, randomization*, and *replication*.

Variations observed in gene expressions may be attributable to differences in the conditions in the samples (e.g., normal and diseased) being compared, or differences in nuisance factors such as array, dye, and spot. Blocking for known nuisance factors avoids systematic bias in estimation. The design we recommend below incorporates array and dye as blocking factors in such a way that that unbiased estimates for differential expressions can be efficiently computed using explicit formulas.

Blocking can also improve the precision of estimation, if within-block variation is small compared to between-block variation. Sample to sample variability and array to array variability can be effectively blocked by hybridizing each replicate sample to a different array. In subsequent sections we will show that, for the design we recommend, formulas for differential expression estimators and their standard errors (SEs) remain the same regardless of whether array is considered a fixed effect or a random effect in modeling the data.

Randomization is the basis for valid statistical inference. Random assignment of replicate samples to arrays and dyes avoids unintended systematic bias due to such nuisance factors. To avoid systematic spot bias, the positions at which the genes are spotted should be completely randomized, separately for each array.

Replication is necessary for the estimation of variability. Replication can be accomplished by having multiple samples, and by spotting each gene more than once on a microarray.

The particular design we recommend, which we call a balanced $2 \times a$ generalized Latin square design, is as follows:

1. Randomly assign treatment labels T_1 and T_2 to treatments 1 and 2
2. Randomly assign array labels 1 through a (a *is* even) to the actual microarrays
3. Randomly assign dye labels 1 and 2 to the Cy5 (red) and Cy3 (green) dyes
4. Assign treatment labels T_1 and T_2 to dyes and arrays according to Table 7.1
5. Randomly spot the lth gene n_l times on each array

TABLE 7.1

Dye\Array	1	2	3	...	$a-1$	a
1	T_1	T_2	T_1	...	T_1	T_2
2	T_2	T_1	T_2	...	T_2	T_1

This design can be thought of as $a/2$ replications of a 2×2 Latin square. As the number of times each gene is spotted remains the same across different arrays, for each 2×2 Latin square, the difference of gene expression averaged across the two arrays are free of array and dye effects. These averaged differences are the least squares estimators for differential expressions from each 2×2 Latin square. With the 2×2 Latin squares being replicates of each other, the covariance structure of these estimators remains the same across the Latin squares regardless of whether the array effect is fixed or random. Consequently, the average of these averaged differences is the least squares estimators of differential expressions, as we prove in the following section.

7.3 MODELING 2-CHANNEL MICROARRAY GENE EXPRESSION DATA

We assume the following model for a gene expression microarray design,

$$\log(z_{ijklm}) = y_{ijklm} = \mu + \alpha_i + \delta_j + \tau_k + \gamma_l + (\delta\gamma)_{jl} + (\tau\gamma)_{kl} + \varepsilon_{ijklm} \tag{7.1}$$

$$\varepsilon_{ijklm} \sim N(0, \sigma^2) \text{ and } \alpha_i \sim N(0, \sigma_A^2)$$

$$\varepsilon_{ijklm}\text{s and } \alpha_i\text{s are all mutually independent}$$

$$i = 1, 2, \ldots, a;\ \ j = 1, 2;\ \ k = 1, 2;\ \ l = 1, 2, \ldots, g$$

$$m = 1, 2, \ldots, n_{il}$$

where y_{ijklm} is the logarithm of the intensity of the observed gene expression level z_{ijklm}, α_i is the effect of the ith randomly selected array, δ_j is the effect of the jth dye, τ_k is the effect of the kth tissue, γ_l is the effect of the lth gene, $(\tau\gamma)_{kl}$ is the effect of the kth tissue and lth gene interaction, $(\delta\gamma)_{jl}$ is the effect of jth dye and lth gene interaction. It then follows that the variance covariance of the observation vector $\boldsymbol{Y}, V(\boldsymbol{Y})$, is

$$V(\boldsymbol{Y}) = I_a \otimes [\sigma^2 I_{2K} + \sigma_A^2 J_{2K}]$$

where notationally I_n is an $n \times n$ identity matrix, 1_n is an $n \times 1$ vector of ones, J_n is an $n \times n$ matrix of ones, $K = \sum_{l=1}^{g} n_{il}$ is the number of spots in each array, and $\otimes$ denotes the Kronecker product of two matrices.

With this model, preprocessing to adjust for array and dye differences amounts to fitting array and dye effects and gene $\times$ dye interaction in the model. Chosen genes are used to calculate normalized expression differentials. These estimated normalized differential expressions form a set of estimable functions. If the correlation structure of estimates of these functions can be derived based on the model, then accurate multiplicity adjustment becomes possible, and one can obtain simultaneous confidence intervals for differential expressions.

Note that, in contrast to this approach, Hsu et al. [9] treated array as a fixed effect in the model. Treating array as a random effect rather than a fixed effect affects the estimation of the array effects and its interaction with other factor effects. But for our design, it does not affect the estimation and testing of the fixed dye, treatment, gene, and treatment by gene interaction effects, as we show in the following section.

7.3.1 Model in Matrix Form

With array being a random factor, the model 7.1 can be written as:

$$\begin{aligned} Y &= 1_{2aK}\mu + X_{\mathrm{A}}\alpha + X_{\mathrm{D}}\delta + X_{\mathrm{T}}\tau + X_{\mathrm{G}}\gamma + X_{\mathrm{DG}}\phi + X_{\mathrm{TG}}\chi + \varepsilon \\ V &= I_a \otimes [\sigma^2 I_{2K} + \sigma_{\mathrm{A}}^2 J_{2K}] \end{aligned} \tag{7.2}$$

where α is a vector of a array effects, δ is a vector of *two* dye effects, τ is a vector of the *two* treatment effects, γ is a vector of the g gene effects, ϕ is a vector of the $2g$ dye $\times$ gene interaction effects, χ is a vector of the $2g$ treatment $\times$ gene interaction effects, and $X_{\mathrm{A}}, X_{\mathrm{D}}, X_{\mathrm{T}}, X_{\mathrm{G}}, X_{\mathrm{DG}}$, and X_{TG} are the design matrices corresponding to $\alpha, \delta, \tau, \gamma, \phi$, and χ respectively.

7.3.2 Reduced Normal Equation and Least Squares Estimates

Let $X_1 = [X_{\mathrm{TG}}], X_2 = [1_{2aK}, X_{\mathrm{A}}, X_{\mathrm{D}}, X_{\mathrm{T}}, X_{\mathrm{G}}, X_{\mathrm{DG}}], X = [X_1, X_2], \varpi_1 = \chi, \varpi_2 = (\mu, \alpha', \delta', \tau', \gamma', \phi', \chi')'$, and $\varpi = [\varpi_1, \varpi_2]$, then model 7.2 can be written as:

$$Y = X_1\varpi_1 + X_2\varpi_2 + \varepsilon$$

A least square estimator $\hat{\varpi}$ of ϖ minimizes $||Y - X\varpi||$, and can be obtained by solving the normal equation:

$$(X'V^{-1}X)\hat{\varpi} = X'V^{-1}y$$

This is the same as solving:

$$X_1'V^{-1}X_1\hat{\varpi}_1 + X_1'V^{-1}X_2\hat{\varpi}_2 = X_1'V^{-1}y \tag{7.3}$$

and

$$X_2'V^{-1}X_1\hat{\varpi}_1 + X_2'V^{-1}X_2\hat{\varpi}_2 = X_2'V^{-1}y \tag{7.4}$$

Eliminating $\hat{\varpi}_2$ by multiplying both sides of Equation 7.4 by $(X_2'V^{-1}X_2)^-$ and subtracting it from Equation 7.3, we find that $\hat{\varpi}_1$ can be obtained by solving

$$\begin{aligned}&[X_1'V^{-1}X_1-(X_1'V^{-1}X_2(X_2'V^{-1}X_2)^-X_2'V^{-1}X_1)]\hat{\varpi}_1\\&\quad=X_1'V^{-1}y-X_1'V^{-1}X_2(X_2'V^{-1}X_2)^-X_2'V^{-1}y\end{aligned}$$

where V^- denotes a generalized inverse of matrix V. That is, the coefficient matrix, C_{TG}, for estimating treatment × gene treatment combinations, eliminating array, dye, treatment, gene, and dye × gene effects, is

$$C_{\mathrm{TG}}=[X_1'V^{-1}X_1-(X_1'V^{-1}X_2(X_2'V^{-1}X_2)^{-1}X_2'V^{-1}X_1)]$$

Note that

$$X_1'V^{-1}X_1=X_{\mathrm{TG}}'V^{-1}X_{\mathrm{TG}'}$$

$$X_2'V^{-1}X_2=\begin{bmatrix}1'V^{-1}1 & 1'V^{-1}X_{\mathrm{A}} & 1'V^{-1}X_{\mathrm{D}} & 1'V^{-1}X_{\mathrm{T}} & 1'V^{-1}X_{\mathrm{G}} & 1'V^{-1}X_{\mathrm{DG}}\\ X_{\mathrm{A}}'V^{-1}1 & X_{\mathrm{A}}'V^{-1}X_{\mathrm{A}} & X_{\mathrm{A}}'V^{-1}X_{\mathrm{D}} & X_{\mathrm{A}}'V^{-1}X_{\mathrm{T}} & X_{\mathrm{A}}'V^{-1}X_{\mathrm{G}} & X_{\mathrm{A}}'V^{-1}X_{\mathrm{DG}}\\ X_{\mathrm{D}}'V^{-1}1 & X_{\mathrm{D}}'V^{-1}X_{\mathrm{A}} & X_{\mathrm{D}}'V^{-1}X_{\mathrm{D}} & X_{\mathrm{D}}'V^{-1}X_{\mathrm{T}} & X_{\mathrm{D}}'V^{-1}X_{\mathrm{G}} & X_{\mathrm{D}}'V^{-1}X_{\mathrm{DG}}\\ X_{\mathrm{T}}'V^{-1}1 & X_{\mathrm{T}}'V^{-1}X_{\mathrm{A}} & X_{\mathrm{T}}'V^{-1}X_{\mathrm{D}} & X_{\mathrm{T}}'V^{-1}X_{\mathrm{T}} & X_{\mathrm{T}}'V^{-1}X_{\mathrm{G}} & X_{\mathrm{T}}'V^{-1}X_{\mathrm{DG}}\\ X_{\mathrm{G}}'V^{-1}1 & X_{\mathrm{G}}'V^{-1}X_{\mathrm{A}} & X_{\mathrm{G}}'V^{-1}X_{\mathrm{D}} & X_{\mathrm{G}}'V^{-1}X_{\mathrm{T}} & X_{\mathrm{G}}'V^{-1}X_{\mathrm{G}} & X_{\mathrm{G}}'V^{-1}X_{\mathrm{DG}}\\ X_{\mathrm{DG}}'V^{-1}1 & X_{\mathrm{DG}}'V^{-1}X_{\mathrm{A}} & X_{\mathrm{DG}}'V^{-1}X_{\mathrm{D}} & X_{\mathrm{DG}}'V^{-1}X_{\mathrm{T}} & X_{\mathrm{DG}}'V^{-1}X_{\mathrm{G}} & X_{\mathrm{DG}}'V^{-1}X_{\mathrm{DG}}\end{bmatrix}$$

and

$$X_1'V^{-1}X_2=\left[X_{\mathrm{TG}}'V^{-1}1,X_{\mathrm{TG}}'V^{-1}X_{\mathrm{A}},X_{\mathrm{TG}}'V^{-1}X_{\mathrm{D}},X_{\mathrm{TG}}'V^{-1}X_{\mathrm{T}},X_{\mathrm{TG}}'V^{-1}X_{\mathrm{G}},X_{\mathrm{TG}}'V^{-1}X_{\mathrm{DG}}\right]$$

In the following section, we will establish that the design we recommend is orthogonal in the sense that all the off-diagonal cross product terms are 0, simplifying the solutions.

7.3.3 Orthogonality Property of Balanced 2 × *a* Generalized Latin Square Designs

Let n_l denote the common number of times the lth gene is spotted on all the arrays and let $K=\sum_{l=1}^{g}n_l$. It can be seen that for a balanced $2\times a$ generalized Latin square design, we have

$$X_A=I_a\otimes 1_{2K}$$

$$X_D=1_a\otimes I_2\otimes 1_K$$

$$X_T = 1_{a/2} \otimes \begin{bmatrix} 1 & 0 \\ 0 & 1 \\ 0 & 1 \\ 1 & 0 \end{bmatrix} \otimes 1_K$$

$$X_G = 1_{2a} \otimes \text{ Diag}(1_{n_1}, 1_{n_2}, \ldots, 1_{n_g}), X_{\text{TG}} = 1_{a/2} \otimes \begin{bmatrix} 1 & 0 \\ 0 & 1 \\ 0 & 1 \\ 1 & 0 \end{bmatrix}$$

$$\otimes \text{Diag}(1_{n_1}, 1_{n_2}, \ldots, 1_{n_g})$$

$$X_{DG} = 1_a \otimes I_2 \otimes \text{Diag}(1_{n_1}, 1_{n_2}, \ldots, 1_{n_g})$$

Note that

$$V^{-1} = I_a \otimes \left[\frac{1}{\sigma^2} \left(I_{2k} - \frac{\sigma_A^2}{\sigma^2 + 2K\sigma_{\text{A}}^2} J_{2K} \right) \right]$$

$$= \frac{1}{\sigma^2} I_{2aK} - \frac{\sigma_A^2}{\sigma^2(\sigma^2 + 2K\sigma_{\text{A}}^2)} I_a \otimes J_{2K}$$

To show the orthogonality property of a balanced $2 \times a$ generalized Latin square designs ($n_{il} = n_l$), we reparameterized model 7.2 to

$$Y = 1_{2ak}\mu^* + X_{\text{A}}^*\alpha + X_{\text{D}}^*\delta + X_{\text{T}}^*\tau + X_{\text{G}}^*\gamma + X_{\text{TG}}^*\chi + X_{\text{DG}}^*\phi + \varepsilon \tag{7.5}$$

where

$$X_{\text{A}}^* = X_{\text{A}} - \frac{1}{a} 1_{2aK} 1_a'$$

$$X_{\text{D}}^* = X_{\text{D}} - \frac{1}{2} 1_{2aK} 1_2'$$

$$X_{\text{T}}^* = X_{\text{T}} - \frac{1}{2} 1_{2aK} 1_2'$$

$$X_{\text{G}}^* = X_{\text{G}} - \frac{1}{K} 1_{2aK} N', \quad \text{where } N' = (n_1, n_2, \ldots, n_g),$$

$$X_{\text{DG}}^* = X_{\text{DG}} - \frac{1}{K} 1_a \otimes I_2 \otimes \begin{bmatrix} n_1 & n_2 & \ldots & n_g \\ n_1 & n_2 & \ldots & n_g \\ \vdots & \vdots & \vdots & \vdots \\ n_1 & n_2 & \ldots & n_g \end{bmatrix}$$

$$- \frac{1}{2} 1_a \otimes J_2 \otimes \text{Diag}(1_{n1}, 1_{n2}, \ldots, 1_{ng}) + \frac{1}{2aK} 1_{2aK} r'$$

and

$$X_{\mathrm{TG}}^{*} = X_{\mathrm{TG}} - \frac{1}{K}1_{a/2} \otimes \begin{bmatrix} 1 & 0 \\ 0 & 1 \\ 0 & 1 \\ 1 & 0 \end{bmatrix} \otimes \begin{bmatrix} n_1 & n_2 & \dots & n_g \\ n_1 & n_2 & \dots & n_g \\ \vdots & \vdots & \vdots & \vdots \\ n_1 & n_2 & \dots & n_g \end{bmatrix}$$
$$- \frac{1}{2}1_a \otimes J_2 \otimes \mathrm{Diag}(1_{n1}, 1_{n2}, \dots, 1_{ng}) + \frac{1}{2aK}1_{2aK}r'$$

where $r' = (an_1, an_2, \dots, an_g, an_1, \dots, an_g)$.

Lemmas 7.4, 7.6, and Theorem 7.1, applicable to model 7.2 with array being a random effect, is similar to Lemmas 7.1 to 7.3 stated in the appendix, applied to a model with array being a fixed effect [10].

Lemma 7.4

(1) $X_{\mathrm{A}}^{*\prime}V^{-1}1_{2aK} = 0$
(2) $X_{\mathrm{D}}^{*\prime}V^{-1}1_{2aK} = 0$
(3) $X_{\mathrm{T}}^{*\prime}V^{-1}1_{2aK} = 0$
(4) $X_{\mathrm{G}}^{*\prime}V^{-1}1_{2aK} = 0$
(5) $X_{\mathrm{TG}}^{*\prime}V^{-1}1_{2aK} = 0$
(6) $X_{\mathrm{DG}}^{*\prime}V^{-1}1_{2aK} = 0$

Proof. Since

$$V^{-1} = \frac{1}{\sigma^2}I_{2aK} - \frac{\sigma_{\mathrm{A}}^2}{\sigma^2\left(\sigma^2 + 2K\sigma_{\mathrm{A}}^2\right)}I_a \otimes J_{2K}$$

$$V^{-1}1_{2aK} = \left(\frac{1}{\sigma^2}I_{2aK} - \frac{\sigma_{\mathrm{A}}^2}{\sigma^2\left(\sigma^2 + 2K\sigma_{\mathrm{A}}^2\right)}I_a \otimes J_{2K}\right)1_{2aK}$$
$$= \frac{1}{\sigma^2}1_{2aK} - \frac{2K\sigma_{\mathrm{A}}^2}{\sigma^2\left(\sigma^2 + 2K\sigma_{\mathrm{A}}^2\right)}1_{2aK}$$

By Lemma 7.1 (1)

$$X_{\mathrm{A}}^{*\prime}V^{-1}1_{2aK} = X_{\mathrm{A}}^{*\prime}V^{-1}1_{2aK} = \frac{1}{\sigma^2}X_{\mathrm{A}}^{*\prime}1_{2aK} - \frac{2K\sigma_{\mathrm{A}}^2}{\sigma^2\left(\sigma^2 + 2K\sigma_{\mathrm{A}}^2\right)}X_{\mathrm{A}}^{*\prime}1_{2aK}$$
$$= \frac{1}{\sigma^2}X_{\mathrm{A}}^{*\prime}1_{2aK} - \frac{2K\sigma_{\mathrm{A}}^2}{\sigma^2\left(\sigma^2 + 2K\sigma_{\mathrm{A}}^2\right)}X_{\mathrm{A}}^{*\prime}1_{2aK}$$
$$= 0 - 0 = 0$$

Results (2) to (6) follow by similar arguments.

By Lemma 7.4, the array, dye, treatment, treatment×gene, dye×gene parameter vectors are orthogonal to the mean parameter in model 7.5. Hence, the least squares estimators of $(\alpha', \delta', \tau', \gamma', \chi', \phi')'$ for the model 7.5 are the same as the least squares estimators of $(\alpha', \delta', \tau', \gamma', \chi', \phi')'$ for the model

$$Y = X^*_{\mathrm{A}}\alpha + X^*_{\mathrm{D}}\delta + X^*_{\mathrm{T}}\tau + X^*_{\mathrm{G}}\gamma + X^*_{\mathrm{TG}}\chi + X^*_{\mathrm{DG}}\phi + \varepsilon \tag{7.6}$$

$$V(y) = I_a \otimes [\sigma^2 I_{2k} + \sigma^2_{\mathrm{A}} J_{2K}] \tag{7.7}$$

The full information matrix for $(\alpha', \delta', \tau', \gamma', \chi', \phi')'$ under the model 7.6 is $X^{*\prime}V^{-1}X^*$ where $X^* = [X^*_{\mathrm{A}}, X^*_{\mathrm{D}}, X^*_{\mathrm{T}}, X^*_{\mathrm{G}}, X^*_{\mathrm{DG}}, X^*_{\mathrm{TG}}]'$.

For orthogonality between treatment×gene and the rest of the parameter vectors, it is sufficient to show that the off-diagonal blocking matrices of $X^{*\prime}V^{-1}X^*$ are all zero matrices. These results follow from Lemmas 7.5 and 7.6 given below. The proof of Lemma 7.5 is given in the appendix, while the proof of Lemma 7.6 follows from Lemmas 7.2 and 7.5.

Lemma 7.5

(1) $V^{-1}X^*_{\mathrm{A}} = \frac{1}{\sigma^2}X^*_{\mathrm{A}}$
(2) $V^{-1}X^*_{\mathrm{D}} = \frac{1}{\sigma^2}X^*_{\mathrm{D}}$
(3) $V^{-1}X^*_{\mathrm{T}} = \frac{1}{\sigma^2}X^*_{\mathrm{T}}$
(4) $V^{-1}X^*_{\mathrm{G}} = \frac{1}{\sigma^2}X^*_{\mathrm{G}}$
(5) $V^{-1}X^*_{\mathrm{DG}} = \frac{1}{\sigma^2}X^*_{\mathrm{DG}}$
(6) $V^{-1}X^*_{\mathrm{TG}} = \frac{1}{\sigma^2}X^*_{\mathrm{TG}}$

Lemma 7.6

(1) $X^{*\prime}_{\mathrm{DG}}V^{-1}X^*_{\mathrm{A}} = 0$
(2) $X^{*\prime}_{\mathrm{DG}}V^{-1}X^*_{\mathrm{D}} = 0$
(3) $X^{*\prime}_{\mathrm{DG}}V^{-1}X^*_{\mathrm{T}} = 0$
(4) $X^{*\prime}_{\mathrm{DG}}V^{-1}X^*_{\mathrm{G}} = 0$
(5) $X^{*\prime}_{\mathrm{DG}}V^{-1}X^*_{\mathrm{TG}} = 0$
(6) $X^{*\prime}_{\mathrm{TG}}V^{-1}X^*_{\mathrm{A}} = 0$
(7) $X^{*\prime}_{\mathrm{TG}}V^{-1}X^*_{\mathrm{D}} = 0$
(8) $X^{*\prime}_{\mathrm{TG}}V^{-1}X^*_{\mathrm{T}} = 0$
(9) $X^{*\prime}_{\mathrm{TG}}V^{-1}X^*_{\mathrm{G}} = 0$
(10) $X^{*\prime}_{\mathrm{G}}V^{-1}X^*_{\mathrm{A}} = 0$
(11) $X^{*\prime}_{\mathrm{G}}V^{-1}X^*_{\mathrm{D}} = 0$
(12) $X^{*\prime}_{\mathrm{G}}V^{-1}X^*_{\mathrm{T}} = 0$

(13) $X_{\mathrm{T}}^{*\prime}V^{-1}X_{\mathrm{A}}^{*} = 0$
(14) $X_{\mathrm{T}}^{*\prime}V^{-1}X_{\mathrm{D}}^{*} = 0$
(15) $X_{\mathrm{D}}^{*\prime}V^{-1}X_{\mathrm{A}}^{*} = 0$

Thus, the full information matrix for $(\alpha', \delta', \tau', \gamma', \chi', \phi')'$ under the model (7.6) is

$$X^{*\prime}V^{-1}X^{*} = \begin{bmatrix} X_{\mathrm{A}}^{*\prime} \\ X_{\mathrm{D}}^{*\prime} \\ X_{\mathrm{G}}^{*\prime} \\ X_{\mathrm{T}}^{*\prime} \\ X_{\mathrm{DG}}^{*\prime} \\ X_{\mathrm{TG}}^{*\prime} \end{bmatrix} V^{-1} \left[X_{\mathrm{A}}^{*}\ X_{\mathrm{D}}^{*}\ X_{\mathrm{G}}^{*}\ X_{\mathrm{T}}^{*}\ X_{\mathrm{DG}}^{*}\ X_{\mathrm{TG}}^{*}\right]$$

$$= \begin{bmatrix} X_{\mathrm{A}}^{*\prime}V^{-1}X_{\mathrm{A}}^{*} & 0 & 0 & 0 & 0 & 0 \\ 0 & X_{\mathrm{D}}^{*\prime}V^{-1}X_{\mathrm{D}}^{*} & 0 & 0 & 0 & 0 \\ 0 & 0 & X_{\mathrm{T}}^{*\prime}V^{-1}X_{\mathrm{T}}^{*} & 0 & 0 & 0 \\ 0 & 0 & 0 & X_{\mathrm{G}}^{*\prime}V^{-1}X_{\mathrm{G}}^{*} & 0 & 0 \\ 0 & 0 & 0 & 0 & X_{\mathrm{DG}}^{*\prime}V^{-1}X_{\mathrm{DG}}^{*} & 0 \\ 0 & 0 & 0 & 0 & 0 & X_{\mathrm{TG}}^{*\prime}V^{-1}X_{\mathrm{TG}}^{*} \end{bmatrix}$$

Therefore, estimates of treatment×gene interaction parameters are based on the sample means of treatment×gene combinations. No adjustment for blocking factors is necessary.

Theorem 7.1 The least squares estimates of $(\delta', \tau', \gamma', \chi', \phi')'$ for the model 7.5 are

(1) $\hat{\delta} = \left(X_{\mathrm{D}}^{*\prime}V^{-1}X_{\mathrm{D}}^{*}\right)^{-} X_{\mathrm{D}}^{*\prime}V^{-1}Y = \begin{bmatrix} \bar{y}_{.1...} & -\bar{y}_{.....} \\ \bar{y}_{.2...} & -\bar{y}_{.....} \end{bmatrix}_{2\times 1}$

(2) $\hat{\tau} = \left(X_{\mathrm{T}}^{*\prime}V^{-1}X_{\mathrm{T}}^{*}\right)^{-} X_{\mathrm{T}}^{*\prime}V^{-1}Y = \begin{bmatrix} \bar{y}_{..1..} & -\bar{y}_{.....} \\ \bar{y}_{..2..} & -\bar{y}_{.....} \end{bmatrix}_{2\times 1}$

(3) $\hat{\gamma} = \left(X_{\mathrm{G}}^{*\prime}V^{-1}X_{\mathrm{G}}^{*}\right)^{-} X_{\mathrm{G}}^{*\prime}V^{-1}Y = \begin{bmatrix} \bar{y}_{...1.} & -\bar{y}_{.....} \\ & \vdots \\ \bar{y}_{...g.} & -\bar{y}_{.....} \end{bmatrix}_{g\times 1}$

(4) $\hat{\chi} = \left(X_{\mathrm{TG}}^{*\prime}V^{-1}X_{\mathrm{TG}}^{*}\right)^{-} X_{\mathrm{TG}}^{*\prime}V^{-1}Y = \begin{bmatrix} \bar{y}_{..11.} & -\bar{y}_{..1..} & -\bar{y}_{...1.} & +\bar{y}_{.....} \\ & & \vdots & \\ \bar{y}_{..1g.} & -\bar{y}_{..1..} & -\bar{y}_{...g.} & +\bar{y}_{.....} \\ \bar{y}_{...21.} & -\bar{y}_{..2..} & -\bar{y}_{...1.} & +\bar{y}_{.....} \\ & & \vdots & \\ \bar{y}_{..2g.} & -\bar{y}_{..2..} & -\bar{y}_{...g.} & +\bar{y}_{.....} \end{bmatrix}$

$$(5)\ \hat{\phi} = \left(X^{*\prime}_{\text{DG}} V^{-1} X^{*}_{\text{DG}}\right)^{-} X^{*\prime}_{\text{DG}} V^{-1} Y = \begin{bmatrix} \bar{y}_{.1.1.} & -\bar{y}_{.1...} & -\bar{y}_{...1.} & +\bar{y}_{.....} \\ & & \vdots & \\ \bar{y}_{.1.g.} & -\bar{y}_{.1...} & -\bar{y}_{...g.} & +\bar{y}_{.....} \\ \bar{y}_{.2.1.} & -\bar{y}_{.2...} & -\bar{y}_{...1.} & +\bar{y}_{.....} \\ & & \vdots & \\ \bar{y}_{.2.g.} & -\bar{y}_{.2...} & -\bar{y}_{...g.} & +\bar{y}_{.....} \end{bmatrix}$$

Proof. The results follow immediately from Lemmas 7.3 and 7.5 in the appendix.

The design we discuss above is an orthogonal design, for which it is known that the generalized least squares estimators are the same as ordinary least squares estimators. As a consequence, all the information on treatment $\times$ gene effects is contained in comparisons within treatment $\times$ gene stratum, whether effects are random or fixed. This simplification in estimation makes possible the derivation of joint distribution of normalized estimators of differential expressions below, which in turn allows for precise multiplicity adjustment.

7.3.4 Estimators of Parameters of Interest

Yang et al. [2] discussed different ways of normalizing cDNA microarray data. They proposed that the average of all genes on the array may be used for normalization when only a small proportion of genes are expected to be differentially expressed, or when there is symmetry in the expression levels of the up/down-regulated genes. Otherwise housekeeping genes, the genes that are present in virtually all cell types [11, p. 231], can be used.

Kerr et al. [5,8] and Kerr and Churchill [6,7] normalized with respect to the average of all genes, as did Hsu et al. [9].

It is assumed that the housekeeping genes do not influence the disease under study as they are present in all cell types. Hence, the expressions for the housekeeping genes are expected to be constant across different tissue types. However, the housekeeping genes are not necessarily constantly expressed across different tissue types and this aspect is often criticized when normalizing with respect to the housekeeping genes. There appears to be some recent success in finding subgroups of housekeeping genes whose *differentials* remain relatively constant across different cell types, using real-time PCR (RT-PCR) which is much more accurate in measuring gene expressions than using microarray [12]. These genes then become good candidates for normalization. In this research, we used the housekeeping genes as the normalization genes.

For each of these normalization schemes, the estimators of differential expressions and their variance–covariance structure are given in the following section.

7.3.4.1 Using the Unweighted Average of Housekeeping Gene Differentials as Control

The parameter of interest is the difference of log intensities between treatments for each gene, compared to the average difference of housekeeping genes (control). Based on the model 7.1, the parameter we wish to estimate is

$$\theta_l = (\tau\gamma)_{1l} - (\tau\gamma)_{2l} - \left[\overline{(\tau\gamma)}^{\mathrm{H}}_{1.} - \overline{(\tau\gamma)}^{\mathrm{H}}_{2.}\right]$$

$$= (\tau\gamma)_{1l} - (\tau\gamma)_{2l} - \left[\frac{\sum_{i=1}^{h}(\tau\gamma)_{1i}}{h} - \frac{\sum_{i=1}^{h}(\tau\gamma)_{2i}}{h}\right], \quad l = h+1, \ldots, g$$

By Gauss–Markov theorem, the best linear unbiased estimator (BLUE) of θ_l is

$$\hat{\theta}_l = \bar{y}_{..1l.} - \bar{y}_{..2l.} - \left[\frac{\sum_{i=1}^{h}\bar{y}_{..1i.}}{h} - \frac{\sum_{i=1}^{h}\bar{y}_{..2i.}}{h}\right], \quad l = h+1,\ h+2, \ldots, g$$

Since

$$\mathrm{var}(\bar{y}_{..kl.}) = \frac{\sigma^2}{an_l} + \frac{\sigma_{\mathrm{A}}^2}{a}, \quad k = 1, 2;\ l = h+1,\ h+2, \ldots, g$$

$$\mathrm{cov}(\bar{y}_{..kl.}, \bar{y}_{..kl'.}) = \frac{\sigma_{\mathrm{A}}^2}{a}, \quad k = 1, 2;\ l \neq l'$$

Thus, the variance of $\hat{\theta}_l$ is

$$\mathrm{var}(\hat{\theta}_l) = \mathrm{var}(\bar{y}_{..1l.} - \bar{y}_{..2l.}) + \mathrm{var}\left[\frac{\sum_{i=1}^{h}\bar{y}_{..1i.} - \sum_{i=1}^{h}\bar{y}_{..2i.}}{h}\right]$$

$$- 2\mathrm{cov}\left(\bar{y}_{..1l.} - \bar{y}_{..2l.} - \left[\frac{\sum_{i=1}^{h}\bar{y}_{..1i.}}{h} - \frac{\sum_{i=1}^{h}\bar{y}_{..2i.}}{h}\right]\right)$$

$$= 2\left(\frac{\sigma^2}{an_l} + \frac{\sigma_{\mathrm{A}}^2}{a} - \frac{\sigma_{\mathrm{A}}^2}{a}\right) + \frac{2}{h^2}\left[\sum_{i=1}^{h}\left(\frac{\sigma^2}{an_i} + \frac{\sigma_{\mathrm{A}}^2}{a} - \frac{\sigma_{\mathrm{A}}^2}{a}\right)\right] - 0$$

$$= \frac{2\sigma^2}{an_l} + \frac{2\sigma^2}{h^2 a}\left[\sum_{i=1}^{h}\frac{1}{n_i}\right], \quad l = h+1, 2, \ldots, g$$

And, the covariance of $\hat{\theta}_l$ and $\hat{\theta}_{l'}$ is

$$\mathrm{cov}(\hat{\theta}_l, \hat{\theta}_{l'}) = \mathrm{cov}\left(\bar{y}_{..1l.} - \bar{y}_{..2l.} - \left[\frac{\sum_{i=1}^{h}\bar{y}_{..1i.}}{h} - \frac{\sum_{i=1}^{h}\bar{y}_{..2i.}}{h}\right], \bar{y}_{1l'} - \bar{y}_{2l'}\right.$$

$$\left. - \left[\frac{\sum_{i=1}^{h}\bar{y}_{..1i.}}{h} - \frac{\sum_{i=1}^{h}\bar{y}_{..2i.}}{h}\right]\right)$$

$$= 0 - 0 - 0 + \operatorname{var}\left[\frac{\sum_{i=1}^{h}\bar{y}_{..1i.} - \sum_{i=1}^{h}\bar{y}_{..2i.}}{h}\right]$$

$$= \frac{2\sigma^2}{ah^2}\left[\sum_{i=1}^{h}\frac{1}{n_i}\right], \quad l \neq l'$$

7.3.4.2 Using the Weighted Average of Housekeeping Gene Differentials as Control

In case the sample sizes for the housekeeping genes are not all equal, we can use the weighted average of housekeeping gene differentials as control. Then, the parameters we wish to estimate are

$$\theta_l = (\tau\gamma)_{1l} - (\tau\gamma)_{2l} - \left[\left(\overline{\tau\gamma}\right)_{1.}^{\mathrm{WH}} - \left(\overline{\tau\gamma}\right)_{2.}^{\mathrm{WH}}\right]$$
$$= (\tau\gamma)_{1l} - (\tau\gamma)_{2l} - \left[\frac{\sum_{i=1}^{h} n_i(\tau\gamma)_{1i}}{\sum_{i=1}^{h} n_i} - \frac{\sum_{i=1}^{h} n_i(\tau\gamma)_{2i}}{\sum_{i=1}^{h} n_i}\right], \quad l = h+1, \ldots, g$$

The least squares estimator of θ_l is

$$\hat{\theta}_l = \bar{y}_{..1l.} - \bar{y}_{..2l.} - \left[\frac{\sum_{i=1}^{h} n_i\bar{y}_{..1i.}}{\sum_{i=1}^{h} n_i} - \frac{\sum_{i=1}^{h} n_i\bar{y}_{..2i.}}{\sum_{i=1}^{h} n_i}\right], \quad l = h+1,\ h+2, \ldots, g$$

The variance of $\hat{\theta}_l$ is

$$\operatorname{var}(\hat{\theta}_l) = \operatorname{var}\left(\bar{y}_{..1l.} - \bar{y}_{..2l.}\right) + \operatorname{var}\left[\frac{\sum_{i=1}^{h} n_i\bar{y}_{..1i.}}{\sum_{i=1}^{h} n_i} - \frac{\sum_{i=1}^{h} n_i\bar{y}_{..2i.}}{\sum_{i=1}^{h} n_i}\right]$$
$$- 2\operatorname{cov}\left(\bar{y}_{..1l.} - \bar{y}_{..2l.} - \left[\frac{\sum_{i=1}^{h} n_i\bar{y}_{..1i.}}{\sum_{i=1}^{h} n_i} - \frac{\sum_{i=1}^{h} n_i\bar{y}_{..2i.}}{\sum_{i=1}^{h} n_i}\right]\right)$$
$$= \frac{2\sigma^2}{an_l} + \frac{2\sigma^2}{a\sum_{i=1}^{h} n_i} - 0$$
$$= \frac{2\sigma^2}{an_l} + \frac{2\sigma^2}{a\sum_{i=1}^{h} n_i}, \quad l = h+1,\ h+2, \ldots, g$$

and the covariance of $\hat{\theta}_l$ and $\hat{\theta}_{l'}$ is

$$\operatorname{cov}(\hat{\theta}_l, \hat{\theta}_{l'}) = \operatorname{cov}\left[\bar{y}_{..1l.} - \bar{y}_{..2l.} - \left[\frac{\sum_{i=1}^{h} n_i\bar{y}_{..1i.}}{\sum_{i=1}^{h} n_i} - \frac{\sum_{i=1}^{h} n_i\bar{y}_{..2i.}}{\sum_{i=1}^{h} n_i}\right], \bar{y}_{1l'} - \bar{y}_{2l'}\right.$$
$$\left. - \left[\frac{\sum_{i=1}^{h} n_i\bar{y}_{..1i.}}{\sum_{i=1}^{h} n_i} - \frac{\sum_{i=1}^{h} n_i\bar{y}_{..2i.}}{\sum_{i=1}^{h} n_i}\right]\right]$$

$$= 0 - 0 - 0 + \text{var}\left[\frac{\sum_{i=1}^{h} n_i \bar{y}_{..1i.}}{\sum_{i=1}^{h} n_i} - \frac{\sum_{i=1}^{h} n_i \bar{y}_{..2i.}}{\sum_{i=1}^{h} n_i}\right]$$

$$= \frac{2\sigma^2}{a\sum_{i=1}^{h} n_i}, \quad l \neq l'$$

7.3.4.3 Using the Unweighted Average of Differentials of All the Gene as Control

In this case, by Gauss–Markov Theorem, the Best Linear Unbiased Estimator (BLUE) of θ_j is

$$\hat{\theta}_l = \bar{y}_{..1l.} - \bar{y}_{..2l.} - \left[\frac{\sum_{i=1}^{g} \bar{y}_{..1i.}}{g} - \frac{\sum_{i=1}^{g} \bar{y}_{..2i.}}{g}\right], \quad j = 1, 2, \ldots, g$$

The variance of $\hat{\theta}_l$ is

$$\begin{aligned}
\text{var}(\hat{\theta}_l) &= \text{var}(\bar{y}_{..l.} - \bar{y}_{..2l.}) + \text{var}\left[\frac{\sum_{i=1}^{g} \bar{y}_{..1i.} - \sum_{i=1}^{g} \bar{y}_{..2i.}}{g}\right] \\
&\quad - 2\text{cov}\left(\bar{y}_{..1l.} - \bar{y}_{..2l.}, \frac{\sum_{i=1}^{g} \bar{y}_{..1i.} - \sum_{i=1}^{g} \bar{y}_{..2i.}}{g}\right) \\
&= \frac{2\sigma^2}{an_l} + \frac{2\sigma^2}{ag^2}\left[\sum_{i=1}^{g} \frac{1}{n_i}\right] - \frac{2}{g}\left[\frac{2\sigma^2}{an_l}\right] \\
&= \frac{2\sigma^2}{an_l}\left(1 - \frac{2}{g}\right) + \frac{2\sigma^2}{ag^2}\left[\sum_{i=1}^{g} \frac{1}{n_i}\right], \quad l = 1, 2, \ldots, g
\end{aligned}$$

and the covariance of $\hat{\theta}_l$ and $\hat{\theta}_{l'}$ is

$$\begin{aligned}
\text{cov}(\hat{\theta}_l, \hat{\theta}_{l'}) &= \text{cov}\left[\bar{y}_{..1l.} - \bar{y}_{..2l.} - \frac{\sum_{i=1}^{g} \bar{y}_{..1i.} - \sum_{i=1}^{g} \bar{y}_{..2i.}}{g}, \bar{y}_{..1l'.} - \bar{y}_{..2l'.}\right. \\
&\quad \left. - \frac{\sum_{i=1}^{g} \bar{y}_{..1i.} - \sum_{i=1}^{g} \bar{y}_{..2i.}}{g}\right] \\
&= -\frac{1}{g}\left(\frac{2\sigma^2}{an_l}\right) - \frac{1}{g}\left(\frac{2\sigma^2}{an_{l'}}\right) + \frac{2\sigma^2}{ag^2}\left[\sum_{i=1}^{g} \frac{1}{n_i}\right], \quad l \neq l'
\end{aligned}$$

For these covariance structures, exact critical values for multiplicity adjustments become possible. For normalization with respect to averages of housekeeping genes, such computations can be done using the factor analytic algorithm of Hsu [13]. For normalization with respect to the average of all genes, the computation can be done using the algorithm in Soong [14].

7.4 ESTIMATION WHEN THE MICROARRAY DESIGN IS NOT ORTHOGONAL

In the previous section, we showed that if the design of the microarray experiment is orthogonal as we recommend, then least squares estimators of differential expressions when the array effect is considered fixed are also the generalized least squares estimators when the array effect is considered random. Orthogonality is ensured if the number of times n_l the lth gene is spotted on each array remains the same across all arrays.

Suppose the number of times a gene is spotted on each array does not remain the same across the arrays, so that the design is nonorthogonal. If array is considered a random factor, then least squares estimators of differential expressions when the array effect is considered fixed are not necessarily the same as the generalized least squares estimators when the array effect is considered random. This is because estimates of treatment × gene contrasts may also be available from comparisons between arrays. If the array effects are large, then the amount of information which may be recovered from inter-array comparisons will be small. On the other hand, if the array effects are small, then there may be substantial gains to be achieved by recovering the inter-array information. Inference based on least squares estimators are easier to construct and compute. In the rest of this section, we examine whether it is worthwhile to use generalized least squares estimators to recover the inter-array information for nonorthogonal microarray designs.

Chapter 8 of John [15] discusses recovery of inter-block information. Section 8.6 in Reference 15, in particular, pertains to the situation where the number of times a gene is spotted on each array differs from array to array, but the *total* number of times each gene is spotted across the arrays is the same for all genes. By identifying "treatment × gene" with "treatment" and each array as a "block" in John's discussion, it can be seen that when array is random, the combined (inter-block and intra-block) treatment × gene ($\tau\gamma_{ij}$) treatment interaction estimators (generalized least squares estimator) and their variances differ from the least squares estimators when array is fixed by factors which are functions of $\phi_1^{-1} = \sigma^2/2K\sigma_A^2$. When $\phi_1^{-1} = 0$, the combined estimators (generalized least squares estimators) of $\tau\gamma_{ij}$ when array is random are the same as the least square estimators of $\tau\gamma_{ij}$ when array is considered a fixed effect. Since the number of spots is typically large and array effects are expected to be significant, ϕ_1^{-1} will be close to zero. Therefore, in this case, inference based on least squares estimators will be a good approximation of inference based on generalized least squares estimators.

7.5 SUMMARY

Good statistical principles such as randomization, replication, and blocking, are applicable not only to 2-channel microarrays but also to oligonucleotide. Currently, due to inconvenience, such principles are rarely applied in their manufacturing. It is our belief, however, that given proper motivation, technology will innovate itself to make these principles conveniently applicable. One such motivation might be the necessity of approval by the Center for Radiologic Research and Health of the Food and Drug Administration (FDA) to market microarrays for diagnostic and prognostic purposes. Given the advantages demonstrated in this chapter, it will in general be easier to meet the requirements in the draft guidance on *Multiplex Testing of DNA Heritable Markers, Mutations, and Expression Patterns* issued by the FDA if these principles are adhered to in microarray experiments.

APPENDIX

The proof of Lemmas 7.1 to 7.3 stated below are provided in Chang et al. [10].

Lemma 7.1

(1) $X_{\mathrm{A}}^{*\prime} 1_{2aK} = 0$
(2) $X_{\mathrm{D}}^{*\prime} 1_{2aK} = 0$
(3) $X_{\mathrm{T}}^{*\prime} 1_{2aK} = 0$
(4) $X_{\mathrm{G}}^{*\prime} 1_{2aK} = 0$
(5) $X_{\mathrm{TG}}^{*\prime} 1_{2aK} = 0$
(6) $X_{\mathrm{DG}}^{*\prime} 1_{2aK} = 0$

Lemma 7.2

(1) $X_{\mathrm{DG}}^{*\prime} X_{\mathrm{A}}^{*} = 0$
(2) $X_{\mathrm{DG}}^{*\prime} X_{\mathrm{D}}^{*} = 0$
(3) $X_{\mathrm{DG}}^{*\prime} X_{\mathrm{T}}^{*} = 0$
(4) $X_{\mathrm{DG}}^{*\prime} X_{\mathrm{G}}^{*} = 0$
(5) $X_{\mathrm{DG}}^{*\prime} X_{\mathrm{TG}}^{*} = 0$
(6) $X_{\mathrm{TG}}^{*\prime} X_{\mathrm{A}}^{*} = 0$
(7) $X_{\mathrm{TG}}^{*\prime} X_{\mathrm{D}}^{*} = 0$
(8) $X_{\mathrm{TG}}^{*\prime} X_{\mathrm{T}}^{*} = 0$
(9) $X_{\mathrm{TG}}^{*\prime} X_{\mathrm{G}}^{*} = 0$
(10) $X_{\mathrm{G}}^{*\prime} X_{\mathrm{A}}^{*} = 0$
(11) $X_{\mathrm{G}}^{*\prime} X_{\mathrm{D}}^{*} = 0$
(12) $X_{\mathrm{G}}^{*\prime} X_{\mathrm{T}}^{*} = 0$
(13) $X_{\mathrm{T}}^{*\prime} X_{\mathrm{A}}^{*} = 0$
(14) $X_{\mathrm{T}}^{*\prime} X_{\mathrm{D}}^{*} = 0$
(15) $X_{\mathrm{D}}^{*\prime} X_{\mathrm{A}}^{*} = 0$

Lemma 7.3

(1) $\hat{\delta} = \left(X_{\mathrm{D}}^{*\prime} X_{\mathrm{D}}^{*}\right)^{-} X_{\mathrm{D}}^{*\prime} Y = \begin{bmatrix} \bar{y}_{.1...} & -\bar{y}_{.....} \\ \bar{y}_{.2...} & -\bar{y}_{.....} \end{bmatrix}_{2\times 1}$

$$(2)\ \hat{\tau} = \left(X_{\mathrm{T}}^{*\prime}X_{\mathrm{T}}^{*}\right)^{-} X_{\mathrm{T}}^{*\prime}Y = \begin{bmatrix} \bar{y}_{..1..} & -\bar{y}_{.....} \\ \bar{y}_{..2..} & -\bar{y}_{.....} \end{bmatrix}_{2\times 1}$$

$$(3)\ \hat{\gamma} = \left(X_{\mathrm{G}}^{*\prime}X_{\mathrm{G}}^{*}\right)^{-} X_{\mathrm{G}}^{*\prime}Y = \begin{bmatrix} \bar{y}_{...1.} & -\bar{y}_{.....} \\ & \vdots \\ \bar{y}_{...g.} & -\bar{y}_{.....} \end{bmatrix}_{g\times 1}$$

$$(4)\ \hat{\chi} = \left(X_{\mathrm{TG}}^{*\prime}X_{\mathrm{TG}}^{*}\right)^{-} X_{\mathrm{TG}}^{*\prime}Y = \begin{bmatrix} \bar{y}_{..11.} & -\bar{y}_{..1..} & -\bar{y}_{...1.} & +\bar{y}_{.....} \\ & & \vdots & \\ \bar{y}_{..1g.} & -\bar{y}_{..1..} & -\bar{y}_{...g.} & +\bar{y}_{.....} \\ \bar{y}_{...21.} & -\bar{y}_{..2..} & -\bar{y}_{...1.} & +\bar{y}_{.....} \\ & & \vdots & \\ \bar{y}_{..2g.} & -\bar{y}_{..2..} & -\bar{y}_{...g.} & +\bar{y}_{.....} \end{bmatrix}_{2g\times 1}$$

$$(5)\ \hat{\phi} = \left(X_{\mathrm{DG}}^{*\prime}X_{\mathrm{DG}}^{*}\right)^{-} X_{\mathrm{DG}}^{*\prime}Y = \begin{bmatrix} \bar{y}_{.1.1.} & -\bar{y}_{.1...} & -\bar{y}_{...1.} & +\bar{y}_{.....} \\ & & \vdots & \\ \bar{y}_{.1.g.} & -\bar{y}_{.1...} & -\bar{y}_{...g.} & +\bar{y}_{.....} \\ \bar{y}_{.2.1.} & -\bar{y}_{.2...} & -\bar{y}_{...1.} & +\bar{y}_{.....} \\ & & \vdots & \\ \bar{y}_{.2.g.} & -\bar{y}_{.2...} & -\bar{y}_{...g.} & +\bar{y}_{.....} \end{bmatrix}_{2g\times 1}$$

Lemma 7.5

$$(1)\ V^{-1}X_{\mathrm{A}}^{*} = \frac{1}{\sigma^2}X_{\mathrm{A}}^{*}$$

$$(2)\ V^{-1}X_{\mathrm{D}}^{*} = \frac{1}{\sigma^2}X_{\mathrm{D}}^{*}$$

$$(3)\ V^{-1}X_{\mathrm{T}}^{*} = \frac{1}{\sigma^2}X_{\mathrm{T}}^{*}$$

$$(4)\ V^{-1}X_{\mathrm{G}}^{*} = \frac{1}{\sigma^2}X_{\mathrm{G}}^{*}$$

$$(5)\ V^{-1}X_{\mathrm{DG}}^{*} = \frac{1}{\sigma^2}X_{\mathrm{DG}}^{*}$$

$$(6)\ V^{-1}X_{\mathrm{TG}}^{*} = \frac{1}{\sigma^2}X_{\mathrm{TG}}^{*}$$

Proof.

$$(1)\quad V^{-1}X_{\mathrm{A}}^{*} = \left(\frac{1}{\sigma^2}I_{2aK} - \frac{\sigma_{\mathrm{A}}^2}{\sigma^2\left(\sigma^2 + 2K\sigma_{\mathrm{A}}^2\right)}I_a \otimes J_{2K}\right)X_{\mathrm{A}}^{*}$$

$$= \frac{1}{\sigma^2}X_{\mathrm{A}}^{*} - \frac{\sigma_{\mathrm{A}}^2}{\sigma^2\left(\sigma^2 + 2K\sigma_{\mathrm{A}}^2\right)}(I_a \otimes J_{2K})X_{\mathrm{A}}^{*}$$

$$= \frac{1}{\sigma^2} X_{\mathrm{A}}^* - \frac{\sigma_{\mathrm{A}}^2}{\sigma^2\left(\sigma^2 + 2K\sigma_{\mathrm{A}}^2\right)} (I_a \otimes J_{2K}) \left(1_a \otimes I_2 \otimes 1_K - \tfrac{1}{2} 1_{2aK} \otimes 1_a'\right)$$

$$= \frac{1}{\sigma^2} X_{\mathrm{A}}^* - \frac{\sigma_{\mathrm{A}}^2}{\sigma^2\left(\sigma^2 + 2K\sigma_{\mathrm{A}}^2\right)} \left(K 1_{2aK\times a} - \tfrac{1}{2} 2K 1_{2aK\times a}\right)$$

$$= \frac{1}{\sigma^2} X_{\mathrm{A}}^*$$

(2) $$V^{-1} X_{\mathrm{D}}^* = \left(\frac{1}{\sigma^2} I_{2aK} - \frac{\sigma_{\mathrm{A}}^2}{\sigma^2\left(\sigma^2 + 2K\sigma_{\mathrm{A}}^2\right)} I_a \otimes J_{2K}\right) X_{\mathrm{D}}^*$$

$$= \left(\frac{1}{\sigma^2} I_{2aK} - \frac{\sigma_{\mathrm{A}}^2}{\sigma^2\left(\sigma^2 + 2K\sigma_{\mathrm{A}}^2\right)} I_a \otimes J_{2K}\right) X_{\mathrm{D}}^*$$

$$= \frac{1}{\sigma^2} X_{\mathrm{D}}^* - \frac{\sigma_{\mathrm{A}}^2}{\sigma^2\left(\sigma^2 + 2K\sigma_{\mathrm{A}}^2\right)} (I_a \otimes J_{2K}) X_{\mathrm{D}}^*$$

$$= \frac{1}{\sigma^2} X_{\mathrm{D}}^* - \frac{\sigma_{\mathrm{A}}^2}{\sigma^2\left(\sigma^2 + 2K\sigma_{\mathrm{A}}^2\right)} (I_a \otimes J_{2K})$$
$$\times \left(1_a \otimes I_2 \otimes 1_K - \tfrac{1}{2} 1_{2aK} \otimes 1_2'\right)$$

$$= \frac{1}{\sigma^2} X_{\mathrm{D}}^* - \frac{\sigma_{\mathrm{A}}^2}{\sigma^2\left(\sigma^2 + 2K\sigma_{\mathrm{A}}^2\right)} \left(K 1_{2aK\times 2} - \tfrac{1}{2} 2K 1_{2aK\times 2}\right)$$

$$= \frac{1}{\sigma^2} X_{\mathrm{D}}^*$$

(3) $$V^{-1} X_{\mathrm{T}}^* = \left(\frac{1}{\sigma^2} I_{2aK} - \frac{\sigma_{\mathrm{A}}^2}{\sigma^2\left(\sigma^2 + 2K\sigma_{\mathrm{A}}^2\right)} I_a \otimes J_{2K}\right) X_{\mathrm{T}}^*$$

$$= \frac{1}{\sigma^2} X_{\mathrm{T}}^* - \frac{\sigma_{\mathrm{A}}^2}{\sigma^2\left(\sigma^2 + 2K\sigma_{\mathrm{A}}^2\right)} (I_a \otimes J_{2K}) X_{\mathrm{T}}^*$$

$$= \frac{1}{\sigma^2} X_{\mathrm{T}}^* - \frac{\sigma_{\mathrm{A}}^2}{\sigma^2\left(\sigma^2 + 2K\sigma_{\mathrm{A}}^2\right)} (I_a \otimes J_{2K}) \left(1_{a/2} \otimes \begin{bmatrix} 1 & 0 \\ 0 & 1 \\ 0 & 1 \\ 1 & 0 \end{bmatrix} \otimes 1_K \right.$$
$$\left. - \frac{1}{2} 1_{2aK} 1_2' \right)$$

$$= \frac{1}{\sigma^2} X_{\mathrm{T}}^* - \frac{\sigma_{\mathrm{A}}^2}{\sigma^2\left(\sigma^2 + 2K\sigma_{\mathrm{A}}^2\right)} \left(K 1_{2aK} 1_2' - K 1_{2aK} 1_2'\right)$$

$$= \frac{1}{\sigma^2} X_{\mathrm{T}}^*$$

$$(4)\quad V^{-1}X_{\mathrm{G}}^{*} = \left(\frac{1}{\sigma^2}I_{2aK} - \frac{\sigma_{\mathrm{A}}^2}{\sigma^2\left(\sigma^2+2K\sigma_{\mathrm{A}}^2\right)}I_a\otimes J_{2K}\right)X_{\mathrm{G}}^{*}$$

$$= \frac{1}{\sigma^2}X_{\mathrm{G}}^{*} - \frac{\sigma_{\mathrm{A}}^2}{\sigma^2\left(\sigma^2+2K\sigma_{\mathrm{A}}^2\right)}(I_a\otimes J_{2K})X_{\mathrm{G}}^{*}$$

$$= \frac{1}{\sigma^2}X_{\mathrm{G}}^{*} - \frac{\sigma_{\mathrm{A}}^2}{\sigma^2\left(\sigma^2+2K\sigma_{\mathrm{A}}^2\right)}(I_a\otimes J_{2K})\left(1_{2a}\otimes \mathrm{Diag}(1_{n1},\cdots,1_{ng})\right.$$

$$\left. -\frac{1}{K}1_{2aK}N'\right)$$

$$= \frac{1}{\sigma^2}X_{\mathrm{G}}^{*} - \frac{-\sigma_{\mathrm{A}}^2}{\sigma^2\left(\sigma^2+2K\sigma_{\mathrm{A}}^2\right)}\times\left\{2\times 1_a\otimes J_2\otimes\begin{bmatrix} n_1 & \cdots & n_g\\ \vdots & \ddots & \vdots\\ n_1 & \cdots & n_g\end{bmatrix}\right.$$

$$\left. -\frac{1}{K}2K1_a\otimes J_2\otimes\begin{bmatrix} n_1 & \cdots & n_g\\ \vdots & \ddots & \vdots\\ n_1 & \cdots & n_g\end{bmatrix}\right\}$$

$$= \frac{1}{\sigma^2}X_{\mathrm{G}}^{*}$$

$$(5)\quad V^{-1}X_{\mathrm{DG}}^{*} = \left(\frac{1}{\sigma^2}I_{2aK} - \frac{\sigma_{\mathrm{A}}^2}{\sigma^2\left(\sigma^2+2K\sigma_{\mathrm{A}}^2\right)}I_a\otimes J_{2K}\right)X_{\mathrm{DG}}^{*}$$

$$= \left(\frac{1}{\sigma^2}I_{2aK} - \frac{\sigma_{\mathrm{A}}^2}{\sigma^2\left(\sigma^2+2K\sigma_{\mathrm{A}}^2\right)}I_a\otimes J_{2K}\right)X_{\mathrm{DG}}^{*}$$

$$= \frac{1}{\sigma^2}X_{\mathrm{DG}}^{*} - \frac{\sigma_{\mathrm{A}}^2}{\sigma^2\left(\sigma^2+2K\sigma_{\mathrm{A}}^2\right)}(I_a\otimes J_{2K})X_{\mathrm{DG}}^{*}$$

$$= \frac{1}{\sigma^2}X_{\mathrm{DG}}^{*} - \frac{-\sigma_{\mathrm{A}}^2}{\sigma^2\left(\sigma^2+2K\sigma_{\mathrm{A}}^2\right)}$$

$$\times\left\{1_a\otimes J_2\otimes\begin{bmatrix} n_1 & \cdots & n_g\\ \vdots & \ddots & \vdots\\ n_1 & \cdots & n_g\end{bmatrix}\right.$$

$$-\frac{1}{K}1_a\otimes J_2\otimes\begin{bmatrix} Kn_1 & \cdots & Kn_g\\ \vdots & \ddots & \vdots\\ Kn_1 & \cdots & Kn_g\end{bmatrix}$$

$$-\frac{1}{2}1_a\otimes J_2\otimes\begin{bmatrix} 2n_1 & \cdots & 2n_g\\ \vdots & \ddots & \vdots\\ 2n_1 & \cdots & 2n_g\end{bmatrix}$$

$$+\frac{1}{2aK}1_a \otimes J_2 \otimes \begin{bmatrix} 2Kan_1 & \cdots & 2Kan_g \\ \vdots & \ddots & \vdots \\ 2Kan_1 & \cdots & 2Kan_g \end{bmatrix} \Bigg\}$$

$$= \frac{1}{\sigma^2} X^*_{\mathrm{DG}}$$

$$\text{(6)} \quad V^{-1}X^*_{\mathrm{TG}} = \left(\frac{1}{\sigma^2} I_{2aK} - \frac{\sigma^2_{\mathrm{A}}}{\sigma^2\left(\sigma^2 + 2K\sigma^2_{\mathrm{A}}\right)} I_a \otimes J_{2K} \right) X^*_{\mathrm{TG}}$$

$$= \frac{1}{\sigma^2} X^*_{\mathrm{TG}} - \frac{\sigma^2_{\mathrm{A}}}{\sigma^2\left(\sigma^2 + 2K\sigma^2_{\mathrm{A}}\right)} (I_a \otimes J_{2K}) X^*_{\mathrm{TG}}$$

$$= \frac{1}{\sigma^2} X^*_{\mathrm{TG}} - \frac{-\sigma^2_{\mathrm{A}}}{\sigma^2\left(\sigma^2 + 2K\sigma^2_{\mathrm{A}}\right)} \Bigg\{ 2 \times 1_{a/2} \otimes J_2 \otimes \begin{bmatrix} n_1 & \cdots & n_g \\ \vdots & \ddots & \vdots \\ n_1 & \cdots & n_g \end{bmatrix}$$

$$- \frac{2}{K} 1_{a/2} \otimes J_2 \otimes \begin{bmatrix} Kn_1 & \cdots & Kn_g \\ \vdots & \ddots & \vdots \\ Kn_1 & \cdots & Kn_g \end{bmatrix}$$

$$- \frac{1}{2} 1_{a/2} \otimes 2J_2 \otimes \begin{bmatrix} n_1 & \cdots & n_g \\ \vdots & \ddots & \vdots \\ n_1 & \cdots & n_g \end{bmatrix}$$

$$+ \frac{1}{2aK} 1_{a/2} \otimes J_2 \otimes \begin{bmatrix} 2Kan_1 & \cdots & 2Kan_g \\ \vdots & \ddots & \vdots \\ 2Kan_1 & \cdots & 2Kan_g \end{bmatrix} \Bigg\}$$

$$= \frac{1}{\sigma^2} X^*_{\mathrm{TG}}$$

REFERENCES

1. S. Dudoit, Y.H. Yang, T.P. Speed, and M.J. Callow. Statistical methods for identifying differentially expressed genes in replicated cDNA microarray experiments. *Statistica Sinica* 12: 111–140, 2002.
2. Y.H. Yang, S. Dudoit, P. Luu, and T.P. Speed. Normalization for cDNA microarray data. In M.L. Bittner, Y. Chen, A.N. Dorsel, and E.R. Dougherty (eds.), *Microarrays: Optical Technologies and Informatics*, Vol. 4266 of Proceedings of SPIE, 2001.
3. G. Grant, E. Manduchi, and C. Stoeckert. Using non-parametric methods in the context of multiple testing to identify differentially expressed genes. In S.M. Lin and K.F. Johnson (eds.), *Methods of Microarray Data Analysis*, Kluwer Academic Publishers, Boston, pp. 37–55, 2002.
4. S. Dudoit, J.P. Shaffer, and J.C. Boldrick. Multiple hypothesis testing in microarray experiments. *Statistical Science* 18: 71–103, 2003.

5. M.K. Kerr, M. Martin, and G.A. Churchill. Analysis of variance for gene expression microarray data. *Journal of Computational Biology* 7: 819–837, 2000.
6. M.K. Kerr, and G.A. Churchill. Experimental design for gene expression microarrays. *Biostatistics* 2: 183–201, 2001.
7. M.K. Kerr, and G.A. Churchill. Statistical design and the analysis of gene expression microarray data. *Genetical Research* 77: 123–128, 2001.
8. M.K. Kerr, C.A. Afshari, L. Bennett, P. Bushel, J. Martinez, N. Walker, and G.A. Churchill. Statistical analysis of a gene expression microarray experiment with replication. *Statistica Sinica* 12: 203–218, 2002.
9. J.C. Hsu, Y.J. Chang, and T. Wang. Simultaneous confidence intervals for differential gene expressions. *Journal of Statistical Planning and Inference*, 2004 (to appear).
10. Y.J. Chang, T. Wang, and J.C. Hsu. Efficient least squares estimation for differential gene expressions. *Proceedings of the American Statistical Association, Biopharmaceutical Section [CD-ROM]* Alexandria, VA: American Statistical Association, 2002.
11. B. Alberts, A. Johnson, J. Lewis, M. Raff, K. Roberts, and P. Walter. *Molecular Biology of the Cell*. New York: Garland Science, 2002.
12. J. Vandesompele, K. De Preter, F. Pattyn, B. Poppe, N. Van Roy, A. De Paepe, and F. Speleman. Accurate normalization of real-time quantitative RT-PCR data by geometric averaging of multiple internal control genes. *Genome Biology* 3: 0034.I–0034.II, 2002.
13. J.C. Hsu. The factor analytic approach to simultaneous inference in the general linear model. *Journal of Graphical and Computational Statistics* 1: 151–168, 1992.
14. W.C. Soong. Exact simultaneous confidence intervals for multiple comparisons with the mean. *Computational Statistics and Data Analysis* 37: 33–47, 2001.
15. J.A. John. *Cyclic Designs*. London: Chapman & Hall, 1987.

8 Overview of Standard Clustering Approaches for Gene Microarray Data Analysis

Elizabeth Garrett-Mayer

CONTENTS

8.1 Introduction 132
8.2 Distance and Similarity Measures 135
8.2.1 Continuous Distance and Similarity Measures 135
8.2.2 Categorical Distance and Similarity Measures 137
8.3 Hierarchical Clustering 138
8.3.1 Up vs. Down 139
8.3.2 Hierarchical Clustering Algorithms 140
8.3.2.1 Agglomerative 140
8.3.2.1.1 Single Linkage (Nearest Neighbor) 140
8.3.2.1.2 Complete Linkage (Furthest Neighbor) 141
8.3.2.1.3 Average Linkage 141
8.3.2.2 Divisive 141
8.3.3 Visualization Methods 142
8.3.3.1 Dendrograms 142
8.3.3.2 Banner Plots 143
8.4 *K*-means and *K*-medoids 143
8.5 Self-Organizing Maps 144
8.6 Cluster Affinity Search Technique 146
8.7 Other Related Methods 147
8.8 Assessing Cluster Fit and Choosing *k* 147
8.8.1 Homogeneity and Separation of Clusters 148
8.8.2 Cluster Silhouettes 148
8.8.3 Weighted Average Discrepant Pairs 150
8.8.4 The Gap Statistic 150
8.8.5 Consensus Trees 151
8.8.6 Other Approaches 152
8.9 Choosing Genes and Samples for Clustering 153

8.10 Computer Packages and Software 155
8.11 Problems and Pitfalls 155
References 156

8.1 INTRODUCTION

Clustering methods have become some of the most prevalent exploratory approaches for evaluating gene expression experiments. Methods such as hierarchical clustering provide a relatively concise visualization of the whole experiment, allowing us to see which genes and samples tend to "cluster" together. From a broader perspective, the goal of cluster analyses is pattern recognition: to identify concise systematic patterns amongst the large amount of data produced by microarray experiments. It must be stressed, however, that clustering methods should be considered exploratory approaches to the analysis of gene expression arrays. In-depth evaluation of the resulting cluster structure or implementation of other hypothesis-driven analyses are critical steps that follow cluster analyses.

When clustering gene expression array data, there are two types of objects which can be clustered (1) genes, of which there are usually between 500 and 50,000 in any given experiment and (2) samples, which usually number somewhere between 6 and 200. *Sample* refers to the person, mouse, cell line, etc. from which the genetic data has been acquired. One goal of clustering genes is to identify which genes tend to behave similarly across the set of samples being studied. We can also identify which samples tend to be similar based on the gene expression patterns.

Figure 8.1 shows an example of a hierarchical cluster analysis of two kinds of lung cancer samples. Garber et al. [1] studied a set of lung cancer samples including 37 adenocarcinomas (adenos) and 10 squamous cell carcinomas (squamous). Two hundred genes out of their set of over 23,000 cDNA clones in the original experiment are shown in Figure 8.1. Notice that there are tree-structures (i.e., dendrograms) on two sides of the figure, displaying the hierarchical clustering of samples and of genes. The genes and samples have been sorted based on this clustering, creating relatively homogeneous areas of red (overexpression) and green (underexpression) in Figure 8.1.

The clustering methods described in this chapter are "unsupervised" methods. What this means is that none of the information about sample phenotype or gene function is used in determining how genes or samples cluster together. This is a sensible approach when we are attempting "class discovery": trying to find new subtypes within a population based on gene expression patterns. One example of the class discovery approach is found in Rosenwald et al. [2] where subtypes of diffuse large B-cell lymphoma were found using hierarchical clustering and another complementary approach. However, in the case where sample phenotype is known and the goal is to find genes which can distinguish phenotypes, it is usually thought to be a more appropriate analytic strategy to use "supervised" approaches, which use phenotypic or gene function information to find patterns. In many cases, though, unsupervised approaches are applied to experiments where phenotype is known and class discovery is not the goal. Although it is not the most appropriate approach for finding genes associated with phenotype, many investigators find the clustering

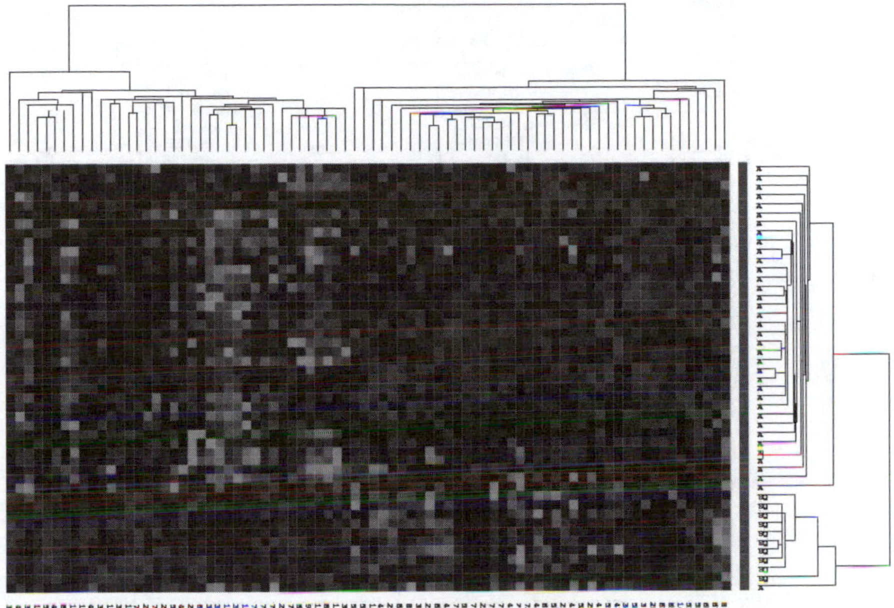

FIGURE 8.1 An example of a hierarchical cluster analysis of two kinds of lung cancer samples (37 adenocarcinomas and 10 squamous cell carcinomas) based on expression data from 200 genes. Data are from Garber et al. *Proceedings of the National Academy of Sciences, USA*, 98: 13784–13789, 2001. The genes and samples have been sorted based on a hierarchical clustering. Red indicates overexpression and green underexpression.

approaches useful in these settings if only for the visual displays they provide of the experiment.

Clustering is used frequently in gene expression experiments, and there are many commonly used procedures for clustering. This can cause some confusion and also preclude results from being reproduced by other investigators. For example, there are many publicly available gene expression datasets for which cluster analysis results have been published, but reproducing the results can prove difficult unless *either* the clustering procedure is described in elaborate detail or *the source code of the software used to perform the clustering is made available.*[*] There are many decisions for the investigator to make when performing a cluster analysis. In Algorithm 8.1, a general procedure for undertaking a hierarchical cluster analysis is outlined. Although it might seem rather straightforward, the choices made can greatly impact the resulting cluster structure. In Figure 8.2, examples of this are shown. The same dataset was clustered using eight slightly different clustering algorithms and eight rather different clustering structures are produced. This suggests that there is significant instability in these cluster results. (The results in Figure 8.2 will be discussed in more detail in later

[*] Commercial Software Code is often proprietary, preventing researchers from understanding exactly how the data are manupulated.

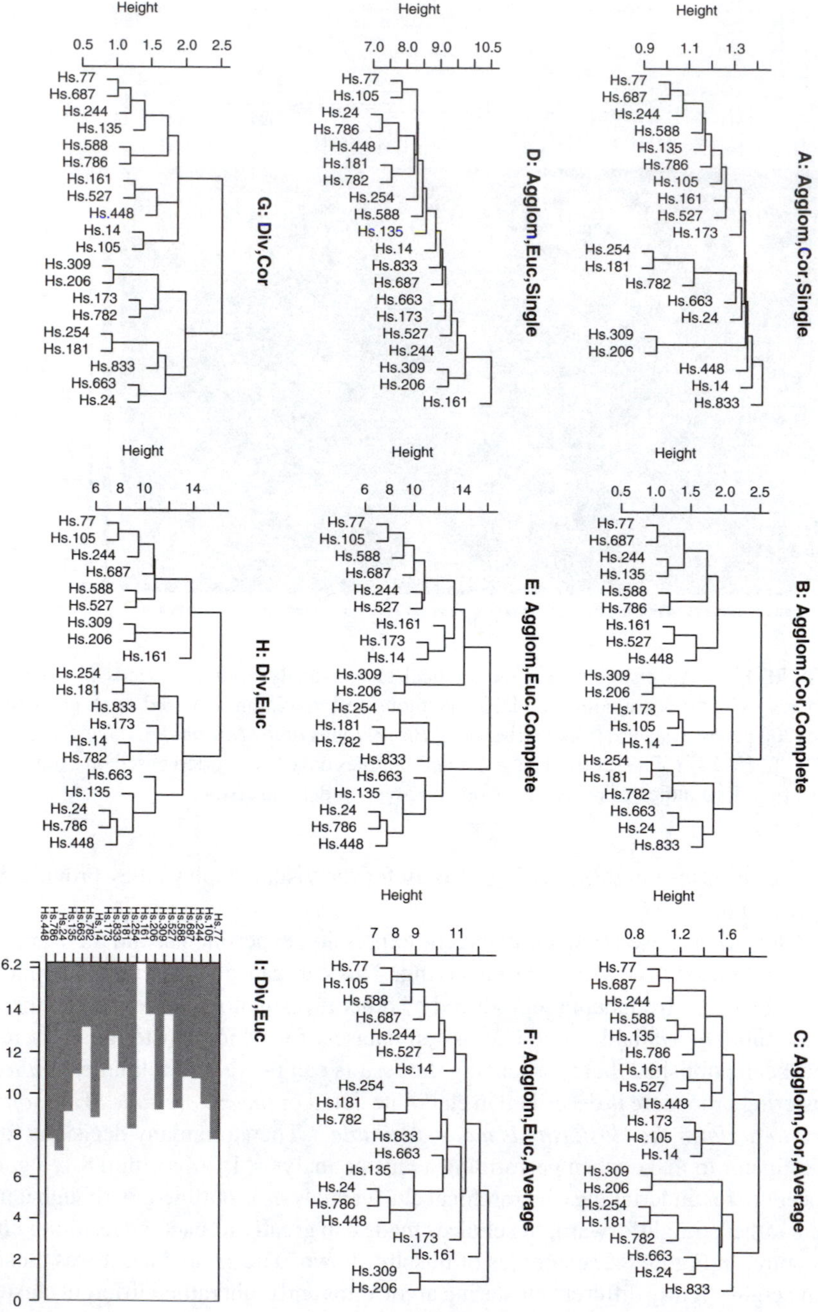

FIGURE 8.2 A–H: Results of cluster analyses performed on one simulated gene expression dataset. Differences in results are due to choice of agglomerative (Agglom) vs. divisive (Div) algorithm, similarity/distance metric (correlation [Corr] vs. Euclidean distance [Euc]), and linkage method (single, complete, average). I: Banner plot of hierarchical clustering using divisive algorithm with Euclidean distance as distance metric.

sections.) As a result, it is important to try several different approaches (e.g., different distance metrics, different linkage methods) and see how sensitive the results are to these choices. Another method for determining robustness of clusters is to consider assessing cluster fit and estimating consensus trees. This will be briefly described in Section 8.8.2.

Algorithm 8.1 [Steps in hierarchical clustering procedure]
1. Select samples to include in the cluster analysis.
2. Select genes to include in the cluster analysis.
3. Choose distance and similarity metric (e.g., Euclidean distance).
4. Choose clustering method (e.g., agglomerative).
5. Choose linkage method (e.g., average).
6. Estimate hierarchical clustering structure.
7. Choose height and number of clusters for interpretation.
8. Evaluate fit of clusters and validate results.

Other authors have described cluster analyses for gene expression applications which are recommended for further reading [3,4].

8.2 DISTANCE AND SIMILARITY MEASURES

A cluster analysis groups objects together based on how close they are to or far they are from each other. Naturally, we need a way to define this distance between or, inversely, similarity of objects. The most commonly used measure of gene expression distance is the Euclidean distance. The most commonly used similarity measure is the correlation. For each pair of items to be measured in the dataset of interest, we calculate the distance (similarity) between them and form the distance (similarity) matrix. If there are N items to be clustered, the distance or similarity matrix will have dimension $N \times N$.

A distance measure $d(x, y)$ is defined as a "metric distance" if it satisfies the following four properties for any distinct objects x, y, and z (1) The distance between x and y is nonnegative: $d(x, y) \geqslant 0$, (2) Distance is measured symmetrically: $d(x, y) = d(y, x)$, (3) The distance between an object and itself is 0: $d(x, x) = 0$, and (4) the distance between two points is not longer than the sum of distances between those points and another point: $d(x, y) \leqslant d(x, z) + d(y, z)$. While these are appealing properties in many analytic settings, all of these properties do not appear to be crucial for distance measures used in cluster analyses. For example, one minus the Pearson correlation is a widely accepted and valid measure of distance used in gene expression analysis. However, it fails both property 1 (i.e., it can take values less than 0) and property 4. Distance based on one minus the Spearman correlation or the uncentered correlation also fails these two distance properties.

8.2.1 Continuous Distance and Similarity Measures

There are a number of choices for determining distance or similarity when the data to be clustered are continuous. Definitions are given below for measuring distance (d)

or similarity (s) between two objects, x and y, where each is measured on K variables. In the case of measuring distance or similarity between genes, consider that x and y are two genes and that there are K samples in the experiment. For measuring distance or similarity between samples, x and y are two samples in the experiment and are measured on K genes. Although other metrics can be used, we will describe several of the commonly used metrics in the gene expression setting and refer the reader to Gordon (1999) for discussion of other continuous variable metrics.

1. *Euclidean distance*: $d(x,y) = \sqrt{\sum_{k=1}^{K} (x_k - y_k)^2}$

This metric distance measure represents a geometric interpretation of distance.

2. *Manhattan*: $d(x,y) = \sum_{k=1}^{K} |x_k - y_k|$

The Euclidean and the Manhattan (a.k.a. city-block distance) distances are both metric distances and are from the Minkowski family of metrics. The Euclidean distance will be more sensitive to outliers than the Manhattan distance, but neither should be considered robust to outliers.

3. *Canberra distance*: $d(x,y) = \sum_{k=1}^{K} \frac{|x_k - y_k|}{|x_k| + |y_k|}$

The Canberra is a metric distance and is a scaled relative of the Euclidean and Manhattan distances. It has a range from 0 to 1.

4. *Correlation*: $s(x,y) = \frac{\sum_{k=1}^{K} (x_k - \bar{x})(y_k - \bar{y})}{\sqrt{\sum_{k=1}^{K} (x_k - \bar{x})^2 \sum_{k=1}^{K} (y_k - \bar{y})^2}}$

The Pearson correlation works well when the relationship between x and y appears strongly linear, but does have the drawback that it is sensitive to outliers. One minus the correlation can be used instead as a distance measure. Many users choose to use the absolute value of the correlation instead, with the idea being that negative correlation and positive correlation both imply similarity, while 0 correlation is the least associated two genes can be.

5. *Uncentered correlation*: $s(x,y) = \frac{\sum_{k=1}^{K} x_k y_k}{\sqrt{\sum_{k=1}^{K} x_k^2 \sum_{k=1}^{K} y_k^2}}$

The uncentered correlation is essentially the Pearson correlation assuming that the means of x and y are both 0. In the case where two genes have a perfectly linear relationship but the average expressions are different, the standard Pearson correlation will be 1, but the uncentered correlation will be less than 1. The main difference with using the uncentered correlation is that, unlike the standard Pearson correlation, it considers that magnitude of the observed values.

6. *Spearman correlation*: $s(x, y) = \dfrac{\sum_{k=1}^{K}(r(x_k) - \bar{r})(r(y_k) - \bar{r})}{\sqrt{\sum_{k=1}^{K}(r(x_k) - \bar{r})^2 \sum_{k=1}^{K}(r(y_k) - \bar{r})^2}}$

The Spearman correlation is the same as the Pearson correlation except that the ranks of the data ($r(x)$ and $r(y)$) are used instead of the observed values of x and y. This estimate is very robust to outliers and deviations from linearity unlike the Pearson correlation. When the relationship between x and y is approximately linear, the Pearson and Spearman correlations will produce very similar results. Just as in Pearson correlation, some people prefer using absolute Spearman correlation.

Euclidean distance and its relatives (Canberra and Manhattan distances) and the uncentered correlation all have the similar property that they consider the magnitude of difference between two genes and not just the similarity of the gene expression patterns across samples. In Figure 8.3A, the relative gene expression of two genes is plotted against time where the mean expression for each gene is denoted by the horizontal lines. The Pearson correlation, uncentered correlation, and the Euclidean distance are 0.74, 0.77, and 3.46, respectively. A similar plot is shown in Figure 8.3B where the same two genes are plotted, except that for the gene indicated by the dashed line, a constant has been added to the expression levels. So, although the "pattern," or trend, of gene expression is identical to that in Figure 8.3A and the Pearson correlation remains unchanged, the uncentered correlation and Euclidean distance change rather drastically: from 0.7 to 0.05 and 13.46 to 15.21, respectively. It can be argued that this is an unfavorable characteristic when trying to find genes which behave similarly. When choosing a distance measure, it is important to keep this property in mind. In general, for gene expression analyses, we usually care more about whether two genes have similar trends in gene expression, but we do not necessarily care that their average expression levels be similar. For this reason, Pearson and Spearman correlations (and their absolute values) may be considered superior to the Euclidean-type distances and the uncentered correlation for gene expression analysis.

8.2.2 Categorical Distance and Similarity Measures

Many measures exist for determining distances between binary and other categorical variables. Although these have been used in the context of gene expression analysis, almost all gene microarray experiments produce continuous gene expression values, and so we will not cover categorical distance measures. The reader should see Gordon [5] for additional discussion of distance measures for categorical data.

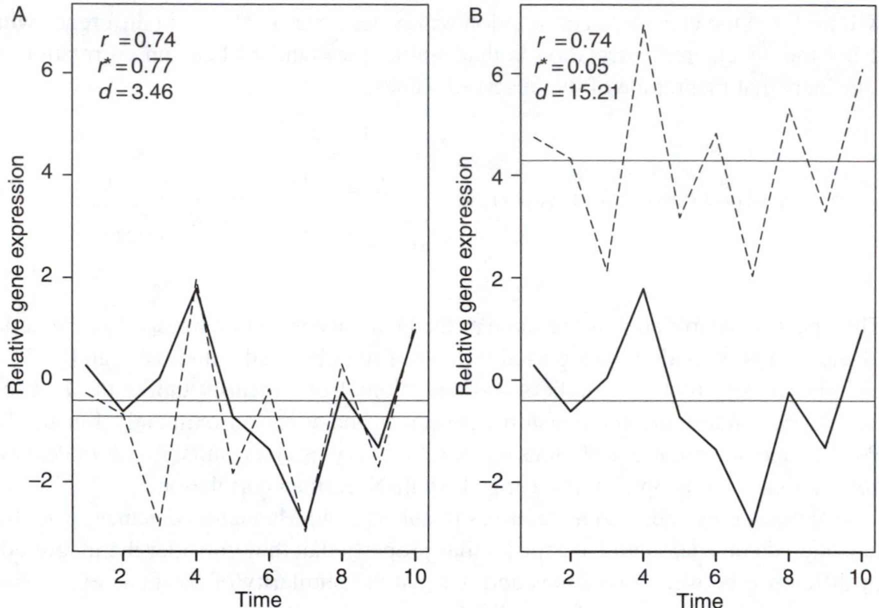

FIGURE 8.3 A comparison of the Pearson correlation, uncentered correlation, and Euclidean distance for two genes. Panel A: Both genes have approximately the same average expression (indicated by horizontal lines) and Pearson and uncentered correlation are almost the same. Panel B: For the gene indicated by the dashed line in Panel A, a constant has been added to the gene expression values. While Pearson correlation between the two genes remains the same, the uncentered correlation and the Euclidean distance between the two genes is drastically changed from Panel A, demonstrating the sensitivity of these measures to differences in average gene expression.

8.3 HIERARCHICAL CLUSTERING

Hierarchical clustering is the most commonly seen method for displaying gene expression analysis results. The general idea is to provide a hierarchy of association between the genes and samples. Many examples of using cluster analysis in gene expression analyses can be found, including those by Blader et al. [6], Spellman et al. [7], Chen et al. [8], Wen et al. [9], and Eisen et al. [10].

The simplest way to understand the mechanics is via a figure. In Figure 8.2, we have examples of "dendrograms" which shows the results of different hierarchical cluster analyses of one dataset. To illustrate the method, we have simulated a very small gene expression dataset with only 20 genes and 30 samples. For each dendrogram, the y-axis represents a measure of distance between two genes. To see how distant two genes are we look to see at what height they are joined. For example, in Figure 8.2A, Hs.309 and Hs.206 are close in distance: the height at which they are joined is approximately 1.0. The genes Hs.161 and Hs.527 are more distant, joined at a height of about 1.25. To determine distance between genes that are not directly joined, we look to see the smallest height at which clusters they belong to are joined.

For example, Hs.24 and Hs.782 are approximately 1.3 units away, while Hs.181 and Hs.782 are 1.1 units apart.

It is not necessarily important to see exactly how far apart two genes are, but instead to identify groupings of genes that look to be relatively close together, yet well separated from other genes in the analysis. In Figure 8.2B, we see two large clusters (joined at 2.5), and then possibly two other large separations (joined at about 2.0 and 1.8). There will be more discussion about how to determine how many "significant" clusters exist in a hierarchical cluster analysis in later sections.

Additional reading of hierarchical clustering can be found in Everitt [11], Gordon [5], and Kaufmann and Rousseeuw [12].

8.3.1 Up vs. Down

There are two "directions" in which we can determine our hierarchy. In "bottom–up," or agglomerative clustering, each object is its own cluster at the beginning of the analysis. The first step is to determine which two objects (e.g., genes) are most similar and merge them into one cluster. We repeat this, merging the closest genes and gene clusters until all genes are in the same cluster. The other option is "top–down," or divisive clustering, where we begin with all genes in the same cluster, and then the cluster is split so that genes within the two new clusters are most similar. This is repeated until all objects are in separate clusters.

Both the agglomerative and divisive approaches are considered "step-wise optimal," meaning that at each step the most optimal merge or split is made. However, this does not imply that the resulting cluster structure is optimal.* This is one reason why different clusters occur when using divisive vs. agglomerative approaches.

It is most common to see the agglomerative approach used in gene expression clustering. However, there is good justification for choosing divisive instead. In most cases, we have many genes and are interested in finding relatively few interesting clusters. If we are interested in the top four clusters, one way to determine this is to "cut" the dendrogram at the height that gives us exactly four clusters. In Figure 8.2B, this would be at approximately 1.8. Using the divisive clustering algorithm, there will be exactly three splits that will determine which genes are in the four clusters. However, in the case of agglomerative clustering, there will be many merges before arriving at the top four clusters. In the example in Figure 8.2A, there are 16 merges before arriving at the top four clusters. The more merges or splits have occurred, the more room for error there is in the clustering (recall that these approaches are only step-wise optimal). If we are interested in the "bottom" of the dendrogram (i.e., where we have many small clusters), then it is sensible to use the agglomerative approach because it has relatively few merges at the bottom. However, when we are interested in the top (i.e., a relatively small number of large clusters), as is most often the case, using the divisive method is more appropriate [5]. One reason that we often see more use of the agglomerative approach may be due to the relative computational intensity of the divisive algorithm. However, with current technology, the divisive approach

* Optimal in the sense of homogenity and separation is described in Section 8.8.1.

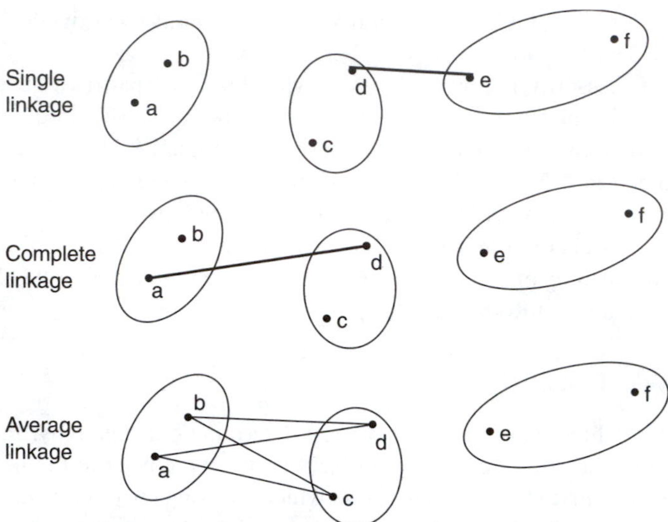

FIGURE 8.4 Three commonly used linkages in agglomerative hierarchical clustering. Lines are drawn connecting points which would be used to determine how the next cluster merging is determined.

poses no significant problems for estimation, even with large gene expression datasets.

8.3.2 Hierarchical Clustering Algorithms

8.3.2.1 Agglomerative

At the onset of the agglomerative algorithm, determining which two genes or samples to join into a single cluster is simple: choose the two objects with the shortest distance (or, the highest similarity) based on the chosen distance or similarity metric. However, once there are clusters containing two or more objects, rules must be defined for determining distance between clusters. The agglomerative method has several different choices of algorithms for defining how to join. The choices are referred to as "linkage" methods. We will discuss the three most common, illustrated in Figure 8.4 where an agglomerative clustering method is being applied where three clusters have already been formed: clusters *ab*, *cd*, and *ef*. The example in Figure 8.4 is rather simplified compared to the gene expression array setting, but is effective for illustrative purposes.

8.3.2.1.1 Single Linkage (Nearest Neighbor)

In single linkage, the distance between clusters is defined as the "distance between the two closest objects in two clusters." For example, in Figure 8.4, in the top panel, the *ef* cluster and the *cd* cluster are joined because the distance between their closest objects (*d* and *e*) is shorter than the other possible combinations (*c* and *b*, and *f* and *b*). This method of clustering appears sensible, but because it ignores the distance between more remote objects when joining clusters, the resulting clusters may not be as homogeneous as might be preferred. The single linkage method will often find nonellipsoid (i.e., long and skinny) shaped clusters in data, unlike other commonly used linkage approaches [13].

8.3.2.1.2 Complete Linkage (Furthest Neighbor)

Complete linkage is the opposite of single linkage: the distance between clusters is defined as the distance between the two furthest objects in two clusters. In Figure 8.4, the middle panel shows that the next step in the clustering would be to join clusters *ab* and *cd*. The distance between a and d is shorter than the distance between c and f, and between f and b. This approach has the benefit that all objects within the resulting cluster are within the distance on which the merging was based (e.g., in Figure 8.4, all objects in the cluster *abcd* are closer than the distance between a and d). Complete linkage usually leads to very compact, spherical clusters.

8.3.2.1.3 Average Linkage

Average linkage uses the distance between all the points across clusters to determine the distance between the clusters. In Figure 8.4, the distance between clusters *ab* and *cd* is $(d(a,c) + d(a,d) + d(b,c) + d(b,d))/4$, which is the average of all pairwise distances between points in the two clusters. This is an attractive option because it incorporates both the closest and the furthest distances when estimating distance, unlike the previous two linkage methods discussed. The clusters from average linkage are usually round or elliptical.

In general, it is sensible to try at least two of these linkage methods to estimate the hierarchical cluster. If the resulting structures appear quite different, this is an indication that clusters may not be stable. For more discussion of the stability of the clustering results, see Section 8.8.

8.3.2.2 Divisive

The divisive algorithm is more computationally intensive than the agglomerative, which is probably why the agglomerative method is so commonly used and the divisive is not despite other obvious advantages of the divisive approach in many settings. Consider the initial step in the agglomerative algorithm: Two of N genes are joined. To determine which two to join, the number of possibilities to consider is $N(N-1)/2$, which grows quadratically in N. In the initial step of the divisive algorithm, the number of possible splits to consider is $2^{N-1} - 1$, which grows exponentially in N. As an example, if you are interested in performing a hierarchical clustering of only ten genes, the number of merges to consider at the first step in the agglomerative algorithm is only 45 as compared to the 511 splits to consider in the divisive algorithm. For 100 genes, these numbers climb to 4950 and 6.34×10^{29} for agglomerative and divisive, respectively. For a naïve estimation approach, the number of splits to consider at each step of the divisive algorithm makes it practically untenable to estimate.

However, faster algorithms have been developed which do not require the estimation of all possible splits at each step. The divisive algorithm by MacNaughton-Smith et al. [14] is relative efficient and has been implemented by Kaufman and Rousseeuw in their clustering programs [12] and in the software packages R and Splus, which is described in Algorithm 8.2:

Algorithm 8.2

1. Compute distance between objects, $d(x_i, x_j)$, for all x_i and x_j.
2. Calculate average distance between each object and other objects:

$$d^*(x_i) = \frac{1}{N-1} \sum_{j:j \neq i} d(x_i, x_j)$$

3. Choose gene which has the largest average distance and place it in its own "splinter" cluster.
4. Calculate distances $d^*(x_i)$ in the large cluster as in Step 2, and also calculate distance between each object in large cluster and gene in the "splinter" cluster.
5. Move objects that are closer (via $d^*(x_i)$) to splinter cluster than the large cluster into the splinter cluster.
6. Repeat Steps 4 and 5 until there are no more reallocations.
7. Divide the cluster with the largest diameter (i.e., the cluster with the largest distance between two objects) using Steps 2 through 6.
8. Repeat until all genes are in single gene clusters.

8.3.3 Visualization Methods

8.3.3.1 Dendrograms

The standard graphical device that shows the result of a hierarchical clustering is the dendrogram, or tree, which was introduced in Figure 8.1. The interpretation of the dendrograms for agglomerative and divisive clustering are generally the same, and when comparing the results of these two algorithms, what we tend to focus on is whether the splits and merges occur in the same order.

An important point to note is the interpretation of the y-axes in these plots. The y-axis on the agglomerative tree represents the distance of two clusters from one another. The y-axis on the divisive tree represents the diameter of the cluster, meaning the largest distance between two genes in the cluster. So, when assessing the difference between the two clustering approaches, it is important to remember that the y-axes are not comparable. In the gene expression setting, we usually do not see the numbers on the y-axis of the plot. Instead, we see the tree paired alongside with the red–green display of gene expression intensity, as in Figure 8.1 where both genes and samples are clustered.

One should be aware of the "nonuniqueness" of the dendrogram result. It is often misconceived that because two genes are next to each other in the dendrogram, they are close together. However, relative distance on the x-axis of a dendrogram means little or nothing. In Figure 8.1, we can see that the cluster analysis almost perfectly distinguishes the 37 lung adenocarcinoma samples from the ten squamous cell lung carcinomas. If we consider the two large clusters, there is one adenocarcinoma that has been allocated into the cluster with the ten squamous cell samples. And, one might be initially surprised to see that this sample is placed furthest from the adenocarcinoma samples. However, at second glance realize that the dendrogram could just have easily been drawn with the stray adeno being placed adjacent to the adeno cluster. That is, it is an arbitrary decision when depicting a split or a merge as to which cluster falls to the right and which to the left of the split or merge. So, imagine that you could lift just the squamous cluster off the page, turn it upside, and put it back on the page in the same place. The resulting dendrogram would be equivalent to the one shown in Figure 8.1. Some people prefer to think of dendrograms as 3-dimensional mobiles, so that the height at which objects are joined remains constant, but the horizontal distance between objects can vary depending on the current state of the mobile.

8.3.3.2 Banner Plots

Banner plots contain the same information as a dendrogram, but give a clearer picture of the state of the clustering structure for a given height of the tree. By drawing a straight (vertical) line along the plot at any level, one can easily tell how many clusters there are and which object belong to which cluster. An example of a banner plot is shown in Figure 8.2I, where divisive clustering was used with Euclidean distance. Figure 8.2H and Figure 8.2I are the same clustering just displayed in different ways.

8.4 *K*-MEANS AND *K*-MEDOIDS

K-means clustering is a nonhierarchical approach to creating spherical clusters of objects which can be used for finding groups of similar genes and samples in gene expression arrays. For example, Tavazoie et al. [53] present a *k*-means clustering to identify transcriptional regulatory sub-networks in yeast. This approach identified novel regulons (sets of co-regulated genes) and their putative *cis*-regulatory elements. Unlike hierarchical clustering where you can obtain any number of clusters by simply cutting the dendrogram at the desired height, in *k*-means clustering, the number of clusters is fixed. Each object is in one and only one cluster, and each cluster contains at least one object. In most estimation procedures, the number of clusters (k) must be chosen in advance. If k is unknown and it needs to be estimated, fitting successive *k*-means cluster analyses can be used to determine the optimal value for k. This is discussed in Section 8.8.

The *k*-means clustering algorithm is based on the idea that there are k centroids which are the centers of the clusters. The centroid is defined as the point in the cluster that minimizes the sum of squared distances between the objects and centroid. In other words, it is the "average" of the cluster. The *k*-means clustering algorithm begins with an initial partitioning of the objects into clusters. The algorithm proceeds as described in Algorithm 8.3. *K*-means clustering is estimated using Euclidean distance. If other distances are preferable to the user, a similar algorithm can be implemented but the algorithm no longer produces clusters for which the centroid is the mean of the cluster.

Algorithm 8.3 *k*-means clustering

1. Create an initial partition of objects into k clusters.
2. Calculate the centroid of each of the k clusters.

3. a. For object i, calculate its distance to each of the centroids.
   b. Allocate object i to cluster with closest centroid.
   c. If object was reallocated, recalculate centroids based on new clusters.
4. Repeat Step 3 for objects $i = 1, \ldots, N$.
5. Repeat Steps 3 and 4 until no reallocations occur.

The initial partitioning is usually based on randomly assigning each of the objects to one of k groups. Another approach is to randomly determine k centroids. It is generally not thought to be favorable to choose an initial partition based on a prediction of how the data will tend to cluster, because the initial partition can influence the final result and hence the results would be biased. As such, it is generally good practice to repeat the k-means clustering algorithm using different initial random partitions and check to see that the resulting clustering structures match. This ensures that resulting clustering solutions are relatively stable and have converged to a reasonable solution. But, like hierarchical clustering, there is no guarantee that the resulting clustering structure is "optimal" in a statistical sense.

k-medoids is an analogous approach, but instead of defining clusters using the average of the objects within the clusters, "representative objects" are found to define each cluster. This is similar to using the median vs. the mean in a univariate setting. One attractive feature of the k-medoid approach is that it is more robust to outliers than k-means. However, by restricting cluster centers to medoids the clustering methods are sensitive to the inclusion of unrelated features and the overall noise level of the data. Note that in the k-medoids algorithm, an initial partition is defined by randomly choosing k objects as the medoids.

Hartigan and Wong [15] give additional helpful discussion of k-means clustering. Kaufmann and Rousseeuw [12] discuss k-medoids clustering and their software application in detail.

8.5 SELF-ORGANIZING MAPS

Self-organizing maps (SOMs), developed by Kohonen [16,17], are primarily a data visualization technique which reduce the dimensionality of high-dimensional data into a two or three dimensional figure. SOMs are similar to k-means and k-medoids approaches in their goals, but have an algorithmically different approach. Tamayo et al. [18] used SOMs to analyze hematopoietic differentiation in human samples and have developed the GENECLUSTER program for implementing SOMs specifically for gene expression data.

The SOM algorithm, although similar to k-means, is more complex than the others described thus far. It is a more modern technique, allowing for thousands of iterations to achieve the resulting cluster structure. The general steps are shown in Algorithm 8.4. The idea is that clusters are represented by nodes. We begin with a two-dimensional grid where the initial nodes of the clusters are represented. This is somewhat arbitrary, but approximately square-shaped grids tend to be preferred (e.g., a 6×5 grid would be preferable to a 15×2 grid in the case of $k = 30$ nodes). At each step in the algorithm, for a particular gene, the node closest to each gene is identified. Then, all nodes reasonably close by are updated by moving them closer to the gene.

This can be seen in Algorithm 8.4: all nodes within radius r of the gene of interest are updated and those further are not updated. There is also a "learning function" which serves the purpose of making large updates at early iterations and smaller updates at later ones.

Algorithm 8.4 An example of a SOM algorithm.

1. Choose a number of clusters, k, and an appropriate initial geometric grid on which to represent the initial nodes of the clusters.
2. Map the nodes into the grid.
3. Randomly order the genes to be clustered. This ordering is determined once and reused as necessary.
4. For the first gene in the ordering, x_I, find its closest node, N_{xI}.
5. Location of node N at iteration t is defined as $q_t(N)$. Update all nodes using the following equation:

   $$q_{t+1}(N) = q_t(N) + \tau(d(N_{x_1}, N), t)(q(x_1) - q_t(N_{x_1}))$$

   where τ is the "learning function." There are numerous choices for the learning function, and one example is:

   $$\tau(d,t) = \begin{cases} \frac{0.02T}{T+100t}, & \text{if } d_t(N, N_{x_i}) \leq r \\ 0, & \text{if } d_t(N, N_{x_i}) > r \end{cases}$$

   where r is a prespecified radius and T is the maximum number of iterations in the algorithm (usually in the tens of thousands or more).
6. Repeat this for each of the genes in the ordering and continue iterating through the list up to T iterations.

A simple example of the beginning of a SOM algorithm is shown in Figure 8.5 where nodes are represented by the black squares. In panel A, the algorithm is initiated with the nodes arranged in a geometric grid. The objects have been randomly ordered with the first object shown in the circle. The top-right node is closest to the first object and is updated using the equation shown in the algorithm and its new location is shown in panel B. No other nodes are within the radius r so no other nodes need to be updated. In panel C, we see the second point used for updating the nodes. The bottom-left node is closest to the point, and when the bottom-left mode shifts towards this point, the bottom-left and bottom-right nodes are within r. So, the bottom-right node is also updated according to the equation in Step 5. Notice that both of these nodes move in the same direction, but the bottom-right node moves a greater distance as demonstrated in Step 5. This process will continue for a large number of iterations, T.

Ripley has shown that, theoretically, SOM and k-means are the same when the radius r is 0 [19,20]. However, there are algorithmic differences that can cause the resulting clustering structures to differ from each other even when r is set to 0 in the SOM approach.

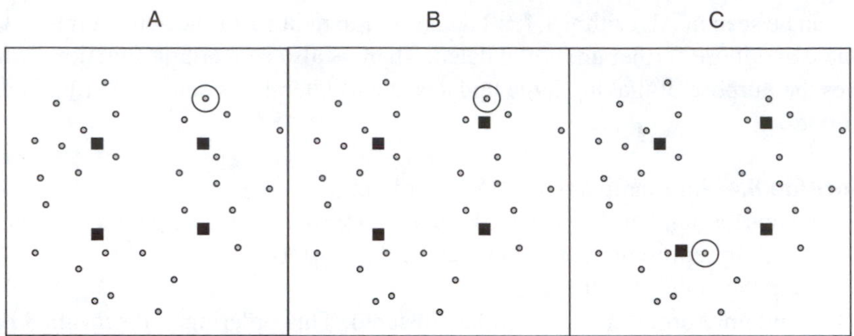

FIGURE 8.5 Beginning of the (SOM) algorithm. Nodes are indicated by black squares and objects by small circles. Panel A shows the initialization and the first object used for node updating, indicated by the circle. Panel B displays the result of the node updating. Panel C shows the second object used for updating and the resulting node updates, where two nodes have changed position.

8.6 CLUSTER AFFINITY SEARCH TECHNIQUE

The Cluster Affinity Search Technique (CAST) algorithm, introduced by Ben-Dor et al. [21], is based on a stochastic model for cluster formation and requires a similarity matrix of genes to be clustered and a threshold which determines the level of similarity required. Clusters are constructed one at a time based on the affinity (i.e., similarity) of genes. Ben-Dor et al. [21] define the affinity of gene x for cluster C as

$$a(x, C) = \sum_{y \in C} s(x, y).$$

One clear benefit of the CAST algorithm is that not all genes are necessarily included in clusters. For example, if there exists a gene g, which has similarity with all genes that is smaller than the threshold, then gene g will not be clustered with any other genes. So, unlike k-means and related algorithms which require a fixed number clusters and assume that all genes belong to exactly one of the clusters, the resulting number of clusters in CAST is purely determined by the threshold and, in most cases, the clusters will exclude a large fraction of the genes under analysis from the resulting clusters. Differences in the number and homogeneity of clusters can be explored by changing the threshold t and comparing results.

Algorithm 8.5 The CAST algorithm.

1. Obtain similarity matrix (S), and choose threshold, t.
2. Choose two genes that are most similar (and not already in a cluster) and define cluster C.
3. Find gene, x, which maximizes $a(x, C)$, and add x to C.
4. Repeat 3 while $a(x, C) < t|C|$.
5. Find gene, w, which is in C and minimizes $a(x, C)$.
6. Repeat 5 while $a(x, C) < t|C|$.

7. Repeat Steps 3 through 6 until no more changes occur.
8. Close cluster C and remove all genes in C from comparison.
9. Begin new cluster by repeating Steps 1 through 7.

8.7 OTHER RELATED METHODS

Principal components analysis (PCA) and multi-dimensional scaling (MDS) are techniques whose main goal is to reduce the dimensionality of the data [22–24]. In both approaches, high-dimensional data is used to identify a smaller set of representative variables that are weighted averages of the original input data. Just as in the clustering methods described above, there are user-specified parameters that may greatly impact the results. There are many examples and references for PCA [25–28] and MDS [29,30] applied to gene expression data. West et al. [31] and Hastie et al. [32] develop PCA to more sophisticated approaches for analyzing gene expression data.

There are many other clustering approaches being developed for specifically for gene expression data analysis. Several others worth mentioning include relevance networks by Butte et al. [33], where gene entropy is used to define large networks of gene associations, neural networks, as described for gene expression analysis by Herrero [34], and probabilistic models, which do not suffer from some of the same drawbacks as some of the standard clustering approaches [35,36].

8.8 ASSESSING CLUSTER FIT AND CHOOSING *K*

One thing that is often not considered when interpreting the results of a cluster analysis is that there can be considerable instability. No information about precision is included in most standard output from the above methods. As a result, clustering results are often overstated and inappropriately represented. To try to guard against this common problem, assessing the cluster result for stability and fit is always an important consideration.

There are several types of stability to consider. The first is how sensitive the clustering result is to the user-specifications within the algorithm. For example, hierarchical clustering results based on Euclidean distance of a gene expression dataset may be quite different from those based on Spearman correlation. Resolving this issue can be done based on careful consideration of the steps in the hierarchical clustering, and exploratory data checking. For example, one can assess whether or not Pearson correlation adequately describes associations, or if a more robust measure (e.g., Spearman correlation) is necessary. The second kind of stability deals with how much noise there is in the data that is contributing to the hierarchical structure.

The approaches discussed can be used to assess clusterings developed using the same method with varying values of k, and also to compare different clustering approaches on the same dataset. Note that most of the approaches discussed below require a fixed number of clusters. In the case of hierarchical clustering, a fixed k can be found be cutting or "pruning" the dendrogram at the height which produces k clusters.

8.8.1 Homogeneity and Separation of Clusters

Chen et al. [37] describe a series of cluster assessment tools and compare k-means, k-medoids, hierarchical, SOM with $r = 0$, and SOM with $r = 1$ clustering approaches. One of the approaches they present developed by Shamir and Sharan [38] estimates homogeneity (H_{ave}) and separation (S_{ave}) of clusters. These are defined as

$$H_{\text{ave}} = \frac{1}{N_g} \sum_{i=1}^{N_g} d(x_i, C(x_i))$$

$$S_{\text{ave}} = \frac{1}{\sum_{i \neq j} N_k N_l} \left(\sum_{i \neq j} N_k N_l d(C_k, C_l) \right)$$

where N_g is the total number of genes, N_k and N_l are the number of genes in clusters k and l, x_i is the vector of information on gene i, C represents a cluster, and $d(,)$ is a measure of distance. Homogeneity measures the compactness of the clusters, while separation measures how far clusters are from each other. Large values of H_{ave} suggest poor fit, while small values of S_{ave} suggest poor fit.

To determine the appropriate number of clusters (k), one can estimate H_{ave} and S_{ave} for various values of k and look for a "leveling off" of the H_{ave} and S_{ave} functions vs. k. This can even be implemented with hierarchical clusters, where k is determined by a particular height cutoff on the dendrogram. For examples, of a comparison of clustering approaches in a gene expression setting using H_{ave} and S_{ave}, see Chen et al. [37].

8.8.2 Cluster Silhouettes

Introduced by Rousseeuw [39], a cluster's silhouette represents how close genes within a cluster are to genes in other clusters. This can be considered a composite measure of both homogeneity and separation of clusters. The silhouette plots provide a graphical depiction of silhouettes so that the user can see which clusters are relatively compact as compared to others [39,40]. An example of a silhouette plot is shown in Figure 8.6. A k-means clustering procedure was used to find three clusters in the left side of the figure and five clusters in the right side of the figure. The silhouettes are shown as horizontal bars: each object has a silhouette where wide silhouettes indicate strong similarity to objects within the cluster and dissimilarity to objects in other clusters. Negative silhouettes suggest a poor allocation of an object to a cluster: a cluster with a negative silhouette is more similar to objects in other clusters than objects in its own cluster.

As a more formal means of evaluating clusters, the silhouette coefficient (SC) can be used [12], which is the average of the width of all silhouettes. SC is defined as

$$\text{SC} = \frac{1}{N_g} \sum_{i=1}^{N_g} \frac{b(i) - a(i)}{\max(a(i), b(i))}$$

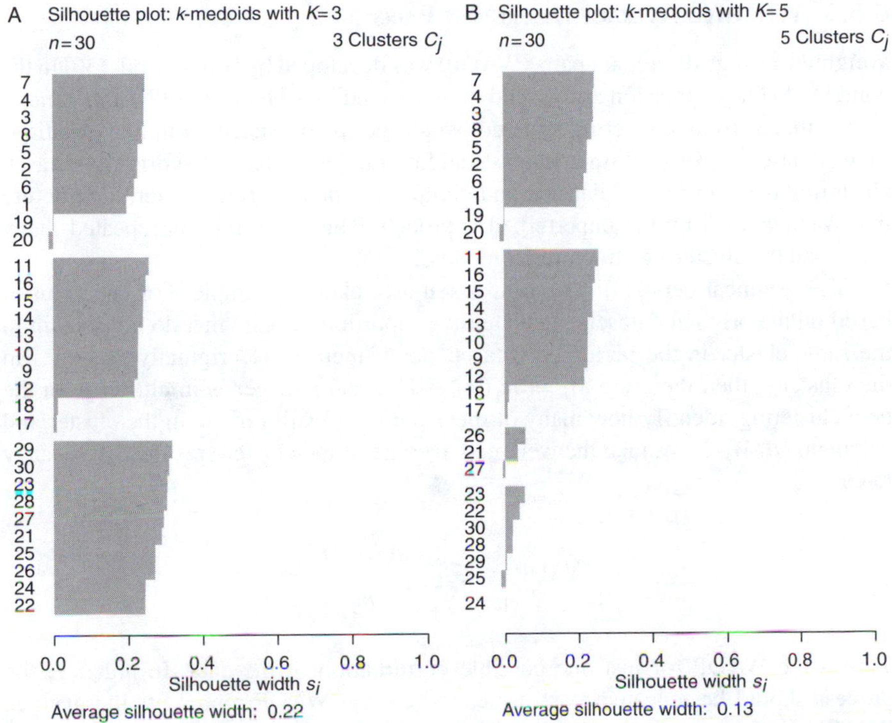

FIGURE 8.6 Silhouette plot. Panel A shows the silhouette plot and silhouette coefficient for a K-medoids clustering where $k = 3$. Panel B shows the silhouette results for a K-medoid clustering of the same data with $k = 5$.

where $a(i)$ is the average distance of gene i to other genes in gene is cluster, $b(i)$ is the average distance of gene i to other genes in the nearest neighbor cluster and N_g is the number of genes. The larger the SC, the better the quality of the clustering structure. In Figure 8.6, the overall SC values are provided at the bottom of each silhouette plot: $\text{SC}_3 = 0.22$ and $\text{SC}_5 = 0.13$ suggesting that the three cluster solution has more homogeneity and separation of clusters than the five cluster solution. The silhouette plot can highlight how clustering procedures are only stepwise optimal: it is not unusual to see objects with negative silhouette widths, suggesting that overall homogeneity and separation were not achieved by the algorithm.

Kaufman and Rousseeuw [12] developed an admittedly subjective scale based on their experience working with the SC in other applications. SC of 0.71 to 1.00 indicates strong structure; 0.51 to 0.70 suggests reasonable structure; 0.26 to 0.50 is indicative of weak or artificial structure and implies another method should be used; less than 0.26 implies no substantial structure in the dataset. However, in gene expression data, these cutoffs seem overly stringent given the amount of noise in the typical gene expression dataset. It may be more useful to use SC to compare cluster results with varying values of k than to expect that SC will be sufficiently high.

8.8.3 Weighted Average Discrepant Pairs

Weighted average discrepant pairs (WADP) was developed by Bittner et al. [30] in the context of gene expression arrays, and implemented by Chen et al. [37]. The general idea is that a strong clustering structure would not be impacted by minor variations in the dataset. As such, Bittner suggests adding random noise to the original dataset, clustering the "perturbed" dataset, and comparing the resulting clustering structure that was obtained by the unperturbed approach. This perturbing is repeated many times and the clustering structures compared.

The technical details of the comparison are relatively simple. For each cluster based on the original dataset, calculate the proportion of pairs that do not remain in the same cluster in the perturbed dataset. So, if there were originally m_j genes in the cluster j, then there are $M_j = m_j(m_j - 1)/2$ pairs of genes in cluster j. In the new clustering, identify how many of these pairs (D_j) still remain in the cluster and calculate D_j/M_j. Now, take the weighted average of the cluster-specific discrepancy rates:

$$\text{WADP} = \frac{\sum_{j=1}^{k} m_j D_j/M_j}{\sum_{j=1}^{k} m_j}$$

This is the WADP for just one possible perturbation of the data. In practice, the process should be repeated many times, where the WADP represents the average WADP based on many perturbations of the original data.

One difficulty with this method is determining how much the data should be perturbed. If too little noise is added, the clustering may appear stable when it should not, and if too much noise is added, a stable cluster result may appear to be unstable. Chen et al. [37] considered just one variance and added Gaussian random noise. Bittner et al. [30] advises using Gaussian noise with standard deviation of the noise determined as the median of the standard deviations of log-ratios as determined for single genes.

8.8.4 The Gap Statistic

Hastie et al. [32] introduced the notion of the "gap" statistic for choosing the number of clusters k in reference to their "gene shaving" technique and then he and colleagues proposed a more general gap statistic that can be used for any clustering result [41]. Their method is based on the idea that the error associated with clustering decreases as the number of clusters increases, but after some number of clusters, k^*, the decrease in error tends to flatten out indicating that additional clusters do not improve the clustering structure.

The gap statistic is based on a measure of homogeneity of clusters:

$$W_k = \sum_{r=1}^{k} \left(\frac{1}{2n_r} \sum_{i,j \in C_r} d_{i,j} \right)$$

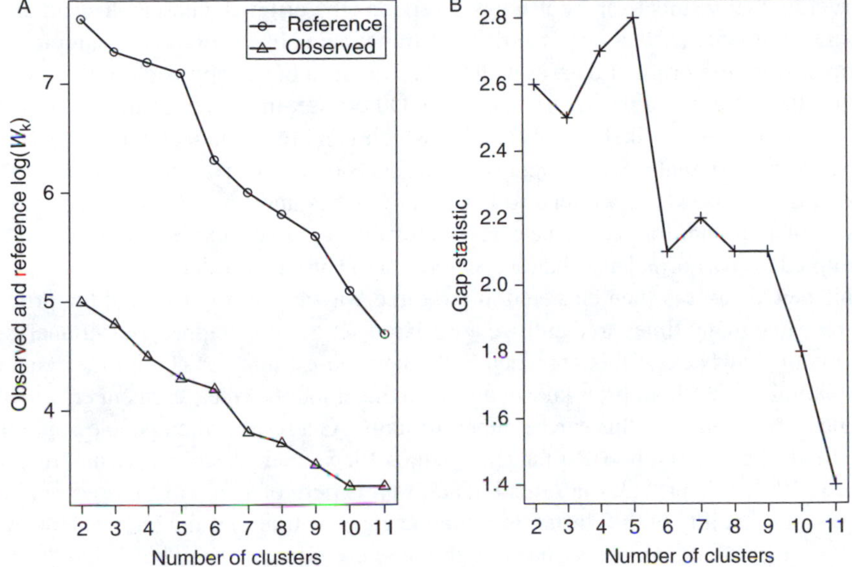

FIGURE 8.7 Example of the gap statistic. Panel A plots the reference and the observed distribution of log(W_k) vs. number of clusters, k. Panel B shows the gap statistic as a function of number of clusters. Based on the gap statistic, five clusters are chosen.

where $d_{i,j}$ is the measure of distance between objects i and j, C_r represents cluster r, and n_r is the number of objects in cluster r. W_k can be estimated for a range of k from 1 to some large number of clusters, k, where larger values of W_k indicate a larger amount of variation within clusters and hence less homogeneity within clusters. The log of the W_k values are then plotted vs. number of clusters.

The null reference distribution of W_k (i.e., $E(W_k)$) is estimated for comparison (see Figure 8.7). For each value of k, the distance between W_k and the reference distribution is calculated and this is defined as the gap statistic:

$$\text{Gap}_n(k) = E_n(\log(W_k)) - \log(W_k)$$

where E_n represents the expected value for a sample size of n. The value k for which the gap statistic is maximized determines the estimated of the number of clusters, k^*. For example, in Figure 8.7, the number of clusters chosen would be five. Although the idea is quite clear, the computation of the reference distribution is somewhat complex. The general idea is similar to the WADP and Consensus Trees (discussed in Section 8.8.5) in that perturbations are used to determine the reference distribution. For details of the estimation of the reference distribution, see Tibshirani et al [41].

8.8.5 Consensus Trees

We briefly describe this approach and refer the reader to Zhang and Zhao [42] for further discussion. The general idea is to compare a clustering structure to the clustering structure that results from bootstrapped versions the original datasets. Instead of the more commonly used nonparametric bootstrap where observations (i.e., samples) are drawn from the original dataset with replacement, a better approach in the clustering setting is to use a parametric approach. Otherwise, in each bootstrapped sample, exact duplicates are likely to be sampled which is more problematic for cluster analysis than other analyses (e.g., regression). This is due to the particular nature of gene expression data where we tend to have many genes and few samples. In parametric bootstrapping, for each gene expression value x_{ij}, a new expression value, $x_{ij}{}^*$ is sampled from a normal distribution, centered at the observed value ($x_{ij}{}^* \sim N(x_{ij}, s_{ij})$). This new dataset is then clustered in the same way as the original, and the process is repeated many times so that there are a large set of bootstrapped cluster analyses. A major drawback of this approach is the need to estimate or choose the standard deviation, s_{ij}. Without prior knowledge as to the amount of measurement error in the gene expression data, this can be rather difficult. As a result, when people apply this method, they often choose a range of values for s_{ij} and create consensus trees for each. While this provides information across a variety of scenarios, it does not help assess the stability of the clustering result because the true s_{ij} is not known. However, if the range of s_{ij} can be estimated with some confidence, then "best" and "worst" case scenario consensus trees can be calculated and compared to the original cluster.

A related approach that is nonparametric, does not require an assumption about the variance of noise in the data, and solves the problem of the sampling exact duplicates is the convex-combination method [43]. In this approach each sampled dataset is generated by selecting two observations (i.e., samples) from the dataset and randomly choosing a weight on the interval [0, 1]. The sampled value is then calculated as the weighted average of the two selected observations. For an example of this approach in gene expression, see Dudoit et al. [44].

Margush and McMorris [45] have defined consensus trees by using a "majority rule." Simply, this means (in the case of hierarchical clustering) that a consensus tree contains all of the nodes that are in at least half of the bootstrapped trees. We can then compare the consensus tree to the original clustering and compare to assess stability by looking at confidence values for each of the nodes. We do not discuss the technicalities of this but refer the reader to the following chapter and other references [42,45].

8.8.6 Other Approaches

Although not discussed in any detail here, the model-based clustering methods used by Fraley and Raftery [36] provide a natural method for model selection (i.e., choosing k): models can be compared using information criteria or other model comparison statistics. Fraley and Raftery advocate the use of the Bayesian information criteria (BIC, [46]) which allows comparison of multiple models and does not restrict comparisons to nested models.

Dudoit and Fridlyand [47] recently developed a resampling method for assessing the number of clusters and cluster stability. Their approach randomly partitions the dataset into two, clusters the data on one part, defines a classifier, and applies the classifier to the other part. This is repeated many times and for varying values of k.

For details of how the results are used to assess the clustering and comparison to other related methods, see Reference 47.

8.9 CHOOSING GENES AND SAMPLES FOR CLUSTERING

In many published gene expression cluster analyses, all genes and all samples from the experiments are included in the clustering. This is not a recommended practice. While it is generally desirable to include as many samples that are representative of the population in which you would like to discovery genetic differences, it is expected that a very large fraction of the genes in the experiment are not at all related to the clustering structure that is sought. Now let us consider what occurs statistically when genes whose expressions are not related to subgroups of samples are included in the clustering. A cluster analysis is a type of latent variable analysis, where the latent variable can be thought of as the true cluster memberships. What we assume is that the correlation (similarity) matrix provided “defines” the underlying latent clustering structure. When genes are included in the analysis that are not related to the cluster structure of interest (e.g., we may be interested in discovering breast cancer subtypes), our results are likely to be quite different than if those genes were excluded.

In Figure 8.8, an example of this is shown using simulated data on 900 genes and 40 samples. The data were generated so that four clusters of samples exist based on gene expression patterns: samples 1–10 are in cluster 1, 11–20 are in cluster 2, 21–30 are in cluster 3, and 31–40 are in cluster 4. These clusters were defined by gene expression in the first 450 genes, while the gene expressions of the other 450 genes were generated randomly. In Figure 8.8A we see an agglomerative hierarchical clustering based on all 900 genes, and in Figure 8.8B we see the same clustering approach, removing the 450 genes that were generated at random. Clearly, the clustering in Figure 8.8B is more consistent with the true underlying clustering of the genes. In Figure 8.8A, we see that the unrelated genes have tainted the analysis, resulting in clusters that are inconsistent with the true clusters. Some may be thinking that it is not possible to know a priori which genes are related to the subgroups of interest. However, it is certainly possible to evaluate the distribution of gene expressions across samples per gene to see which genes show evidence of differential expression. This is not the same as identifying which genes have large variance: what we are interested in genes that appear to show possible multi-modalities. For example, we would be interested in a gene whose expression values form a bimodal distribution so that a proportion of the genes tend to have one expression level and the remaining have another expression level. One approach to detecting this is by Parmigiani et al. [48] where mixture models are used to identify genes exhibiting differential expression. Another simpler approach could be to simply rank genes based on their evidence of nonnormality. Genes showing evidence of nonnormality will have more promise of separating genes into clusters. It has been argued that in normally distributed genes, the variation in expression levels is due to measurement errors and not biologic variability [48].

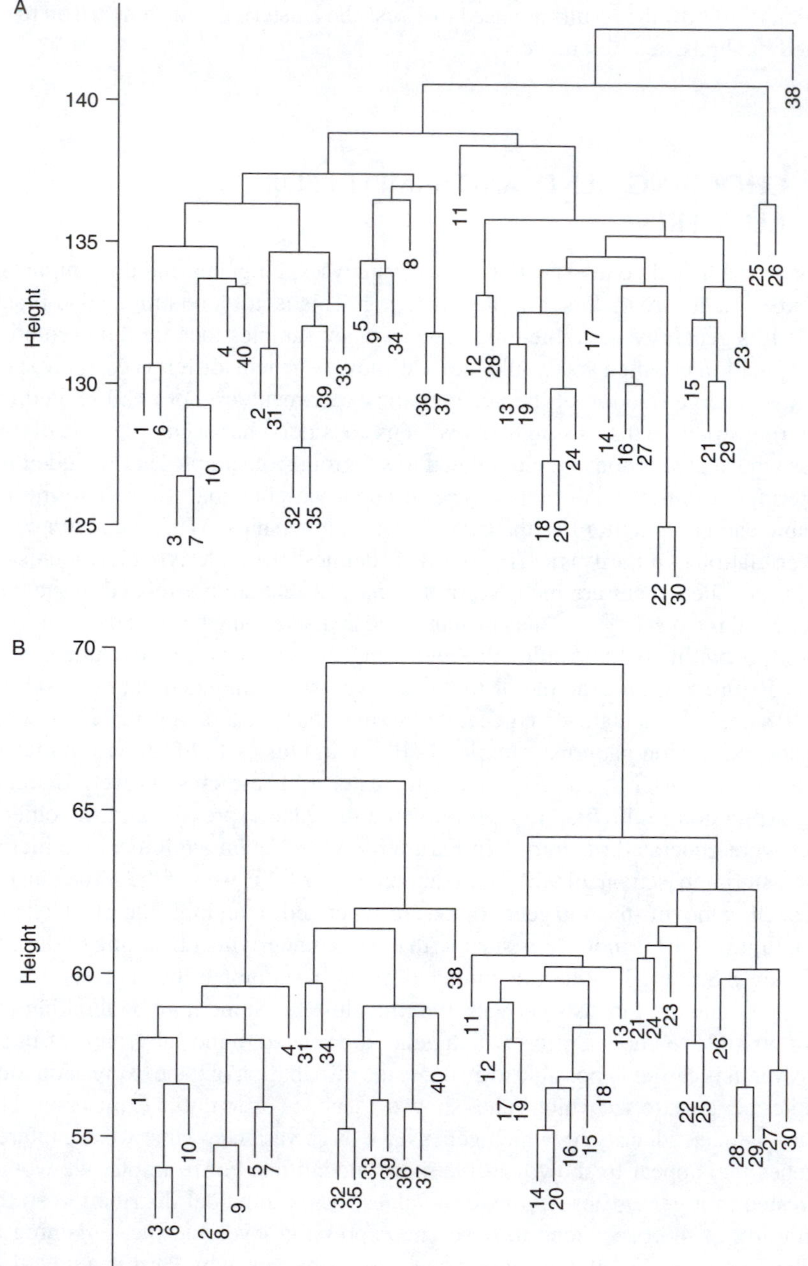

FIGURE 8.8 Hierarchical clustering where four clusters truly exist (where four groups are designated by samples, 1–10, 11–20, 21–30, and 31–40) and clusters are based on the gene expression of 450 genes. Panel A: The hierarchical cluster is based on the 450 relevant genes plus an additional 450 genes whose expressions were generated randomly. Panel B: The hierarchical cluster is based on only the 450 relevant genes.

8.10 COMPUTER PACKAGES AND SOFTWARE

There are many commercial and freely available software packages for performing a variety of cluster analyses. The list here is not comprehensive by any means, but provides a few tools to the reader. Eisen has developed clustering and visualization tools called Cluster and TreeView which are specifically geared for gene expression data (these were used to create Figure 8.1) and can be downloaded at http://rana.lbl.gov [10]. Cluster can perform hierarchical, *k*-means, self-organizing maps, and principal components analysis. However, for hierarchical clustering, only agglomerative clustering can be performed via Cluster. R and Splus have several functions for hierarchical clustering (diana and agnes) which were developed by Kaufmann and Rousseeuw in their programs DIANA and AGNES [12]. The R software package (freely available at http://www.r-project.org) contains libraries and functions for multi-dimensional scaling with the mva and MASS libraries, SOM functions can be found in the library GeneSOM, and model-based clustering using MCLUST [36, 49]. Tamayo et al. [18] GENECLUSTER can also be used for SOM. Other clustering tools include Expression Profiler [50], MeV by TIGR which can be used for data mining and visualization [51], and Genesis described by Sturn et al. [52] which is available at http://genome.tugraz.at. A useful resource is a website from the Classification Society of America which maintains a list of classification software at http://www.pitt.edu/~csna/software.html.

8.11 PROBLEMS AND PITFALLS

Many statisticians who analyze gene expression data criticize clustering methods in this setting. This criticism is not unfounded: the majority of clustering results in gene expression analyses are poorly constructed, not properly validated, and overinterpreted. It must be recognized that different clustering methods can produce drastically different results, many cluster analyses are quite unstable, and cluster analyses are a purely exploratory method.

In Figure 8.2, one dataset has been clustered using different approaches: agglomerative vs. divisive; single, average, and complete linkage; and using correlation and Euclidean distances. Notice the differences among the results. This is not an uncommon result: different options within a clustering method can produce different results, leading to inconsistent inferences. Similarly, while transformations may be appropriate (e.g., many analysts choose to take a log transform of gene expression data when the data is provided to them in ratio format), imposing transformations on the data will undoubtedly affect the resulting cluster structure. As a result, we stress the importance of understanding the limitations of cluster analyses and their validation.

An additional consideration is that the clustering result can only be as good as the data on which the clustering is performed. It is important to make sure that the data have been preprocessed and normalized (see Chapter 2 and Chapter 3 on normalization and quality control). Much effort should be used on these initial steps before embarking on any kind of analyses of gene expression data. Without careful inspection and appropriate adjustment (e.g., adjustment of spatial artifacts, removal of bad spots, etc.) of all arrays under consideration, spurious results are likely to arise in clustering or any other kind of data analytic strategy for microarray data.

REFERENCES

1. M.E. Garber, O.G. Troyanskaya, K. Schluens, S. Petersen, Z. Thaesler, M. Pacyna-Gengelbach, M. van de Rijn, G.D. Rosen, C.M. Perou, R.I. Whyte, R.B. Altman, P.O. Brown, D. Botstein, and I. Petersen, Diversity of gene expression in adenocarcinoma of the lung. *Proceedings of the National Academy of Sciences, USA*, 98: 13784–13789, 2001.
2. A. Rosenwald, G. Wright, A. Wiestner, W.C. Chan, J.M. Connors, E. Campo, R.D. Gascoyne, T.M. Grogan, H.K. Muller-Hermelink, E.B. Smeland, M. Chiorazzi, J.M. Giltnane, E.M. Hurt, H. Zhao, A.L.S. Henrickson, L. Yang, J. Powell, W.H. Wilson, E.S. Jaffe, R. Simon, R.D. Klausner, E. Montserrat, F. Bosch, T.C. Greiner, D.D. Weisenburger, W.G. Sanger, B.J. Dave, J.C. Lynch, J. Vose, J.O. Armitage, R.I. Fisher, T.P. Miller, M. LeBlanc, G. Ott, S. Kvaloy, H. Holte, J. Delabie, L.M. Staudt, and L.L.M.P. Project, The use of molecular profiling to predict survival after chemotherapy for diffuse large-B-cell lymphoma. *New England Journal of Medicine* 346: 1937–1947, 2002.
3. J. Quackenbush, Computational analysis of microarray data. *Nature* 2: 418–427, 2001.
4. R. Tibshirani, T. Hastie, M. Eisen, D. Ross, D. Botstein, and P.O. Brown, Clustering methods for the analysis of DNA microarray data. Department of Statistics, Stanford, CA: Stanford University, 1999.
5. A.D. Gordon, *Classification*, 2nd ed. New York: Chapman & Hall/CRC, 1999.
6. I.J. Blader, I.D. Manger, and J.C. Boothroyd, Microarray analysis reveals previously unknown changes in toxoplasma gondii-infected human cells. *Journal of Biological Chemistry* 276: 24223–24231, 2001.
7. P.T. Spellman, G. Sherlock, M.Q. Zhang, V.R. Iyer, K. Anders, M.B. Eisen, P.O. Brown, D. Botstein, and B. Futcher, Comprehensive identification of cell cycle-regulated genes of the yeast *Saccharomyces cerevisiae* by microarray hybridization. *Molecular Biology of the Cell* 9: 3273–3297, 1998.
8. X. Chen, S.T. Cheung, S. So, S.T. Fan, C. Barry, J. Higgins, K.M. Lai, J. Ji, S. Dudoit, I.O. Ng, M. Van De Rijn, D. Botstein, and P.O. Brown, Gene expression patterns in human liver cancers. *Molecular Biology of the Cell* 13: 1929–1939, 2002.
9. X. Wen, S. Fuhrman, G.S. Michaels, D.B. Carr, S. Smith, J.L. Barker, and R. Somogyi, Large-scale temporal gene expression mapping of central nervous system development. *Proceedings of the National Academy of Sciences, USA* 95: 334–339, 1998.
10. M.B. Eisen, P.T. Spellman, P.O. Brown, and D. Botstein, Cluster analysis and display of genome-wide expression patterns. *Proceedings of the National Academy of Sciences, USA* 95: 14863–14868, 1998.
11. B.S. Everitt, S. Landau, and M. Leese, *Cluster Analysis*, 4th ed. London: Edward Arnold, 2001.
12. L. Kaufman and P.J. Rousseeuw, *Finding Groups in Data: An Introduction to Cluster Analysis*, Wiley Series in Probability and Mathematical Statistics. New York: John Wiley & Sons, Inc., 1990.
13. R.A. Johnson and D. Wichern, *Applied Multivariate Statistical Analysis*, 4th ed. Upper Saddle River, NJ: Prentice Hall, 1998.
14. P. MacNaughton-Smith, W. Williams, M.B. Dale, and L.G. Mockett, Dissimilarity analysis: a new technique of hierarchical sub-division. *Nature* 202: 1034–1035, 1964.

15. J.A. Hartigan and M.A. Wong, A. *k*-means clustering algorithm. *Applied Statistics* 28: 100–108, 1979.
16. T. Kohonen, Analysis of a simple self-organizing process. *Biological Cybernetics*, 43: 59–69, 1982.
17. T. Kohonen, *Self-Organizing Maps*. Berlin: Springer-Verlag, 1995.
18. P. Tamayo, D. Slonim, J. Mesirov, Q. Zhu, S. Kitareewan, E. Dmitrovsky, E.S. Lander, and T.R. Golub, Interpreting patterns of gene expression with self-organizing maps: methods and application to hematopoietic differentiation. *Proceedings of the National Academy of Sciences, USA* 96: 2907–2912, 1999.
19. B.D. Ripley, *Pattern Recognition and Neural Networks*. Cambridge: Cambridge University Press, 1996.
20. B. Kosko, *Neural Networks and Fuzzy Systems: A Dynamical Systems Approach to Machine Intelligence*. Englewood Cliffs, NJ: Prentice Hall, 1992.
21. A. Ben-Dor and Z. Yakhini. Clustering gene expression patterns. In *Proceedings of the 3rd Annual International Conference on Computational Molecular Biology (RECOMB99)*. Lyon, France, 1999.
22. S.K. Kachigan, *Multivariate Statistical Analysis*, 2nd ed. New York: Radius Press, 1991.
23. B.S. Everitt and G. Dunn, *Applied Multivariate Data Analysis*, 2nd ed. New York: Oxford University Press, Inc., 2001.
24. J.B. Kruskal, Multidimensional scaling by optimizing goodness of fit to a non-metric hypotheses. *Psychometrika* 29: 1–27, 1964.
25. K.Y. Yeung, and W.L. Ruzzo, Principal component analysis for clustering gene expression data. *Bioinformatics* 17: 763–774, 2001.
26. J, Quackenbush, Computational analysis of microarray data. *Nature Reviews in Genetics*, 2: 418–427, 2001.
27. S. Knudsen, *A Biologist's Guide to Analysis of DNA Microarray Data*. New York: John Wiley & Sons, 2002.
28. S. Raychaudhuri, J.M. Stuart, and R.B. Altman. Principal components analysis to summarize microarray experiments: application to sporulation time series. In *Proceedings of the 5th Pacific Symposium on Biocomputing*, Waikiki, Hawaii, 2000.
29. J. Khan, R. Simon, M. Bittner, Y. Chen, S. Leighton, T. Pohida, P.D. Smith, Y. Jiang, G.C. Gooden, J.M. Trent, and P. Meltzer, Gene expression profiling of alveolar rhabdomyosarcoma with cDNA microarrays. *Cancer Research* 58: 5008–5013, 1998.
30. M. Bittner, P. Meltzer, Y. Chen, Y. Jiang, E. Seftor, M. Hendrix, M Radmacher, R. Simon, Z. Yakhini, A. Ben-Dor, N. Sampas, E. Dougherty, E. Wang, F. Marincola, C. Gooden, J. Lueders, A. Glatfelter, P. Pollock, J. Carpten, E. Gillanders, D. Leja, K. Dietrich, C. Beaudry, M. Berens, D. Alberts, and V. Sondak, Molecular classification of cutaneous malignant melanoma by gene expression profiling. *Nature* 406: 536–540, 2000.
31. M. West, C. Blanchette, H. Dressman, E. Huang, S. Ishida, R. Spang, H. Zuzan, J.R. Marks, and J.R. Nevins, Predicting the clinical status of human breast cancer using gene expression profiles. *Proceedings of the National Academy of Sciences, USA* 98: 11462–11467, 2001.
32. T. Hastie, R. Tibshirani, M. Eisen, and P. Brown, *Gene Shaving: A New Class of Clustering Methods for Expression Arrays*. Stanford, CA: Stanford University, 2000.
33. A. Butte, P. Tamayo, D. Slonim, T.R. Golub, and I.S. Kohane, Discovering functional relationships between RNA expression and chemotherapeutic susceptibility using relevance networks. *Proceedings of the National Academy of Sciences, USA* 97: 12182–12186, 2000.

34. J. Herrero, A. Valencia, and J. Dopazo, A hierarchical unsupervised growing neural network for clustering gene expression patterns. *Bioinformatics* 17: 126–136, 2001.
35. E. Segal, B. Taskar, A. Gasch, N. Friedman, and D. Koller, Rich probabilistic models for gene expression. *Bioinformatics*, 17: S243–S252, 2002.
36. C. Fraley and A.E. Raftery, How many clusters? Which clustering method? Answers via model-based cluster analysis. *Computer Journal* 41: 578–588, 1998.
37. G. Chen, S.A. Jaradat, and Banerjee, Evaluation and comparison of clustering algorithms in analyzing ES cell gene expression data. *Statistica Sinica* 12: 241–262, 2002.
38. R. Shamir and R. Sharan, *Algorithmic Approaches to Clustering Gene Expression Data*. Current Topics in Computational Molecular Biology, Boston: MIT Press, 2002.
39. P.J. Rousseeuw, *Silhouettes:* a graphical aid to the interpretation and validation of cluster analysis. *Journal of Computation and Applied Mathematics* 20: 53–65, 1987.
40. J. Vilo, A. Brazma, I. Jonassen, A. Robinson, and E. Ukkonenen, Mining for putative regulatory elements in the yeast genome using gene expression data. *Intelligent Systems for Molecular Biology* 8: 384–394, 2000.
41. R. Tibshirani, G. Walther, and T. Hastie, Estimating the number of clusters in a dataset via the Gap statistic. *Journal of the Royal Statistical Society, Series B* 63: 411–423, 2000.
42. K. Zhang and H. Zhao, Assessing reliability of gene clusters from gene expression data. *Functional and Integrative Genomics* 1: 156–173, 2000.
43. L. Breiman, *Using Convex Pseudo-Data to Increase Prediction Accuracy.* Berkeley: Statistics Department, U.C. Berkeley, 1998.
44. S. Dudoit, J. Fridlyand, and T.P. Speed, Comparison of discrimination methods for the classification of tumors using gene expression data. *Journal of the American Statistical Association* 97: 77–87, 2002.
45. T. Margush and F.R. McMorris, Consensus n-trees. *Bulletin of Mathematical Biology* 43: 239–244, 1981.
46. G. Schwarz, Estimating the dimension of a model. *The Annals of Statistics* 6: 461–464, 1978.
47. S. Dudoit and J. Fridlyand, A prediction-based resampling method for estimating the number of clusters in a dataset. *Genome Biology* 3: 0036.1–0036.21, 2002.
48. G. Parmigiani, E. Garrett, R. Anbazhagan, and E. Gabrielson, A statistical framework for expression-based molecular classification in cancer. *Journal of the Royal Statistical Society, Series B* (*with discussion*) 64: 717–736, 2002.
49. C. Fraley and A.E. Raftery, MCLUST: software for model-based cluster analysis. *Journal of Classification* 16: 297–306, 1999.
50. J. Vilo, M. Kapushesky, P. Kemmeren, U. Sarkans, and A. Brazma, Expression profiler. In *The Analysis of Gene Expression Data: Methods and Software*, G. Parmigiani, E. Garrett, R. Irizarry, and S. Zeger (eds.). New York, Springer-Verlag, 2002.
51. A.I. Saeed, V. Sharov, J. White, J. Li, W. Liang, N. Bhagabati, J. Braisted, M. Klapa, T. Currier, M. Thiagarajan, A. Sturn, M. Snuffin, A. Rezantsev, D. Popov, A. Ryltsov, E. Kostukovich, I. Borisovsky, Z. Liu, A. Vinsavich, V. Trush, and J. Quackenbush, TM4: a free, open-source system for microarray data management and analysis. *Biotechniques* 34: 374–378, 2003.
52. A. Sturn, J. Quackenbush, and Z. Trajanoski 1, Genesis: cluster analysis of microarray data. *Bioinformatics* 18: 207–208, 2002.
53. S. Tavajoie, J.D. Hughes, B.J. Campbell, R.J. Cho, and G.M. Church, Systematic determination of genetic network architecture. *Nature Genetics* 22: 281–285, 1999.

9 Cluster Stability

Bernard S. Gorman and Kui Zhang

CONTENTS

9.1 Cluster Stability 159
9.2 Defining Stability 160
9.3 A Brief Overview of Clustering 161
9.4 Choice Points That Influence Stability and Instability 162
9.4.1 Pitfalls at Step 1: Obtaining High-Quality Data 162
9.4.2 Pitfalls at Steps 2 and 3: Selection of Genes and Arrays 164
9.4.3 Pitfalls at Steps 4 and 5: Data Normalization and Data Cleaning .. 165
9.4.4 Pitfalls at Step 6: Choices of Similarity Metrics 166
9.4.5 Pitfalls at Step 7: Choice of a Clustering Algorithm 167
9.5 A General Approach for Detecting Stable Cluster Solutions 171
References 173

9.1 CLUSTER STABILITY

Cluster analysis has provided a set of methods that has been very useful for exploring gene expression patterns from microarray data. The goal of such analysis is to construct classes of genes or classes of samples such that observations within a class are more similar to each other than they are to observations in different classes according to their expression levels. There are several reasons for interest in cluster analysis of microarray data. First, there is evidence that many functionally related genes have similar expression patterns [1,2]. By grouping genes in a coordinated manner according to their expression under multiple conditions, we may be able to reveal the function of those genes which were previously unknown. Second, a class of genes with similar expression pattern may reveal much about regulatory mechanisms. The common regulatory elements (e.g., motifs) identified in a class of genes would greatly facilitate our understanding of genetic networks [3,4]. Third, it provides a more reliable and precise way to distinguish different subtypes of tumors (e.g., breast cancers), which are not achievable by standard microscopic or molecular approaches, by classifying the samples on the basis of their gene expression levels [5–8]. The new subtype of tumors can also be identified. Eventually, such classifications can lead the advancement for successful prognosis, diagnosis, and therapeutics of diseases. Fourth, given that a microarray can potentially contain expression of tens of thousands of genes over several to hundreds of samples, by grouping either genes or samples, or both simultaneously [1,9], clustering analysis potentially provides an effective way to reduce the complexity of data for easy organization, visualization, and interpretation.

A variety of clustering algorithms have been applied to analyze microarray data to partition genes or samples into mutually exclusive classes. These methods include hierarchical clustering algorithms [1,6], *k*-means [10,11], self-organizing maps [12–15], the support vector machine [16,17], model-based clustering [18,19], and other algorithms [20–22]. The detailed description of each algorithm is beyond the scope of this paper. Thus, we use the hierarchical algorithm as an example to illustrate the common features existing in many clustering algorithms. Researchers can refer the corresponding literature for other algorithms they are interested in.

However, one might ask if the resulting clusters, whether from hierarchical methods or any other methods, are artificial or real. By now, many microarray researchers have become aware of the fact that many clustering methods are available and the different methods generally generate different clusters [58]. Furthermore, large variation exists in microarray data, resulting in different clusters with perturbations in the same data set. Our purpose in this chapter on cluster stability is to explore potential sources of artifacts and to suggest methodological and statistical strategies that might allow us to detect genuine clusters. We shall argue that each step of the clustering process may introduce artifacts that compromise stability. On a more positive note, we will suggest ways to meet these challenges.

The concept of "stability" falls under the general rubric of "reliability." Although many use the words "reliability" and "validity" in the same breath, permit us make a distinction between these concepts. "Reliability" refers to the consistency and precision of measurement. "Validity," however, refers to the ability of our measured variables to display functional relationships to other measured variables. Reliability is a necessary but not a sufficient condition for establishing validity. It is quite possible for an accurately measured variable to have no valid relationships to any other variable measure of interest.

We do not view cluster analyses as confirmatory methods that establish validity but, rather, as exploratory descriptive methods that have the potential to suggest fruitful hypotheses and to provide more concise displays of complex data. For our purpose, we want to establish that we have calculated clusters that are reflections of the true underlying nature of the microarray data. Thus, we do not necessarily require the clustering results to relate to other entities as long as they can capture most important of characteristics or patterns of the data at hand. Therefore, we will leave the questions of whether the clusters are meaningful in as much as they might discriminate between patients and controls or are related to interesting biological substrates to others in this volume.

9.2 DEFINING STABILITY

It should be noted at the outset that different researchers vary considerably in their definitions of cluster stability. For example, McShane et al. [23] noted that "Clustering algorithms always detect clusters, even in random data and it is imperative to conduct some statistical assessments of the strength of evidence for any clustering and to examine the reproducibility of individual clusters." Similarly, Smolkin and Ghosh [24] stated that "it is important to separate the clusters which arise due to random chance from those which represent 'true' clusters." Zhang and Zhao [25] focused on instability due to measurement error and the inclusion of aberrant genes or specimens. They stated that, "we need to determine whether this cluster is a real cluster or a superficial one resulting from random variations in gene expression measurements." Roth et al. [26] focused on the replicability of cluster analysis solutions due to variation in arrays define and stability as "the variability of solutions which are computed from different datasets sampled on the same source."

Although no two definitions of stability are identical, we can see two emerging themes. The first theme is that a stable solution is one in which the results will not change much with small modification of specific methods. The second theme concerns the idea that results will not change much with small perturbation on the data. Both themes are interdependent. If we have data of poor quality, then we may have large variations in of cluster results, regardless of the clustering method used. Conversely, even with the best available data, we might insert methodological artifacts.

Let us first look at the some general steps of cluster analysis, which we will illustrate with an extremely simple dataset. At each step, we will note variations that could affect the stability of our cluster analysis solutions.

9.3 A BRIEF OVERVIEW OF CLUSTERING

In order to examine potential sources of instability in microarray cluster analyses, let us analyze a simple dataset and describe some of the steps and decisions that researchers need to perform in a cluster analysis. We have constructed a small dataset, shown in Table 9.1 and Figure 9.1, containing six rows, representing hypothetical genes and four columns, representing four specimens or "arrays."

As is typical in microarray clustering literature, the rows of our matrix are genes. Thus, our example has genes G1 through G6. The columns are typically labeled as "arrays," which are individual specimens from individual microarray chips or repeated measures from the same individuals. The arrays may come from different groups. In our example, arrays P1 and P2 are "patients" and C1 and C2 are hypothetical

TABLE 9.1
The Gene Expression Value for a Simple Dataset with Six Genes and Four Arrays

Gene	P1	P2	C1	C2
G1	10	10	4	3
G2	11	9	3	4
G3	3	2	9	8
G4	3	3	7	9
G5	5	5	4	5
G6	4	4	3	4

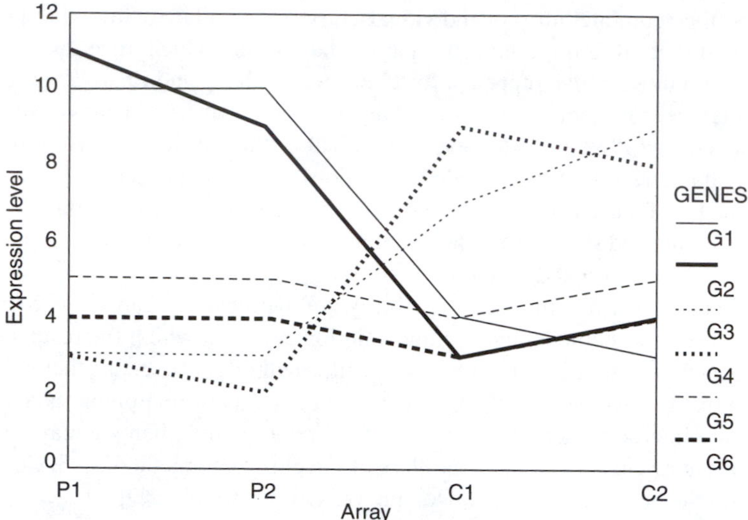

FIGURE 9.1 The gene profiles for a simple data set with six genes and four arrays.

"controls." To simplify matters, let us use the most common form of microarray, the complimentary DNA chip (cDNA) and let the cell entries be relative intensities of gene expression levels. Typically, experimental arrays are dyed red and controls are dyed green and one compares computes relative intensities of red to green. These levels are often normalized against a reference value and since ratios are used, the ratios are often transformed to logarithms.

Hopefully, our readers can see a pattern in our simple dataset. Genes G1 and G2 are highly expressed in the patients and genes G5 and G6 are highly expressed in the controls. Genes G3 and G4 have similar profiles over arrays but show no strong pattern of differential expression between patients and controls.

9.4 CHOICE POINTS THAT INFLUENCE STABILITY AND INSTABILITY

Figure 9.2 presents a flow chart of the typical steps that one might use in cluster analysis. We believe that the each step presents some challenges that may affect the quality of the resulting cluster analyses.

9.4.1 Pitfalls at Step 1: Obtaining High-Quality Data

For a long time, data analysts have subscribed to the dictum "garbage in; garbage out." If our measurements are imprecise, our ability to draw conclusions will be badly compromised. The practice of cluster analyses provides no exception to this principle.

Step 1: Obtaining high-quality data

Step 2: Selection of genes

Step 3: Selection of arrays

Step 4: Data cleaning

Step 5: Normalization and standardization

Step 6: Choice of a similarity metric

Step 7: Choice of a clustering algorithm

FIGURE 9.2 The flow chart of the typical steps that one might use in cluster analysis.

Referring to Step 1 of our cluster analysis procedures, we are going to assume that the data in the cells in Table 9.1 were measured accurately. As we will see, this is not always the case. In classical psychometric test theory [27] every measure, X, has a "true" component (t) and an "error" component (e). Thus, $X = t + e$. Other fields might call t the "signal" and e "noise." The equation $X = t + e$ has two unknowns. Therefore, one can estimate these components but cannot simply obtain them through subtraction. It could also be said that the term "error" is somewhat misleading in as much as "error" can consist of systematic but unintended sources of variance that interfere with our intended measurement plan as well as random sources of variance. For our purposes, let us define "error" as unwanted sources of variance. As such, there are many of them and we will describe procedures that attempt to assess these sources of variance and attempt to separate them from "true," underlying data structures.

An observed intensity ratio for a gene in a given array contains both the "true" value as well as "noise" components. Many authors have commented on sources of "noise" inherent to the process of manufacturing the cDNA array (e.g., [28–30]).

There are many sources of noise. For a good example, of how we can separate data into components that we wish to analyze aside from those that we consider as noise we turn to the work of Kerr et al. [31], who demonstrated that an observed score, Y, can have many sources of variation. These authors expressed their model of variation notion in an equation:

$$Y = \mu + \mathrm{G} + \mathrm{A} + \mathrm{T} + \mathrm{D} + \mathrm{AG} + \mathrm{AD} + \mathrm{DG} + \mathrm{TG} + \mathrm{TA} + \mathrm{TD} + \mathrm{AGT} + \mathrm{AGD} + \mathrm{GTD} + \mathrm{TDAG} + e$$

where μ is the grand mean; the overall average value of relative intensity levels over all genes and arrays in our matrix. G represents gene main effects. That is, they reflect the fact that some genes have higher values than others. In our matrices, these could be seen as marginals for each row. Array main effects are represented by A. "Arrays" are the data columns — typically "experiments" or "subjects" or "cases," or time points within the same case. The differences among genes may be due to true differences between patients and controls or to a time course but it may be also due to artifacts such as the use of different lab conditions, different calibration of instruments, instrument drift over time, etc.

Dye main effects are symbolized by D. For example, red dyes may have stronger signals than green dyes regardless of true pattern differential gene expression. T represents treatment main effects. These may represent differences between patient and controls or among phases in repeated measurements within the same specimens. There are the interactions among the factors above. For example, there are gene by array interactions (GA), gene by dye interactions (GD), array by dye interactions (AD), treatment by dye (TD), treatment by gene (TG), treatment by array (TA), and three-way and four-way interaction of genes, arrays, treatments, and dye. Of these, the component due to by gene array interaction (GA) is the one we most probably want. It is the "true" score, hopefully corrected for dye effects, main effects of arrays and genes, and other interaction terms. Finally, there is a random error (residual) component (e). This is assumed to be randomly and normally distributed.

Kerr et al. [31] employ an analysis of variance (ANOVA) approach to assess the sources of variance. Because there are typically too many unknown terms in the linear equation, their solution utilizes an iterative bootstrapping approach. Interaction components that have confidence intervals that include zero are dropped from further analyses. Estimates of the GA interaction components are then used for further analysis. While it is unlikely that our simple dataset will need such elaborate preprocessing, given "real world" noise, the reliability of other datasets will probably benefit from such "precleaning" strategy. Other ANOVA models were discussed by Wu et al. [32], Cui and Churchill [33], Pavlides and Noble [34], and Emptage et al. [35].

9.4.2 Pitfalls at Steps 2 and 3: Selection of Genes and Arrays

Next, referring to Steps 2 and 3, let us also assume that all six genes in our example are the genes of interest and that the four arrays, consisting of two patients and two controls are also accurately designated. Hopefully, their data were obtained by the same instruments, at roughly the same time period, and calibrated to the same level of accuracy. Also, one might hope the red or green dyes used to provide relative intensity measures of patients vs. controls have the same level of reflectance.

Given the current state of microarray technology, thousands of genes can be deposited on a microarray. Ideally, however, researchers wish to find discrete subsets of genes that show differential expression. Several authors preselect interesting genes by bypassing a preliminary clustering step and immediately perform t-tests, nonparametric tests, or fold level tests on each gene to detect those genes that differentiate between patients and controls or between different time points. They then eliminate subsets of genes that do not seem to discriminate among patients and controls and then conduct further cluster analyses on those remaining genes that demonstrate differential expression.

Given the possibility of reducing computational complexity from thousands of genes to perhaps dozens or hundreds of genes, preselection seems like an advantage. However, one might wonder if this step biases results or if it is a legitimate and prudent way to clean the data. Given so many genes, are some genes retained or deleted merely on chance effects? We leave the assessment of the probabilities of chance effects to other authors in this volume but it bears mention here. Furthermore, the preselection of genes on a gene-by-gene basis may ignore the possibility that whole groups of genes may operate systematically to produce an effect. Prescreening may obscure the multivariate complexity of such systems.

While we are wary of simple gene preselection, a stronger strategy might be found in deliberately including marker genes that we know to have or strongly suspect will show differential expression. This approach is suggested in the "fishing expedition" approach discussed by Zhang et al. [36].

Just as gene preselection presents problems, so does the selection of our arrays. For example, a clinical study, should we select patients vs. controls or only patients? In a repeated measures strategy, which time points should be sampled?

We constructed our simple example. But in less artificial situations, we might ask if a few aberrant arrays, whether real or error-prone, could affect the outcome. We continue to address the issue of finding and working around aberrant arrays in later parts of this chapter.

9.4.3 Pitfalls at Steps 4 and 5: Data Normalization and Data Cleaning

All of our expression levels in the cells of Table 9.1, ranged from 3 to 10. This scale seemed reasonable, but should we have rescaled the data? We could have made other choices. For example, we might have transformed the data, as is commonly done, by transforming the data to their logarithms. We might have wished to control for differences in overall expression levels for each array by subtracting array's mean from each data value in the area. Likewise, we could have subtracted row means from data values to compensate for differences in each gene's overall expression level. We might use standardized scores, in which the mean is subtracted from data values and the difference is divided by the standard deviation. We might have compared all values, whether experimental or control, patient or normal, as has Chu et al. [37], to a ratio of (red signal level — red background level)/ (green signal — green background) levels? Should we attempt other normalizations as suggested by Quackenbush [38]? As can be seen, we can make many data transformation choices. However, it also should be said that the strategy previously mentioned by Kerr et al. [31], in which we correct for many artifacts, seems to be an attractive alternative to normalization.

Perhaps a more serious set of problem can be caused by the presence of outliers. If we had some gene with values out of range, say, with much lower or much higher expression levels, we might wish to ask whether these "outliers" were due to gene that legitimately fit within separate clusters, or whether the data were the gene was inaccurately recorded. Perhaps the inaccuracy was due to an instrumentation error or perhaps it was caused by a simple clerical error. In some cases, there may be wide variability.

We caution against automatic outlier rejection. Instead, we suggest two strategies. The first is to simply perform a cluster analysis in which we will extract many small clusters. Outliers typically appear as clusters with only one or two members and with very high or very low data values. Researchers should examine these clusters and make judgments about their legitimacy. The second strategy will be discussed later in this chapter. However to peek ahead, we will recommend a resampling strategy to is to resampling strategy in which we will obtain consensus of the presence of more typical, robust clusters and obtain low consensus on the presence of atypical clusters.

9.4.4 Pitfalls at Step 6: Choices of Similarity Metrics

For our example, we are going to employ a commonly-used measure of similarity for each pair of genes, the Squared Euclidean distance, D^2. Euclidian distance is defined as the square root of the sum of squared differences over each array. Thus, for genes G1 and G2, the sum of squared differences is $(10-11)^2+(10-9)^2+(4-3)^2+(3-4)^2$ or 4 and, computing the square root, gives us 2.0. The matrix of all pairwise differences is presented in Table 9.2. A small D^2 value indicates high similarity, while a high D^2 value indicates increasing dissimilarity.

It should be noted that our use of the squared Euclidean distance represents only one out of hundreds of choices of similarity and dissimilarity metrics. In fact, many others could have been used (e.g., [39]). Simple visual inspection of Table 9.2 shows that the greatest pairwise similarity can be found between gene pairs G1 and G2, G3 and G4, and G5 and G6.

Even a simple dataset can give widely divergent results that can be attributed to differences among similarity metrics. Profiles of genes can vary in overall level, in scatter, and in shape. For example, let us consider three objects with three attributes in Table 9.3.

By differences in *level*, we mean that the values in the cells can be consistently larger than those in other genes. Gene 3s level is higher than that gene 1. By *scatter*, we mean that the deviations of relatively intensity of values of the gene ay differ considerably from those around the mean expression level for the whole gene. Genes 1, 2, and 3 has virtually no scatter but gene 5 displays a small amount of scatter. By *shape* we mean that genes will have their own unique profiles over arrays. Genes 1, 2, and 4 have flat profiles over arrays but gene 3 has a higher elevation in array C.

TABLE 9.2
The Squared Euclidean Distances for Genes in Table 9.1

Gene	Squared Euclidean distance among genes					
	1: G1	2:G2	3:G3	4:G4	5:G5	6:G6
1:G1	0	4	163	143	54	74
2:G2	4	0	165	141	54	74
3:G3	163	165	0	6	47	57
4:G4	143	141	6	0	33	43
5:G5	54	54	47	33	0	4
6:G6	74	74	57	43	4	0

TABLE 9.3
The Gene Expression Level for a Simple Dataset with Four Genes and Three Arrays

	Arrays		
Gene	A	B	C
G1	1	1	1
G2	2	2	2
G3	3	3	5
G4	2	2	2

TABLE 9.4
The Euclidean Distances for Genes in Table 9.3

Genes	Euclidean Distance			
	1:G1	2:G2	3:G3	4:G4
1:G1	0.000	1.732	4.899	1.732
2:G2	1.732	0.000	3.317	0.000
3:G3	4.899	3.317	0.000	3.317
4:G4	1.732	0.000	3.317	0.000

Table 9.4 presents the Euclidean distances among these four genes. A metric that considers differences in both level and shape, such as the Euclidean Distance or Squared Euclidean Distance, would consider gene 3 to be very dissimilar to the others.

A metric, such as the Pearson product–moment correlation coefficient, however, would be sensitive to shape but would ignore levels, so that genes 1, 3, and 4 would be perfectly correlated but gene 3 would have a lower correlation. Like correlation coefficients, vector cosine measures would show a slight difference between gene 3 and the others but would show genes 1, 2, and 4 to be identical. Standardization within columns by subtracting column means and dividing by column standard deviations would not affect the present dataset, but it could affect others.

These are just a few of the similarity metrics; there are literally hundreds [40,41]. Of course, the problem might be to decide which of these metrics is appropriate for genetic data in a given study.

9.4.5 Pitfalls at Step 7: Choice of a Clustering Algorithm

For Step 7, we chose to use a popular method known as average linkage hierarchical cluster analysis. In this method, genes are successively merged into clusters and the

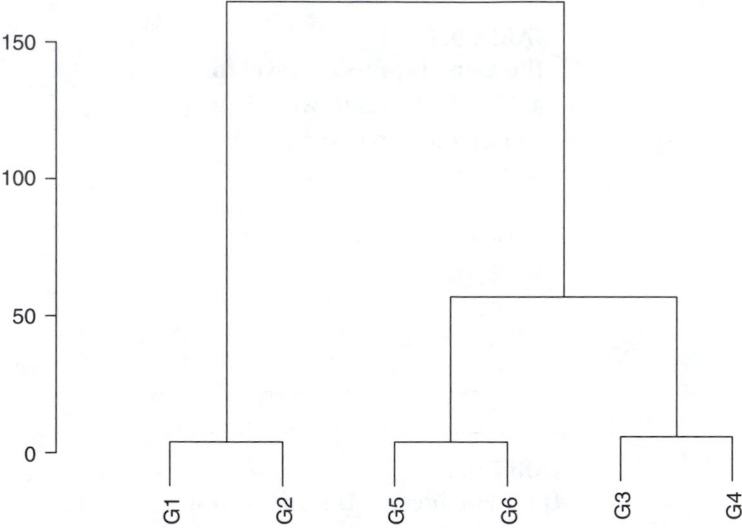

FIGURE 9.3 The dendrogram generated by the hierarchical clustering algorithm with average linage between groups.

subsequent clusters are then merged with other clusters to form a branching tree structure. The joining points or the nodes of the resulting solution have the values of the average similarity of each member of the cluster to those of the other members. The analysis produces a dendrogram — a nested, tree structure, in which early branching represents genes that are very close (similar) and later branching represent increasing distances (dissimilarity). Figure 9.3 presents a dendrogram for the clustering of our six genes from Table 9.1.

Thus, in our example, genes 1 and 2 are joined at a distance value of 4, genes 5 and 6 at a distance of 4, and genes 3 and 4 at a distance of 6. Then, genes 3 and 4 are merged with genes 5 and 6 to form a larger cluster whose similarity value is 45, the average squared distance between the {G3, G4} cluster and the {G5, G6} cluster. Note, however, that other hierarchical cluster merging rules could be used, and we could use a large variety of nonhierarchical cluster procedures.

In our example, we have clustered genes but the previous steps could have also been used to cluster arrays and some researchers report both gene and array clustering.

As mentioned previously, we used all six genes and four arrays for our analyses. However, we could have used a subset of the arrays to examine whether the same general pattern of gene clustering was obtained when some arrays were omitted. Additionally, we could have held some genes out of our analyses to examine whether we could obtain the same general pattern of array clustering. If our solutions changed drastically, then we might suspect that some genes or arrays are unusually influential for forming cluster results.

Notice that a hierarchical cluster analysis forms a dendrogram. There is nothing inherent in the dendrogram to tell us how many clusters we have. In fact, we could say that we have as many as six clusters, each containing a single gene, as few as one cluster that contains all of the genes. However, by examining the pattern of large jumps in distances between clusters as they merge and given the artificial nature of our data set, it probably would be reasonable to say that there are three clusters. However, one might entertain the possibly of two clusters: a cluster consisting of genes 1 and 2 and a second cluster consisting of genes 3, 4, 5, and 6. Although we leave decisions about the validity of a cluster to other contributors in this volume, we might consider the {A, B} and the {E, F} clusters to be important because they discriminate between patients and controls. The {C, D} cluster might also be important for its lack of discrimination between patients and controls.

There are numerous methods for clustering. In fact, Lance and Williams [42] flexible hierarchical clustering algorithm, alone, can provide an infinite variety of clustering strategies. Hierarchical agglomerative fusion rules may impose structures that may not reflect the "true" data structure [43]. For example, single linkage cluster analysis [44] favors long, stringy clusters rather than compact, spherical ones. These solutions may distort the solution, but then again, they may be correct if the structure is really "stringy." Single linkage methods may be very susceptible to a presence of even a few outliers. Complete linkage [44] and Ward's [45] methods favor very compact clusters that are well separated from neighboring clusters. However, they may unnaturally dissect long, stringy, elliptical clusters. Averaging methods, such as the one used here, fall somewhere in between the single and complete linkage properties.

We constructed another simple data set to illustrate how differences in well-known clustering algorithms can give the radically different solutions with the same dataset. Table 9.5 has five observations (genes and two columns [arrays or experiments]).

A plot of the genes in the two-dimensional space can be seen in Figure 9.4. As can be seen in Figure 9.4, there is only one long stringy cluster. Any other attempt to partition this dataset into smaller cluster would be a "dissection" rather than a true cluster analysis. Single linkage cluster analysis produces the nested structure (E,(D,(C,(A, B)))) or its mirror image. However, complete linkage produces (((A, B),C),(D, E)) or ((A, B),(C,(D, E))). Even in this simple example, we can see that algorithms may impose structures. With larger, more complex microarray data sets, the problem can only be greatly magnified.

TABLE 9.5
The Gene Expression Value for a Simple Dataset with Five Genes and Two Arrays

Gene	*X*	*Y*
A	1	1
B	2	2
C	3	3
D	4	4
E	5	5

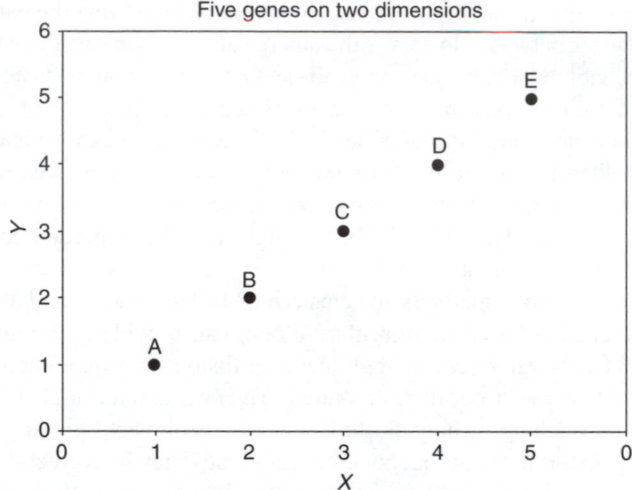

FIGURE 9.4 The plot of five genes in the two-dimensional space.

We can suggest several ways to manage the problem of detecting solutions that have been caused more by the clustering algorithms than by the data itself. One possibility is to perform several different clusterings of the same data using different methods. Then, one could obtain matching index, like the cophenetic correlation [39], or a discrepancy function like "stress" [46] or the Shepard plot in multidimensional scaling that relates the distances created by the cluster solution to distance in the original data [47]. We would then pick the solution that shows the closest similarity to the original data.

Some strategies for confronting the fact that different clustering models are differentially sensitive to cluster shapes might be found in the use of model-based clustering, as seen in the works of Banfield and Raftery [48] and Yeung et al. [19]. Although computer-intensive, these methods permit clusters to vary considerably in their shapes and seem to retrieve extremely complex data structures.

Another possibility will be covered later in this chapter. We will perform multiple analyses of the same data with different clustering algorithms and we will then use a consensus algorithm to find those clusters that have the same memberships across a variety of clustering schemes.

It should also be noted that hierarchical methods are called "greedy" algorithms. Once mergers of single genes or clusters of genes have been made, very few hierarchical algorithms can reassign cluster members to other clusters. What may be locally optimal may not be globally optimal. Nonhierarchical methods, such as k-means methods attempt to find the optimal solution that partitions the dataset into k separate groups. Nearly all k-means solutions have procedures for reassigning cluster members to other clusters where they might fit better. k-means methods often ask researchers to know how many k groups exist beforehand. However, if the researcher lacked such knowledge, it probably would be useful to abandon parsimony and extract many rather than few clusters. In this way, the clustering program could detect potential outliers but also reassign members to other clusters.

Mixture models [49] and fuzzy clustering schemes [10,50] that assign grades of membership to clusters also might free us from forcing genes into mutually exclusive clusters. In this way, a gene can be seen as a member of several clusters. If we do, however, wish to make tighter gene classifications, we might delete genes that have memberships among too many clusters.

9.5 A GENERAL APPROACH FOR DETECTING STABLE CLUSTER SOLUTIONS

By now, it should be obvious that even in our tiny, constructed dataset a multitude of factors can affect cluster stability. The problems will be magnified enormously in the "real world" in which we will have thousands of genes per array. We would now like to take the bold step of suggesting an approach that will attempt to detect stable clusters in the presence of data noise, selections of genes and arrays, different clustering algorithms, and potential outlier genes and arrays. The key to this approach will combine two strategies: resampling and consensus gathering. We will attempt to deal with outlier and selection problems by obtaining thousands of random samples (with replacement) from sets of genes and arrays and we will perform cluster analyses on each subsample. We will then use consensus methods to detect those gene clusters and array clusters that consistently emerge across the subsamples. We will also use consensus methods to detect those clusters that consistently appear regardless of different clustering methods. Finally, following Zhang and Zhao [25] and Davidson et al. [51], we shall vary "noise" and seek consensus of clusters that stand out against such noise.

Again, we will illustrate this general procedure by the hierarchical clustering algorithms based on the simple dataset presented in the Table 9.1. First, we use the bootstrap method to resample many data sets. Bootstrap methods [52] have been widely used in microarray data analysis [53], including cluster analysis [54]. In the commonly used nonparametric bootstrap, the expression value of a gene is obtained across replicated samples with replacement. In the parametric bootstrap, the new expression value of a gene is sampled from a normal distribution $N(x, \sigma^2)$ with the observed expression value x as mean. The variance, σ^2, can be estimated simply from replications [25] or obtained from other methods (e.g., [54]). In general, the nonparametric method is more reliable but it requires a relatively large number of samples to assure that the resampled data are not identical with the original ones. The parametric bootstrap can virtually generate an infinite number of new datasets which are conducive to clustering analysis, but it is needed to estimate the variation which can be rather difficult in some circumstances. This method may also produce data that are not normally distributed. In our example, we employ the parametric bootstrap and simply assume the variance for each gene is proportional to its expression value [25], allowing us to assess the reliability of clusters across a range of variation values. All data sets are then clustered by the same methods. In the second step, we apply the consensus tree defined by Margush and McMorris [55] to the original tree as

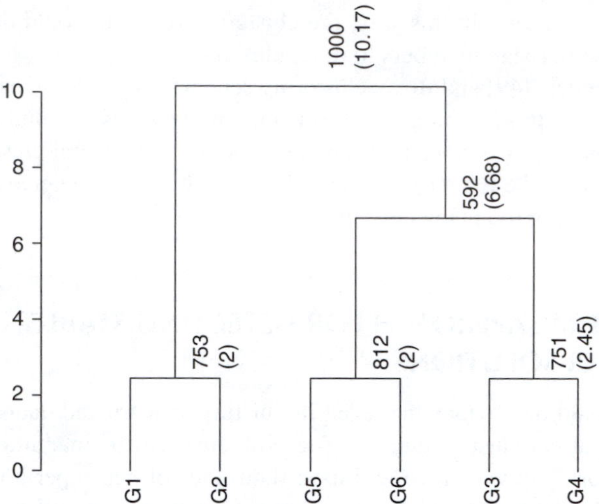

FIGURE 9.5 The original tree with the confidence value and the branch length at each node.

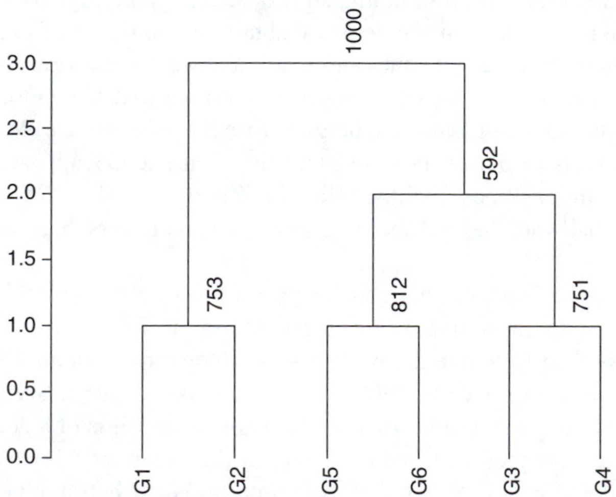

FIGURE 9.6 The consensus tree with the confidence value at each branch. The set of trees are constructed from resampled data sets.

well as the bootstrapped trees. We can establish the "confidence value" for branches of the original tree by comparing it with the bootstrapped trees. We can also derive a consensus tree that contains all the branches that are in at least half of all trees. Figure 9.5 and Figure 9.6 show the original tree and the consensus tree having the times of occurrence in the bootstrapped trees at each branch, respectively. The numbers in parenthesis in Figure 9.5 are the length of branches. Here, the 1000 new datasets are generated according to the normal distribution $N(x, 0.25x)$ and then clustered by the hierarchical algorithm using average linkage and the Squared Euclidean distance.

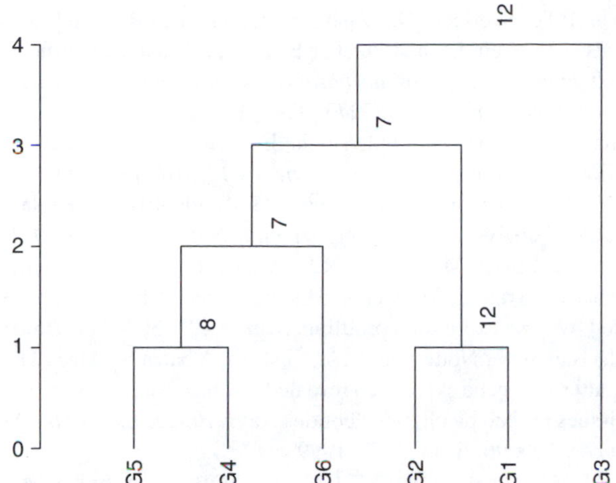

FIGURE 9.7 The consensus tree with the confidence value at each branch. The set of trees are constructed from resampled data sets.

Although the topological structure of the original tree and the consensus tree are same for this simple example, we can still notice that G5 and G6 are more likely to be in a group than the other two pairs (G1 and G2, G3 and G4).

As we stated before, large variations in cluster memberships can come from the perturbation of data as well as modifications of methods and variations in similarity metrics. The consensus rules are applicable to assess the stability of method. Figure 9.7 shows the consensus tree constructed from a set of trees using different clustering methods (single linkage, average linkage, and complete linkage) and different similarity metrics (Squared Euclidean distance, Pearson correlation coefficient, sum of absolute difference, and maximum difference across samples). We can find that only G1 and G2 are still in a group for all 12 methods, while the relationship between the other two pairs (G3 and G4, G5 and G6) are broken by applying different methods. We make no pretence that these methods are suitable for this data set, only that it provides an illustration for our method.

We illustrate our general strategy for assessing the reliability of clusters by a simple example. Nevertheless, we can vary the data and the methods simultaneously to seek consensus of clusters. It is also worth noting that the majority consensus rule proposed by Margush and McMorris [55] is only suitable for the hierarchical tree and is only one of commonly used consensus rules. The researchers could refer to other sources to seek a more suitable consensus rule for their projects (e.g., [56]) and consensus classifications obtained by k-means or other clustering methods (e.g., [57]).

REFERENCES

1. Eisen M.B., Spellman P.T., Brown P.O., and Botstein D. Cluster analysis and display of genome-wide expression patterns. *Proceedings of the National Academy of Sciences, USA* 95: 14863–14868, 1998.

2. Spellman P.T., Sherlock G., Zhang M.Q., Iyer V.R., Anders K., Eisen M.B., Brown P.O., Botstein D., and Futcher B. Comprehensive identification of cell cycle-regulated genes of the yeast *Saccharomyces cerevisiae* by microarray hybridization. *Molecular Biology and Cell* 9: 3273–3297, 1998.
3. Bussemaker H.J., Li H., and Siggia E.D. Regulatory element detection using correlation with expression. *Nature Genetics* 27: 167–171, 2001.
4. Pilpel Y., Sudarsanam P., and Church G.M. Identifying regulatory networks by combinatorial analysis of promoter elements. *Nature Genetics* 29: 153–169, 2001.
5. Alizadeh A.A., Eisen M.B., Davis R.E., Ma C., Lossos I.S., Rosenwald A., Boldrick J.C., Sabet H., Tran T., Yu X et al. Distinct types of diffuse large B-cell lymphoma identified by gene expression profiling. *Nature* 403: 503–511, 2000.
6. Alon U., Barkai N., Notterman D.A., Gish K., Ybarra S., Mack D., and Levine A.J. Broad patterns of gene expression revealed by clustering analysis of tumor and normal colon tissues probed by oligonucleotide arrays. *Proceedings of the National Academy of Sciences, USA* 96: 6745–6750, 1999.
7. Bittner M., Meltzer P., Chen Y., JIang Y., Seftor E., Hendrix M., Radmacher M., Simon R., Yakhini Z., Ben-Dor A et al. Molecular classification of cutaneous malignant melanoma by gene expression profiling. *Nature* 406: 536–540, 2000.
8. Golub T.R., Slonim D.K., Tamayo P., Huard C., Gaasenbeek M., Mesirov J.P., Coller H., Loh M.L., Downing J.R., Caligiuri M.A., Bloomfield C.D., and Lander E.S. Molecular classification of cancer: class discovery and class prediction by gene expression monitoring. *Science* 286: 531–537, 1999.
9. Hastie T., Tibshirani R., Eisen M.B., Alizadeh A., Levy R., Staudt L., Chan W.C., Botstein D., and Brown P. "Gene shaving" as a method for identifying distinct sets of genes with similar expression patterns. *Genome Biology* 1: 1–21, 2000.
10. Dembele D. and Kastner P. Fuzzy C-means method for clustering microarray data. *Bioinformatics* 19: 973–980, 2003.
11. Tavazoie S., Hughes J.D., Campbell M.J., Cho R.J., and Church G.M. Systematic determination of genetic network architecture. *Nature Genetics* 22: 281–285, 1999.
12. Herrero J. and Dopazo J. Combining hierarchical clustering and self-organizing maps for exploratory analysis of gene expression patterns. *Journal of Proteome Research* 1: 467–470, 2002.
13. Tamayo P., Slonim D., Mesirov J., Zhu Q., Kitareewan S., Dmitrovsky E., Lander E.S., and Golub T.R. Interpreting patterns of gene expression with self-organizing maps: methods and application to hematopoietic differentiation. *Proceedings of the National Academy of Sciences, USA* 96: 2907–2912, 1999.
14. Torkkola K., Gardner R.M., Kaysser-Kranich T., and Ma C. Self-organizing maps in mining gene expression data. *Information Sciences* 139: 79–96, 2001.
15. Toronen P., Kolehmainen M., Wong C., and Castren E. Analysis of gene expression data using self-organizing maps. *FEBS Letters* 451: 142–146, 1999.
16. Brown M.P., Grundy W.N., Lin D., Cristianini N., Sugnet C.W., Furey T.S., Ares M Jr., and Haussler D. Knowledge based analysis of microarray gene expression data by using support vector machines. *Proceedings of the National Academy of Sciences, USA* 97: 262–267, 2000.
17. Lee Y. and Lee C.K. Classification of multiple cancer types by tip multicategory support vector machines using gene expression data. *Bioinformatics* 19: 1132–1139, 2003.
18. Pan W., Lin J., and Le C.T. Model-based cluster analysis of microarray gene-expression data. *Genome Biology* 3: 1–10, 2002.

19. Yeung K.Y., Fraley C., Murua A., Raftery A.E., and Ruzzo W.L. Model-based clustering and data transformations for gene expression data. *Bioinformatics* 17: 977–987, 2001.
20. Ben-Dor A. and Yakhini Z. Clustering gene expression patterns. *Journal of Computational Biology* 6: 281–297, 1999.
21. Dudoit S. and Fridly J. Bagging to improve the accuracy of a clustering procedure. *Bioinformatics* 19: 1090–1099, 2003.
22. Zhang H.P., Yu C.Y., Singer B., and Xiong M.M. Recursive partitioning for tumor classification with gene expression microarray data. *Proceedings of the National Academy of Sciences, USA* 98: 6730–6735, 2001.
23. McShane L.M., Radmacher M.D., Freidlin B., Yu R., Li M.C., and Simon R. Methods of assessing reproducibility of clustering patterns observed in analysis of microarray data. *Bioinformatics* 18: 1462–1469, 2002.
24. Smolkin M. and Ghosh D. Cluster stability scores for microarray data in cancer studies. *BMC Bioinformatics* 4: 36, 2003.
25. Zhang K. and Zhao H. Assessing reliability of gene clusters from gene expression data. *Functional and Integrative Genomics* 1: 156–173, 2000.
26. Roth V., Braun M.L., Lange T., and Buhmann J.M. Stability-based model order selection in clustering with applications to gene expression data. *Lecture Notes in Computer Science*, Vol. 2415, Springer-Verlag, Berlin, pp. 607–612, 2002.
27. Torgerson, W.S. *Theory and Methods of Scaling*. John Wiley & Sons, New York, 1958.
28. Bakay M., Chen Y.W., Borup R., Zhao P., Nagaraju K., and Hoffman E.P. Sources of variability and effect of experimental approach on expression profiling data interpretation. *BMC Bioinformatics* 3: 4, 2002.
29. Holloway A.J., van Laar R.K., Tothill R.W., and Bowtell DDL. Options available — from start to finish — for obtaining data from DNA microarrays II. *Nature Genetics* 32: 482–489, 2002.
30. Simon R., Radmacher M.D., Dobbin L., and McShane L.M. Pitfalls in the use of DNA microarray data for diagnostic and prognostic classification. *Journal of National Cancer Institute* 95: 14–18, 2003.
31. Kerr M.K., Martin M., and Churchill G.A. Analysis of variance for gene expression microarray data. *Journal of Computational Biology* 7: 819–837, 2000.
32. Wu H., Kerr K.M., and Churchill G.A. MAANOVA: a software package for the analysis of spotted cDNA microarray experiments. In *the Analysis of Gene Expression Data: Methods and Software*, 2002.
33. Cui X. and Churchill G.A. Statistical tests for differential expression in cDNA microarray experiments. *Genome Biology* 4: 210, 2003.
34. Pavlidis P. and Noble W.S. Analysis of strain and regional variation in gene expression in mouse brain. *Genome Biology* 2: 1–15, 2001.
35. Emptage M.R., Hudson-Curtis B., and Sen K. Treatment of microarray experiments as split-plot designs. *USA Journal of Biopharmaceutical Statistics* 13: 159–178, 2003.
36. Zhang Z., Page G.P., and Zhang H. Fishing expedition — a supervised approach to extract patterns from a compendium of expression profiles. In S.M Lin and K.F Johnson (eds.), *Methods of Microarray Data Analysis: Papers from CAMDA '01 Methods of Microarray Data Analysis II*. Kluwer Academic Publishers, Dordrecht, pp. 86–97, 2002.
37. Chu T.M., Weir B., and Wolfinger R. A systematic statistical linear modeling approach to oligonucleotide array experiments. *Mathematical Biosciences* 176: 35–51, 2002.

38. Quackenbush J. Microarray data normalization and transformation. *Nature Genetics* 32(Suppl): 496–501, 2002.
39. Sneath P.H.A. and Sokal R.R. *Numerical Taxonomy*. Freeman, San Francisco, CA, 1973.
40. SAS Institute, Inc. *SAS Distance Macro*. SAS Institute, Cary, SC., 2003.
41. SPSS, Inc. *SPSS Base System*. SAS Institute, Chicago, IL, 2003.
42. Lance G.N and Williams W.T.A. General theory of classificatory sorting strategies. I. Hierarchical systems. *Computer Journal* 9: 373–380, 1967.
43. Jardine N. and Sibson R. *Mathematical Taxonomy*. John Wiley & Sons, Inc, New York, 1971.
44. Johnson S.C. Hierarchical clustering schemes. *Psychometrika* 32: 241–254, 1967.
45. Ward J.H. Hierarchical grouping to optimize an objective function. *Journal of the American Statistical Association* 58: 236–244, 1963.
46. Chen C.H and Chen J.A. Interactive diagnostic plots for multidimensional scaling with applications in psychosis disorder data analysis. *Statistica Sinica* 10: 665–691, 2000.
47. Shepard R.N. The analysis of proximities: multidimensional scaling with an unknown distance function. I. *Psychometrika* 27: 125–140, 1962.
48. Banfield J.D. and Raftery A.E. Model-based Gaussian and non-Gaussian clustering. *Biometrics* 49: 803–821, 1993.
49. McLachlan G.J. and Peel D. *Finite Mixture Models*. John Wiley & Sons, New York, 2000.
50. Höppner F., Klawonn F., Kruse R., and Runkler T. *Fuzzy Cluster Analysis*, Wiley, Chichester, UK, 1999.
51. Davidson G.B., Wylie B.N, and Boyack K. Cluster stability and the use of noise in interpretation of clustering. In *Proceedings of the 7th IEEE Symposium on Information Visualization (INFOVIS 2001)*, San Diego, 2001.
52. Efron B. and Tibshirani R.J. *An Introduction to the Bootstrap*. Chapman & Hall, New York, 1993.
53. Allison D.B., Gadbury G., Heo M., Fernandez J., Lee C.-K., Prolla T.A., and Weindruch R. A mixture model approach for the analysis of microarray gene expression data. *Computational Statistics and Data Analysis* 39: 1–20, 2002.
54. Kerr M.K. and Churchill A. Bootstrapping cluster analysis: assessing the reliability of conclusions from microarray experiments. *Proceedings of the National Academy of Sciences, USA* 98: 8961–8965, 2000.
55. Margush T. and McMorris F.R. Consensus n-trees. *Bulletin of Mathematical Biology* 43: 239–244, 1983.
56. Adams E. N-trees as nestings: complexity, similarity and consensus. *Journal of Classification* 3: 299–317, 1986.
57. Gordon A.D. and Vichi M. Partitions of partitions. *Journal of Classification* 15: 265–285, 1998.
58. Goldstein D., Ghosh D., and Conlon E. Statistical issues in the clustering of gene expression data. *Statistica Sinica* 12: 219–241, 2002.

10 Dimensionality Reduction and Discrimination

Jeanne Kowalski and Zhen Zhang

CONTENTS

10.1 Introduction 177
10.2 Dimension Reduction 179
10.2.1 The UMSA Learning Algorithm 181
10.2.2 Supervised Component Analysis Using UMSA 182
10.3 Discrimination 184
10.3.1 U-Statistics 184
10.3.2 Hypothesis Testing: Two, Few-Sample Groups 185
10.3.3 Hypothesis Testing: Several, Single-Sample Groups 188
10.4 Conclusion 193
References 194

10.1 INTRODUCTION

With microarrays and other high-throughput genomic profiling technologies that allow for the simultaneous analysis of expression levels of thousands or even tens of thousands of genes, we are able to expand our ability to characterize and understand disease processes at the molecular level and the heterogeneity surrounding them. As advances in biotechnology continue to support this remarkable expansion, however, the need for extracting and synthesizing information from the volumes of expression data has created an equally challenging research area in the development of corresponding statistical methods for their analysis. Within the context of biomedical and clinical research, a common objective of microarray data analysis is the selection of genes, from among the thousands profiled, that characterize groups of similar phenotype based on a small number of samples.

Cluster analysis (CA) (Everitt, 1992) is a technique used to identify samples with similar genetic patterns. The application of CA to gene expression data is based on the biologic premise that genes displaying similar expression patterns may be coregulated and share a common function or contribute to a common pathway. CA is a technique used to identify samples (or genes) with similar intensities based on a metric that is irrespective of phenotype and thus requires secondary analyses to describe samples (genes) that characterize the clusters formed and the differences among them. A related approach, recursive partitioning (RP) (Breiman, 1984), is a technique used to identify subsets of genes that explain most of the variability in some (continuous) phenotypic response, but the actual phenotypic differences between subsets is not apparent and similar to CA, and therefore requires further exploratory analyses for this purpose. A pertinent issue related to both CA and RP is the subjective defining of the number of sample clusters, with no reference to defined population clusters. Since CA is primarily a descriptive rather than an inferential tool, it has enticed the development of many ad hoc approaches toward statistical inference. Instead of conditioning upon samples of similar phenotypes, linear discriminant analysis (LDA) (Fisher et al., 1922) is a tool used to identify (linear) combinations of variables (gene intensities) that best predicts phenotypic group membership. Unlike the dependent observations that may be formed by the use of intensity pair distances, such as in CA, LDA is similar to RP in that it requires the use of independent observations. With LDA, the observations within each group are typically assumed to belong to a multivariate normal distribution that is characterized by different means and a similar covariance matrix that can be estimated by pooled samples. These assumptions, when they are met, guarantee that the LDA will be the optimal predictive model with an error rate approaching the lowest possible rate (Bayes error rate) as the number of training sample increases. However, restrictions such as independence, distributional assumptions, and equal second (and higher) moments, altogether, preclude the usefulness of LDA and its derivatives as data analytic tools for studying differential molecular heterogeneity among groups, since it is plausible that the two groups differ in terms of higher moments, such as skewness and kurtosis. This latter consideration is especially important when working with genomic data of a complex, high-dimensional structure. Support vector machines (SVM) is a supervised data modeling and classification tool (Vapnik, 1998). With its structural risk minimization learning algorithm, SVM circumvents the need for explicit estimation of data distributions and has become a popular tool for microarray analysis for its ability to handle high-dimensional data (Brown et al., 2000; Furey et al., 2000). As a supervised classification model, SVM can be used to identify linear or nonlinear (with the use of kernel functions) combinations of the variables (genes) to best predict phenotypic group memberships and provide the basis for gene selection (Weston et al., 2000).

It is often clinically desirable to be able to simultaneously query the changes in expression patterns of tens of thousands of genes. However, when the expression levels of these genes are regarded as variables, the individual array experiments, or observations, become data points in an extremely high-dimensional space. The convergence of any estimator to the true value of a smooth function defined in such a space where $p \gg n$, will be very slow, as is often reflected in the expression of "the curse of dimensionality." The availability of clinical samples, and the cost and effort associated with microarray experiments dictate that for the time being, the majority of large-scale genomic expression analysis studies will not have a statistically sufficient number of observations to allow for a "good" estimate of a function of the genes that identifies, for example, an altered expression patterns associated with a specific tumor type. In general, this is a pretty dim scenario from all perspectives.

Fortunately, however, we may reasonably hope that in many cases, there are really "a few things that matter" and thus, the function of interest is expected to be constant along most dimensions of the space. This opens up the possibility of conducting statistical analyses in a meaningful and novel way, and in particular, motivates the potential for an inference framework.

Despite the numerous clustering and discriminating algorithms available, there remains a lack of a formal framework wherein to conduct analysis of data from microarray experiments that is flexible especially in terms of analytical assumptions. In some cases, microarray analysis involves the combining of CA and LDA features into a unified framework. For example, the high dimensionality of the problem may be approached by the construction of a single composite measure that summarizes information among the dimensions into a single statistic, or perhaps through a few dimensions, such as in principal components analysis, followed by the characterization of genomic differences based upon this composite measure, such as in LDA. We focus upon some recent, separate developments in the areas of dimension reduction and discrimination. For the former, we highlight a supervised component analysis approach in which the dimensions of the reduced space represent directions along which the groups are best separated. In the latter case, we describe two novel approaches to nonparametric inference for high-dimensional comparisons involving two or more groups, based on very few samples from within each group. By presenting these particular research efforts, we provide a summary of approaches that extends traditional cluster and linear discriminant methods to address current, as well as future challenging analyses settings presented by the continued and more creative use of microarray technology, apart from the initial cancerous vs. noncancerous comparisons.

10.2 DIMENSION REDUCTION

In this section, we first describe the unified maximum separability analysis (UMSA) learning algorithm and the rationale behind it. We then present a supervised component analysis approach based on UMSA for dimension reduction (Zhang et al., 2001, 2002).

For a two-class data set of n observations from two classes and G variables, represented as a set of column vectors $z_1, z_2, \ldots, z_n$ in a G-dimensional space with class label $c_1, c_2, \ldots, c_n$, where $c_i \in \{-1, +1\}$, the solution of a linear classification function may be represented as a unit projection vector d in the p-dimensional space. For LDA, $d = \Sigma^{-1}(M_h - M_k)/\|\Sigma^{-1}(M_h - M_k)\|$. In this solution, the class means and the pooled covariance matrix are estimated using all the available samples. On the other hand, for the linear SVM (optimal soft margin classifier), the unit projection is obtained by solving the following optimization problem:

$$\text{Minimize} \quad \frac{1}{2}\nu \cdot \nu + C\sum_{i=1}^{n} \xi_i$$

$$\text{Subject to} \quad c_i(\nu \cdot z_i + b) \geq 1 - \xi_i, \quad i = 1, 2, \ldots, n$$

and then $d = \nu/\|\nu\|$. As it turns out, the SVM solution $\nu = \sum c_i \alpha_i z_i$ is a linear combination of the training samples. However, in general, only a subset of the samples located along the class boundary (the support vectors) will have $\alpha_i \neq 0$. The remaining interior samples in the class distribution in the p-dimensional space do not contribute to the final solution at all. In the above learning algorithm, note that the constant C limits on the maximum influence an individual data point may have (i.e., the "softness").

The LDA and SVM are just examples of the two different approaches to supervised learning for the derivation of classification functions. In the traditional statistical approach, exemplified by the LDA, the classifiers are constructed based on the estimated conditional distributions of the classes of samples. During the estimation process, the samples are treated equally regardless of their relative location to the class boundaries. In the second approach, exemplified by the SVM, the classifier is determined through an empirical risk minimization (ERM) procedure for a given class of classifiers. The empirical risk functions are typically associated with the classification errors, consequently, the resultant classifiers are mostly defined by the samples that are close to the class boundaries and have little to do with the interior samples that are unlikely to contribute to the classification errors.

For microarray data, the sample size is often much too small for the asymptotic behavior of the learning algorithm of a traditional statistical classifier or an ERM-based classifier to take full effect. The efficiency in utilizing information from the limited number of samples therefore becomes a critical issue in microarray data analysis. One of the arguments for the ERM approach has been that the estimate of data distribution is a more general problem than the construction of a classifier and is an often ill-posed problem (Vapnik, 1995). The interiors of the conditional distributions that are on the faraway sides of two classes should have little to do with the accurate estimation of classification boundaries. However, the reliance of ERM-based methods on boundary points (e.g., the support vectors in SVM) has its own shortcomings for small sample problems. In general, the subset of boundary points of a small sample set is an even smaller set. The final solution of ERM may lack robustness. For example, for small sample problems, SVM is very sensitive to sample labeling errors even with the use of soft margin. However, for microarray data from clinical samples, labeling errors happen frequently due to reasons such as imperfections in current diagnostic methods (e.g., false negative biopsy results) or sample contamination (e.g., the mixture of noncancer cells among cancer cells dissected from tissue specimens).

Intuitively, a more reasonable approach would be to combine the distribution estimate-based methods and the ERM-based methods so that the samples close to the class boundaries would contribute more to the accurate estimation of the classification boundaries and at the same time information on the overall data distributions would help to achieve a more robust solution. The UMSA learning algorithm was developed based on this concept. From a pure computational point of view, the ERM-based optimal soft margin classifier learning algorithm in SVM has the advantage of being very efficient in handling high-dimensional data and hence particularly suitable for microarray data. In UMSA, we modified this learning algorithm to allow for the introduction of additional data distribution information into the ERM process.

10.2.1 The UMSA Learning Algorithm

The UMSA is a two-step process. In the first step, a traditional statistical classifier $f(\cdot)$ is constructed using any one of the many possible models that are based on the estimated conditional distributions. The simplest case, of course, would be the LDA, which is often a sensible choice for microarray dataset of a very small size. Furthermore, unsupervised dimension reduction methods such as PCA/SVD may have to be applied before a statistical classifier can be constructed if the model requires a stable estimate of the covariance matrix. The following description clearly shows that some loss of information due to the dimension reduction is acceptable. In the second step, the optimal soft-margin classifier learning algorithm in SVM is modified to incorporate information about the individual samples extracted in the first step:

$$\text{Minimize} \quad \frac{1}{2}\nu \cdot \nu + \sum_{i=1}^{n} p_i \xi_i$$

$$\text{Subject to} \quad c_i(\nu \cdot z_i + b) \geq 1 - \xi_i, \quad i = 1, 2, \ldots, n$$

where $p_1, p_2, \ldots, p_n$ are n positive constants to individually constrain the maximum contribution of the corresponding samples and are different from the single constant C used in SVM. These constraints quantify a sample's "trustworthiness" to be used as a "support vector" in the solution ν of the above optimization problem. Typically, $p_i = \phi(x_i, f) > 0$ is related to the level of disagreement of a sample x_i to the statistical classifier $f(\cdot)$. Let this level of disagreement be δ_i, a positive decreasing function such as the following may be used to compute p_i:

$$p_i = \phi(\delta) = C \cdot \mathrm{e}^{-\delta_i^2/\sigma^2}$$

where $C > p_i > 0$. The parameter σ modulates the differentiating effect from δ_i. For a very large σ relative to the spread of δ_i, all p_is would essentially become a constant close to C. The UMSA algorithm is then equivalent to the optimal soft-margin classifier learning algorithm in SVM. On the other hand, a very small σ would make the p_is for those samples that have a high level of disagreement with the statistical classifier small so that they would not affect the final solution. The determination of parameter C is similar to that in SVM and have been extensively discussed in the literature (Cortes and Vapnik, 1995; Lin et al., 2002; Sollich, 2002). The choice of parameter σ requires some knowledge of the data, such as in-class variability and labeling error rate. When UMSA is used for dimension reduction as to be described next, the separation between classes of samples may be visualized, which could be used for the interactive adjustment of both parameters.

The solution of the UMSA learning algorithm has exactly the same form as that from SVM. With the use of kernel functions, UMSA can be used to derive nonlinear classifiers.

10.2.2 Supervised Component Analysis Using UMSA

The objective of unsupervised dimension reduction methods is to retain information as much as possible without concerning the type of information to be saved. PCA/SVD, for instance, projects the data to a new space where the dimensions represent the directions along which the data demonstrate the largest variation. For differential analysis of microarray data, the objective, however, is to preserve the separability between different classes of samples, for which a supervised dimension reduction method would be more appropriate. The UMSA-based component analysis algorithm is a recursive process to identify directions that are mutually orthogonal and along which two classes of samples are optimally separated by the linear version of UMSA with a given set of parameters C and σ:

```
UMSA component analysis for a two-class dataset with
    G variables and n samples
    inputs:
        UMSA parameters C and σ,
        number of components q ≤ min(G, n) to be used;
        data z1, z2, ..., zn; and
        class labels c1, c2, ..., cn, ci ∈ {−1, +1}.
    initialization:
        component set D ← {};
        k ← 1.
    while k ≤ q
     1. applying UMSA(σ, C) on z1, z2, ..., zn and c1,c2,...,cn;
     2. dk ← υ/‖ν‖; D ← D ∪ {dk};
     3. zi ← zi − (Zi^T dk)dk, i = 1, 2, ..., n;
     4. k ← k + 1.
    return D.
```

During each of the iterations, a direction d_k in the current $G - (k - 1)$ dimensional space is identified by the UMSA algorithm. All the samples are then projected onto a $G - k$ subspace perpendicular to d_k. For most practical problems, $q = 3$ is sufficient to extract most of the separability information from the original variables. The separability between the classes of the projected data can be viewed in interactive 3D display. Similar to PCA/SVD, the coefficients in the projection vector d_ks may be viewed as a measure of the contribution of the original variables toward the separation of classes of samples along the axes of the new component space. In addition to component analysis, the UMSA algorithm has also been incorporated into other variable selection procedures.

The UMSA component analysis procedure has been implemented as a module in the software package ProPeak (3Z Informatics) using the Java/Netbeans platform (www.netbeans.org/). However, it is fairly easy to implement UMSA using an existing SVM procedure. One needs only to replace SVMs constant C with UMSA's individualized p_is. In the current version of ProPeak, the individualized p_is are calculated by the following steps:

1. Application of SVD to reduce the dimension of data to equal the number of nonnegligible eigenvalues in SVD (this is to allow for a relatively stable estimate of the pooled covariance matrix to be used in LDA).
2. Construction of LDA to project all data points onto a 1D space. Selection of a cutoff between the two group means in proportion to the group standard deviations.
3. If a projected sample point is on the correct side of the LDA cutoff with a sufficient margin, δ_i equals to the distance between the point and the cutoff then multiplied by a constant (100.0 in ProPeak), otherwise δ_i equals to a minimum positive value (0.1 in ProPeak).
4. $p_i = C \cdot \mathrm{e}^{-\delta_i^2/\sigma^2}$.

As an example to illustrate the effect of UMSA-based component analysis and its difference from PCA/SVD, we analyzed the gene expression data of ovarian carcinomas by Schaner et al. (2003) available from Stanford University at genome-www.stanford.edu/ovarian_cancer/ (in file "suppTable9.xls" spreadsheet "1_22_02samimputed"). The analysis was to identify genes that were differentially expressed between samples from 9 patients with histological grade I/II serous papillary ovarian cancer and 10 patients with grade III cancer. The imputed dataset included 3053 genes. Figure 10.1a shows the 19 samples plotted in the first two-component space from SVD. As expected, the components represent directions along which the 19 samples have the largest variance in the original 3053-dimensional space. Figure 10.1b show a similar plot of the samples in the first two-component space

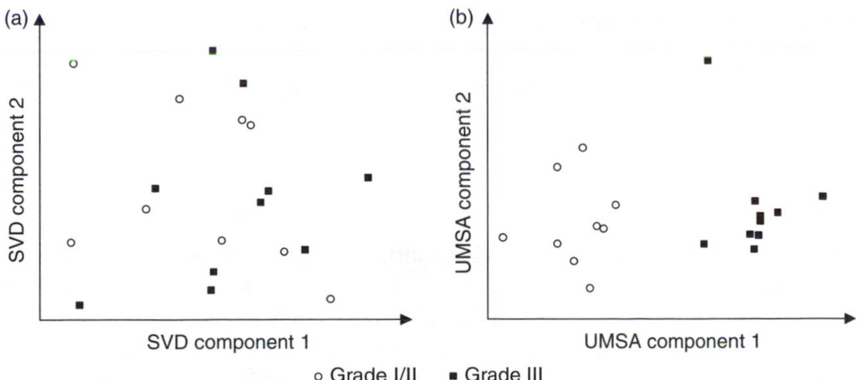

FIGURE 10.1 Comparison of (a) unsupervised component analysis using SVD and (b) supervised component analysis using UMSA ($C = 5.0$ and $\sigma = 0.02$) of a gene expression data set from Schaner et al. (2003) available at Stanford University at genome-www.stanford.edu/ovarian_cancer/ (in file "suppTable9.xls" spreadsheet "1_22_02samimputed"). Note that along the first component extracted by UMSA, the separation between two groups (9 patients with histological grade I/II serous papillary ovarian cancer vs. 10 patients with grade III cancer) is preserved.

from UMSA (exported from ProPeak, $C = 5.0, \sigma = 0.02$). In this case, the direction of the components retains the separation between the two groups. In fact, the top 160 genes that had the highest weights in the first component included all the 30 genes identified in the original paper (Schaner et al., 2003). Furthermore, the top 30 genes with the highest weights in the first UMSA component actually provided a better group clustering result than the 30 genes selected in the original paper. An explanation for this is that the weight in UMSA component is based on a gene's contribution in the collective effort of all genes to separate the two groups while in SAM or PAM type analyses, the contribution of a gene is assessed more or less independently.

10.3 DISCRIMINATION

By comparing and characterizing genomic heterogeneity among defined groups, within a nonparametric framework, in this section, we introduce some recently developed methods for nonparametric analyses of high-dimensional data, within the context of a supervised microarray analysis. In particular, we discuss two nonparametric, inference approaches for comparisons of two or more groups, based on a few, or as little as a single sample from within each group. We first focus upon the setting of two groups, each with a few samples from within each group. In this context, we discuss a distance-based approach to analysis that requires the construction of a composite measure of genomic heterogeneity within and between each group to formulate hypotheses. We then extend this setting to address analysis in the extreme case scenario of comparing very high-dimensional data among three or more groups, based on a single sample from within each group. This scenario often arises in an effort to examine hypothesized molecular pathways. For example, within an immunology context, a goal may be to compare intensities obtained from cell samples singly exposed to several distinct but related experimental conditions. We focus upon the formulation and test of hypotheses to facilitate valid statistical inference without requiring intensities to follow some analytic distribution, and without specifying a correlation structure among genes.

10.3.1 U-STATISTICS

Statistics defined by a sum of dependent random variables or vectors from multiple subjects are introduced in this section. One example of such a class of statistics is the sample variance, where each term contains the sample mean that is defined based on all observations, resulting in a statistic defined by a sum of dependent terms. U-statistics (Serfling, 1980, Chapter 5) are a class of statistics specific to the study of the asymptotic behavior of a sum of dependent terms. Typically, an introduction to U-statistics involves showing that the sample variance may be alternatively expressed in terms of a symmetric expression about pairs of subjects. In some cases, however, the statistic itself is already expressed in terms of pairs of subjects, such as the Mann–Whitney–Wilcoxon Rank sum test and distance measures. While the notion of U-statistics is not new, their application to address timely, high-dimensional analyses, such as microarrays, represents a novel area with extensions to existing theory.

In the following sections, we discuss two novel approaches for nonparametric inference within the context of comparing several groups with high-dimensional data from either a few or very few samples. To facilitate hypothesis testing in both cases, statistics are constructed that are defined by a sum of dependent random variables from multiple subjects. Thus, the definition and asymptotic properties of U-statistics are applied to examine the large sample behavior of such a class of statistics, since asymptotic theory for independent sums of random variables no longer applies.

10.3.2 Hypothesis Testing: Two, Few-Sample Groups

In this section, we highlight recent work developed for the analysis of gene sequence diversity to facilitate comparisons (Kowalski et al., 2002) and characterizations (Kowalski, 2001) of heterogeneity between two groups, within a microarray context. This approach may be viewed as nonparametric inference for discriminant analysis in the sense that groups are conditioned upon and the differences between them, in terms of the degree of heterogeneity, formally compared through hypothesis testing. Similar to principal components analysis, the dimension of the problem is first reduced by collapsing information contained in all genes into a single composite measure of heterogeneity to facilitate hypothesis testing.

Let G (indexed by g) denote the number of cDNA experimental genes arrayed, including known human (or murine) genes and expressed sequence tags (ESTs). Suppose that the n subjects are classified into two (distinct) phenotypic groups, h and k, with sample sizes n_h and n_k, respectively. Denote by $\mathbf{z}_j = (z_{1j}, \ldots, z_{Gj})^{\mathrm{T}}$ the G-dimensional vector of intensities obtained from G genes. Let $\mathbf{z}_h$ refer to intensities from a reference standard (control) tissue that is to be compared against intensities from a target (comparative) tissue, $\mathbf{z}_k$. Because intensities from all G genes of interest are typically obtained from S objects (chips, filters, slides, etc.), we may further partition each intensity vector according to the set of genes arrayed on each object. For example, $\mathbf{z}_j$ may be further partitioned as $\mathbf{z}_j = (\mathbf{z}_{j1}^{\mathrm{T}}, \ldots, \mathbf{z}_{js}^{\mathrm{T}})^{\mathrm{T}}$, where $\mathbf{z}_{js}^{\mathrm{T}} = (z_{j[G_{(s-1)}+1]}, \ldots, z_{j[G_{(s-1)}+G_s]})$ for $1 \leq s \leq S$, such that $G_0 = 0$ and $G = \sum_{s=1}^{S} G_s$.

A composite measure of heterogeneity within the reference, within the comparison condition, and between the reference and comparison conditions, among all genes, are respectively defined by

$$
\begin{aligned}
d^2{}_{ii'\cdot,hh}(\mathbf{z}_{hi}, \mathbf{z}_{hi'}\cdot; \alpha) &= (\mathbf{z}_{hi} - \mathbf{z}_{hi'})^{\mathrm{T}} W(\alpha)(\mathbf{z}_{hi} - \mathbf{z}_{hi'}), \quad 1 \leq i < i' \leq n \\
d^2{}_{jj'\cdot,kk}(\mathbf{z}_{kj}, \mathbf{z}_{kj'}\cdot; \alpha) &= (\mathbf{z}_{kj} - \mathbf{z}_{kj'})^{\mathrm{T}} W(\alpha)(\mathbf{z}_{kj} - \mathbf{z}_{kj'}), \quad 1 \leq j < j' \leq m \\
d^2{}_{ij\cdot,hk}(\mathbf{z}_{hi}, \mathbf{z}_{kj}\cdot; \alpha) &= (\mathbf{z}_{hi} - \mathbf{z}_{kj})^{\mathrm{T}} W(\alpha)(\mathbf{z}_{hi} - \mathbf{z}_{kj}), \quad 1 \leq i \leq n,\ l \leq j \leq m
\end{aligned}
\tag{10.1}
$$

where W denotes a symmetric, nonnegative definite $G \times G$ weighting matrix, characterized by a $n_\alpha \times 1$ parameter vector α.

Using the above distances, we form matrices and their corresponding distributions to formulate hypotheses of equal degrees of heterogeneity within and between conditions, and then characterize observed differences in heterogeneity through estimates of individual gene contributions to the test statistic. To this end, consider the following genomic distance matrices, conditional on each phenotypic classification:

Phenotypic group	Reference (h)	Comparison (k)
Reference (h)	$D_{hh} = \left[d_{ii'\cdot,hh} = d_{hh\cdot}\,(\mathbf{z}_{hi}, \mathbf{z}_{hi'})\right]$	$D_{hk} = \left[d_{ij\cdot,hk} = d_{ij\cdot}\,(\mathbf{z}_{hi}, \mathbf{z}_{kj})\right]$
Comparison (k)	$D_{kh} = \left[d_{ji\cdot,kh} = d_{ji\cdot}\,(\mathbf{z}_{ki}, \mathbf{z}_{hj})\right]$	$D_{kk} = \left[d_{jj'\cdot,kk} = d_{jj'\cdot}\,(\mathbf{z}_{kj}, \mathbf{z}_{kj'})\right]$

The within-condition matrices, D_{hh} and D_{kk}, are symmetric with zero on the diagonal, while the between-condition distance matrices, D_{hk} for $h \neq k$ is nonsymmetric with a nonzero diagonal. The distribution of each of these stochastic distance matrices, D_{hk}, provides information for testing gene region heterogeneity within and between phenotypic groups.

Let $F_{hk}(d) = \Pr(d_{ij\cdot,hk'} \leq d)$ denote the cumulative distribution function (cdf) based on D_{hk}, where d is a mass point among the elements of D_{hk}. Let $\{d_{\min}, \ldots, d_{\max}\}$ denote the ordered distinct mass points, such that $F_{hk}(d_{\min}) = 0$ and $F_{hk}(d_{\max}) = 1$, respectively. Since $d_{ij\cdot,hk}$ is a discrete random variable, the set, $\{d_{\min}, \ldots, d_{\max}\}$ is finite. For a given distance, $d, F_{hk}(d)$ refers to the within group ($h = k$) and between group ($h \neq k$) distance distributions. For a given sample, a nonparametric estimator of $F_{hk}(d)$ is defined through an empirical cdf (ecdf), as given below

$$\hat{F}_{hk}(d) = \begin{cases} \binom{n_h}{2}^{-1} \sum_{(i,j)\in C_h} I_{\{d_{ij\cdot,hh} \leq d\}} & \text{for } h = k \\ \prod_{b=h,k} \binom{n_b}{1}^{-1} \sum_{i,j} I_{\{d_{ij\cdot,hh} \leq d\}} & \text{for } h \neq k \end{cases}$$

where $\binom{n}{k}$ is the binomial coefficient, C_h denotes the set of all distinct (i,j) pairs with $1 \leq i, j \leq n_h$, and $I_{\{x \leq d\}}$ is a binary indicator with the value 1 if $x \leq d$ and 0 otherwise. For notational convenience, we have suppressed the dependency of $\hat{F}_{hk}$ on α. By applying the theory of U-statistics, $\hat{F}_{hk}(d)$ has been shown to be consistent and asymptotically normal (Kowalski et al., 2002). Regardless of whether α is treated as known values (e.g., W is an identity matrix) or estimates (e.g., $W_2 = \hat{\Sigma}_{\bar{\mathbf{d}}}$), an estimate of the asymptotic variance of $\hat{F}_{hk}(d)$ may be readily calculated. To eliminate the dependency of these statistics on d, let $\hat{\mathbf{F}}_{hk} = \left(\hat{\mathrm{F}}_{hk}(d_{\min}), \ldots, \hat{\mathrm{F}}_{hk}(d_{\max})\right)^{\mathrm{T}}$ denote an estimator of $\mathbf{F}_{hk} = (\mathrm{F}_{hk}(d_{\min}), \ldots, \mathrm{F}_{hk}(d_{\max}))^{\mathrm{T}}$. The vector statistic, $\hat{\mathbf{F}}_{hk}$, is well defined, since $d_{ij\cdot,hk}$ is a discrete random variable and in addition, it may also be used to characterize within ($h = k$) and between ($h \neq k$) group differences.

In contrast to $\mathbf{F}_{hk}$, a probability "density" function (pdf) may be defined based on the first-order differences, $f_{hk}(d) = F_{hk}(d + 1) - F_{hk}(d)$. By substitution of the respective ecdfs into this expression, that is, $\hat{f}_{hk}(d) = \hat{F}_{hk}(d + 1) - \hat{F}_{hk}(d)$, we obtain the vector statistic based on the density function, $\hat{\mathbf{f}}_{hk}$. Even though the density statistics are equivalent in terms of statistical inference, they are better for visually displaying within and between group heterogeneity.

Let $F_{rr}(d) = \Pr(d_{rr\cdot,ii'} \leq d)$ denote the cumulative distribution function (cdf) for the composite measure within reference condition samples, where $\Pr(\cdot)$ denotes a probability measure. We similarly denote a within comparison condition and between reference and comparison condition cdf's by $F_{aa}(d)$ and $F_{ra}(d)$, respectively. A nonparametric estimator (empirical cdf) for each underlying population cdf is defined by

$$\hat{F}_{rr}(d) = \binom{I}{2}^{-1} \sum_{i<k} I_{\{d_{rr,jk} \leq d\}}; \quad \hat{F}_{aa}(d) = \binom{J}{2}^{-1} \sum_{j<m} I_{\{d_{aa,mj} \leq d\}}$$

$$\hat{F}_{ra}(d) = (IJ)^{-1} \sum_{k,j} I_{\{d_{ra,kj} \leq d\}} \tag{10.2}$$

where $\binom{I}{2}$ is the binomial coefficient, and $I_{\{a \leq s\}}$ is a binary indicator with the value 1 if $a \leq s$ and 0 otherwise. By applying the theory of U-statistics, the empirical cdfs are shown to be consistent and asymptotically normal (Kowalski, 2001). For visual comparisons of the degree of intensity closeness, it may be preferable to construct density statistics, $\hat{f}_{rr}$, $\hat{f}_{aa}$, and $\hat{f}_{ra}$, through successive subtraction of empirical cdfs.

We construct the following hypotheses to formally compare heterogeneity within and between groups:

$$H_{01}\colon \mathbf{F}_{hh} = \mathbf{F}_{kk}, \quad H_{02}\colon \mathbf{F}_{hk} = \mathbf{F}_{hh}, \quad H_{03}\colon \mathbf{F}_{hk} = \mathbf{F}_{kk} \tag{10.3}$$

The first hypothesis, H_{01}, tests equality in the degree of gene region heterogeneity between the two groups, while $H_{02}(H_{03})$ examines such equality in the combined group against each of the other groups. In comparison to the other two, H_{01} tests for differential within-group heterogeneity between the reference and comparison tissue groups. For example, if the kth group is more homogenous than the hth group, then $d_{ij\cdot,kk}$ will be smaller, on average, relative to $d_{ij\cdot,hh}$ and thus, constitute a difference in their distributions. However, since the distance measure $d_{ij\cdot,kk}$ only discriminates between different types of heterogeneity, but not heterogeneity of the same type, the hypothesis H_{01} does not test for a group difference when the same degree of within-group heterogeneity is characterized by different genes between the two groups. In such cases, H_{02} and H_{03} may be used to test for such homogeneity within groups but heterogeneity between them, since the cross-group distances will be larger than the within-group ones. The scenarios described for rejection of various hypotheses and their corresponding distinct interpretations are summarized in Figure 10.2.

To test each hypothesis in Equation 10.2, define the following difference statistics:

$$\hat{\delta}_j^2 = \hat{\Delta}_j^{\mathrm{T}} W_{3j} \hat{\Delta}_j, \quad \hat{\Delta}_1 = \hat{\mathbf{F}}_{hh} - \hat{\mathbf{F}}_{kk}, \quad \hat{\Delta}_2 = \hat{\mathbf{F}}_{hk} - \hat{\mathbf{F}}_{hh},$$

$$\hat{\Delta}_3 = \hat{\mathbf{F}}_{hh} - \hat{\mathbf{F}}_{kk}, \quad j = 1, 2, 3 \tag{10.4}$$

where W_{3j} denotes a symmetric, nonnegative $n_d \times n_d$ weighting matrix. In the simplest case, for $W_{3j} = I_{n_d}, \hat{\delta}_j^2$ is the squared Euclidean distance. Each of the difference

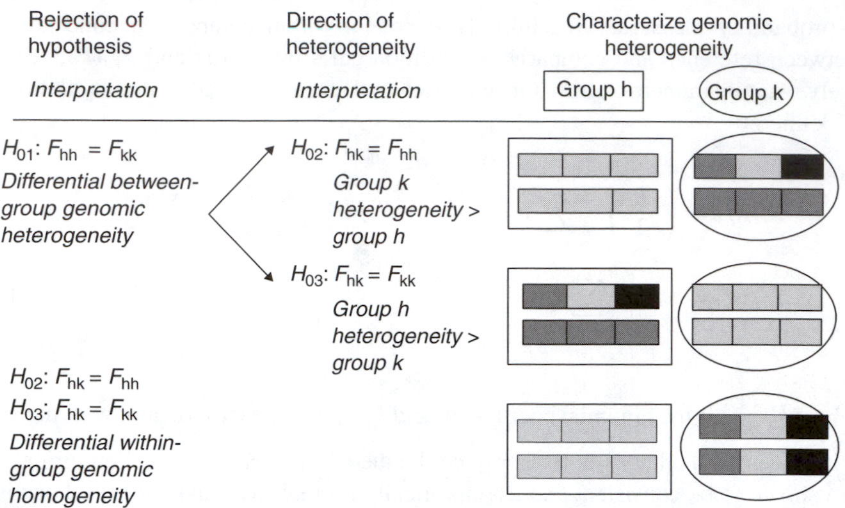

FIGURE 10.2 Overall schema for hypothesis testing based upon the distance approach of Section 10.3.2 for comparison of genomic heterogeneity between two groups, h and k. In this figure, for illustration, assume a genome defined by three genes, the different colors indicate different levels of intensity, and a pair of subjects from within each group. In the first scenario, rejection of hypotheses, H_{01} and H_{02}, is characterized by differences in the degree of genomic heterogeneity between the two groups, where the heterogeneity between the sample pair in group h is smaller relative to the heterogeneity between samples in group k. A similar situation is depicted by the rejection of hypotheses, H_{01} and H_{03}, but with greater heterogeneity exhibited by group k relative to group h. The third scenario involves the rejection of hypotheses, H_{02} and H_{03} (and failure to reject H_{01}), in which the degree of heterogeneity between samples is similar within each group, but characterized differently between groups.

statistics, $\hat{\Delta}_j$ have been shown to be asymptotically normal (Kowalski et al., 2000), in which case, $\hat{\delta}_j^2$ is asymptotically χ^2 under H_{0j}, provided that a consistent estimate of the inverse (or generalized inverse) $\sum_{\hat{\Delta}_j}^{-}$ of the asymptotic variance of $\hat{\Delta}_j$ is used as the weighting matrix for each W_{3j}. Inferences may also be based on a Monte Carlo approximation to the permutation distribution of $\hat{\delta}_j^2$ under each null hypothesis (Fisher, 1936; Efron and Tibshirani, 1993). Note that the hypotheses in Equation 10.3 may be equivalently expressed in terms of pdfs, with the epdf statistics used for testing them. By letting $W_{3j} = \sum_{\hat{\Delta}_j}^{-}$, enables both asymptotic and exact inference methods to be used in obtaining p-values for each of the respective hypothesis, where for the exact methods, a Monte Carlo approximation may be implemented.

10.3.3 Hypothesis Testing: Several, Single-Sample Groups

A central application of high-throughput technologies is the characterization of expression differences among cells. The initial focus of microarray experiments has been to conduct two group comparisons among genes expressed in target (pathological) cells vs. (normal) reference cells. The information from such experiments, however, is limited in the sense that they only provide partial information toward the ultimate goal of identifying the roles of genes and molecular pathways in directing normal and pathological cellular function. For this reason, there has been a shift in the design focus of microarray experiments to incorporating more than two tissue comparisons. Along these lines, the cost from the number of samples and replication are often traded for the cost of adding more than two groups to be compared among thousands of genes. This particular use of microarray technology presents a major challenge ahead for analysis, in terms of comparing several single sample groups, each of very high dimension.

One approach to facilitate comparisons among several conditions is to separately examine intensities between each pair of two phenotypes. In this case, differentially expressed candidate genes are selected within each phenotypic pair, according to some method, and the results between all two-group comparisons intersected to define a set of candidate genes as differentially expressed among all condition comparisons. A major drawback to this approach is that with additional groups (phenotypes), the number of comparisons greatly increases, thus embedding a large multiple testing problem within an already large multiple testing problem, due to the number of genes examined.

In this section, we discuss a novel statistical approach (Kowalski and Powell, 2003) to facilitate comparisons among microarray data obtained from samples singly exposed to several experimental conditions or clinical tissues. In this setting, the goal is to compare gene transcriptions from several conditions to select genes that are simultaneously over- or underexpressed, among the several groups and thousands of genes, beyond experimental variation. We formulate a global hypothesis for some general criteria involving several group comparisons, among thousands of genes. Depending upon whether independent or dependent experimental errors are assumed, individual genes or pairs of two genes (gene pairs) are the respective units of analysis used in hypothesis testing. Upon rejecting the global hypothesis, we estimate the cardinality of a set of candidate genes (or gene pairs) as characterizing the criteria of interest, and with this estimate, we characterize the potential elements of the candidate set by describing relationships within and between genes and comparison pairs, using singular value decomposition and inner product concepts, in combination.

Similar to the notation of the previous section, let $\mathbf{z}_1$ denote the vector of intensities (reference) that is to be compared to intensities from three conditions (phenotypic groups), $\mathbf{z}_2$, $\mathbf{z}_3$, and $\mathbf{z}_4$. For clarity of discussion, suppose that $E = \{\mathbf{z}_1 > \mathbf{z}_2, \mathbf{z}_1 > \mathbf{z}_3, \mathbf{z}_1 > \mathbf{z}_4\}$ denote the event of interest, that is, the event that the reference condition, $\mathbf{z}_1$, is simultaneously overexpressed relative to three other conditions. We consider a permutation approach as an alternative to permuting columns of phenotypes, such as in the significance analysis of microarrays (SAM) algorithm (Tusher et al., 2001), in which we permute phenotypes (columns) within each gene (row), among all genes. In this way, we view intensities as observations from possibly distinct distributions among genes. With this approach, we construct a sampling distribution based on $(J!)^G$ values (as opposed to $J!$ values). Although appealing, this way of permuting the data assumes independence among genes. Since the genes arrayed are from the same tissue, they are generally not independent. One way around this problem is to define a (linear) mapping from a dependent to an independent space, and conduct hypothesis testing in this latter space. Along these lines, we subtract intensities from contiguous genes (with one overlapping gene between pairs) to define a between-gene pair intensity difference within each phenotype, denoted by $\Delta_{zj} = \left(\mathbf{z}_{j(-h)} - \mathbf{z}_{j(-k)}\right)$ for $1 \leq j \leq J$, where for a jth column vector of Z, $\mathbf{z}_{j(-g)}$ denotes removing its gth row and thus, Δ_{zj} is a vector of length $L = (G-1)$. We view the subtraction of intensities between genes as a way of removing potential dependencies among them so that the proposed row-based permutation approach may be used as a way of generating the sampling variability necessary for valid inference. By using gene pairs as the unit of analysis (indexed by l) we express the hypothesis of interest as

$$H_0\text{: } \Pr[K\Delta_{zl} \geq 0] = \Pr\left[K\left(\pi^{c(l)}(\Delta_{zl})\right) \geq 0\right] \quad \text{for some } 1 \leq l \leq L, \pi^{c(l)} \in \pi_J^{c(l)} \text{ vs.} \tag{10.5}$$

$$H_a\text{: } \Pr[K\Delta_{zl} \geq 0] > \Pr\left[K\left(\pi^{c(l)}(\Delta_{zl})\right) \geq 0\right] \quad \text{for some } l \text{ and } \pi^{c(l)}$$

where $\pi_J^{c(l)}$ denotes the set of all distinct $J!$ permutations of the integer set $\{1, \ldots, J\}$, within an lth row of Δ_{zj}. Thus, $\pi^{c(l)}(\Delta_{zl})$ denotes one permutation of columns within an lth row of the observed gene pair difference matrix, $\Delta\mathbf{z}$.

Among all G genes, there is a total of $G!$ gene reassignments (permutations) and a total of $(J!)^G$ permutations of the J columns (phenotypes) within each gth gene (row). Suppose we sample M from among the $G!$ permutations, and K from the $(J!)^G$ permutations of columns to test the global hypothesis in Equation 10.4 of equal events of greater vs. less or no difference in intensity, simultaneously, among all genes arrayed and condition comparisons of interest. Below, we provide an outline of the general approach used to test the hypothesis in Equation 10.4, assuming dependent experimental errors among genes and thus, gene pairs are the units of analysis. Note that the assumption of independent experimental errors is viewed as a special case. For more details on the statistical principles underlying this approach, we refer the reader to the papers of Kowalski and Powell (2003) and Kowalski et al. (2004).

Within an mth permutation of the G genes, we propose the following approach to testing the hypothesis in Equation 10.5:

1. Define the observed data matrix as $Z^{(m)} = (\mathbf{z}_1^{(m)}, \ldots, \mathbf{z}_J^{(m)})$, where $\mathbf{z}_j^{(m)} = (\mathbf{z}_{g(m)1}, \ldots, \mathbf{z}_{G(m)j})^{\mathrm{T}}$ for $1 \leq j \leq J$.
2. Calculate a between gene pair difference within a jth column by

$$\mathbf{z}_{j(-1)}^{(m)} = \left(z_{2(m)1,\ldots,}z_{G(m)j}\right)^{\mathrm{T}}, \quad \mathbf{z}_{j(-G)}^{(m)} = \left(z_{1(m)j,\ldots,}z_{G-1(m)j}\right)^{\mathrm{T}},$$
$$\Delta_{zj}^{(m)} = \left(\mathbf{z}_{j(-1)}^{(m)} - \mathbf{z}_{j(-G)}^{(m)}\right)$$

3. Calculate a between condition pair difference by

$$\delta_1^{(m)} = \left(\delta_{12}^{(m)}, \ldots, \delta_{1J}^{(m)}\right)^{\mathrm{T}}, \quad \delta_{1j}^{(m)} = \left(\Delta_{z1}^{(m)} - \Delta_{zj}^{(m)}\right) \quad \text{for } 2 \le j \le J$$

4. Among the $L = (G-1)$ gene pairs formed, we calculate a test statistic based on redefined observed data as follows:

$$\hat{\gamma}^{(m)} = \frac{1}{L}\sum_l n_l^{(m)}, \quad n_1^{(m)} = \prod_{j=2,3,4} I_{\left\{\delta_{1jl}^{(m)} > 0\right\}}$$

5. For a kth permutation of the J columns within an mth permutation of the G genes, we define the following reassignment of the data:

$$Z^{k(m)} = \left(\mathbf{z}_1^{k(m)}, \ldots, \mathbf{z}_J^{k(m)}\right), \quad \mathbf{z}_j^{k(m)} = \left(\mathbf{z}_{gl}^{k(m)}, \ldots, \mathbf{z}_{Gj}^{k(m)}\right)^{\mathrm{T}} \quad \text{for } 1 \le j \le J$$

6. Calculate between gene pair and between condition pair differences for each kth permutation, denoted by $\delta_1^{k(m)} = \left(\delta_{12}^{k(m)}, \ldots, \delta_{1J}^{k(m)}\right)^{\mathrm{T}}$.
7. Calculate test statistics, $\hat{\gamma}^{k(m)} = \frac{1}{L}\sum_l n_l^k(m)$, where $n_l^k\ (m) = \prod_{j=2,3,4} I_{\left\{\delta_{1ji}^{k(m)} > 0\right\}}$.
8. Redo Steps 5–7 above for each kth permutation of columns within a gene for $1 \le k \le K$.
9. Estimate a p-value by $\hat{p}^{(m)} = \frac{1}{K}\sum_k I_{\{\gamma^{k(m)} \ge \gamma^{(m)}\}}$.
10. Redo Steps 1–9 for each mth permutation of genes for $1 \le m \le M$.
11. Calculate average (over M permutations of genes), p-value estimate by $\bar{\hat{p}} = \frac{1}{M}\sum_m \hat{p}^{(m)}$.

For testing the same global hypothesis under the assumption of independent experimental errors requires the following modifications to the above algorithm: (1) $M = 1$ is based on the assignment of genes from the observed data; (2) step 2 is no longer necessary; (3) redefine between condition pair differences in step 3 by $\delta_1^{(m)} = \left(\delta_{12}^{(m)}, \ldots, \delta_{1J}^{(m)}\right)^{\mathrm{T}}, \delta_{1j}^{(m)} = \left(\mathbf{z}_1^{(m)} - \mathbf{z}_j^{(m)}\right)$, for $2 \le j \le J$; and (4) in step 4, L indexes individual genes from 1 to G. In either case, we apply some upper threshold, $p_0 > 0$, to $\bar{\hat{p}}$, such that if $\bar{\hat{p}} \le p_0$, then we reject H_0.

Upon rejecting the null hypothesis, we estimate a number of gene pairs (or individual genes) as characterizing the criteria of interest. Denote an estimate of this number by $\overline{\Delta}_n = K^{-1}\sum_m \Delta_n^{(m)}$, where

$$n_{\text{obs}}^{(m)} = \sum_l n_l^{(m)}, \quad n_{\text{exp}}^{(m)} = \frac{1}{K}\sum_k \sum_l n_l^{k(m)}, \quad \Delta_n^{(m)} = \left(n_{\text{obs}}^{(m)} - n_{\text{exp}}^{(m)}\right),$$

$$\hat{\sigma}_{\Delta n}^2 = \frac{1}{M}\sum_m \left(\Delta_n^{(m)} - \overline{\Delta}_n\right)^2$$

with $\overline{\Delta}_n = M^{-1} \sum_m \Delta_n^{(m)}$. Thus, $\overline{\Delta}_n$ provides an estimate of the number of gene pairs, on average, observed to meet the criteria of interest, upon permuting the data. Based on this permutation approach, we also obtain a variance estimate for $\overline{\Delta}_n$, denoted by $\hat{\sigma}^2_{\Delta n}$, under dependence, whereas under a working independent (experimental error) assumption, $M = 1$ and thus, Δ_n is an estimate of the number of candidate individual genes.

Next, we characterize candidate genes based upon an algorithm that utilizes singular value decomposition and inner product concepts to select a candidate set, the cardinality of which is estimated from the hypothesis test. For details on the algorithm and the statistical principles underlying it, see Kowalski et al. (2004). Briefly, a nonparametric closeness measure among genes, between phenotypic pairs is constructed based on the inner product. This measure enables an examination of intensity relations between genes, in terms of comparing their intensity functionals, across genes, between each phenotypic pair. In this case, intensity vectors for a pair of conditions are considered the same if their intensity functionals are parallel. An estimate of the effect of each gene on the degree of similarity between intensity functionals, within each phenotypic pair, is defined by removing a gth gene at a time. If intensities between each pair of phenotypic differences, such as $(\mathbf{r} - \mathbf{a})$ vs. $(\mathbf{r} - \mathbf{b})$, are close to each other, then the inner product measure will be close to one, indicating a strong degree of similarity in intensities among genes, between the two phenotypes. Otherwise, this measure will deviate from one, with smaller values indicating a lesser degree of similarity. In comparison to regression, this approach does not require that the intensity vectors be rescaled, that a liner relationship hold among genes, or that the errors among genes are independent, such as in a regression setting. Perhaps a most important distinguishing feature of the use of the inner product is that it is defined based on the notion of closeness that exploits an affine property for functionals, rather than defined using a traditional, distance-based notion of closeness.

Among all phenotypic difference pairs formed, we then construct an inner product matrix, which is decomposed into a rank one matrix, and that represents the error in approximation. Based on this decomposition, we construct summary statistics that provides a two-dimensional measure for each gene; one that estimates a within-gene effect that is common among all phenotypic comparisons, and a second statistic that estimates a genes' deviation from this common effect. We therefore, define a desirable gene as one that is relative to all other genes, exhibits both a large, common effect, among all comparison pairs, and a small deviation from this common fit. Another way to think of the utility of both measures is to view them in an analogous regression setting, where one obtains a parameter estimate reflecting the effect of some covariate on a response, as in our common estimated effect, and then examines the overall fit of a model to the data, as in our deviation statistic. The selection of candidate genes is then made by defining a norm that combines information from a two-dimensional ranking (Kowalski et al., 2004).

In addition to the above algorithm, similar to the distance-based approach discussed, we characterize comparison groups further by partitioning the selected candidate gene set according to differential gene effects within each phenotypic pair, relative to a common effect among all pairs. To this end, we refer to a gene as dominant triple if, on the basis of some defined measure, a genes' effect within each pair is similar across all three-condition pairs. Similarly, we refer to a gene as dominant double or dominant single, if their estimated measure of effect is similar among two of the three condition pairs, or a single condition pair effect is much larger than others, respectively. In the case of gene pairs, we also characterize genes according to individual gene effects within a gene pair. To this end, we refer to a gene pair as dominant single if one gene in the pair more closely resembles the overall, estimated common effect among all conditions for that pair. We denote a pair in which both genes appear to equally contribute to an overall effect among all conditions, as dominant double. In this case, two scenarios may arise; one in which effects from both genes are positive (in the same direction) and another in which the estimated effect of one gene is positive and the other is negative (in opposing directions), in the same degree. We differentiate these two situations by referring to a dominant double gene pair as either the same or opposing directions on overall effect among all condition pairs. The partition of genes into such groups enables an initial focus upon a smaller set of genes that do not simply depict several criteria, but the various ways in which the condition pairs are observed to meet it. For example, by focusing upon the dominant double group, we gain insight into a set of genes that may characterize one hypothesized pathway, apart from another.

10.4 CONCLUSION

In this chapter, we present two approaches toward achieving dimension reduction for the analysis of microarray data, a nonparametric hypothesis test method for multiple-group comparisons and a computational method that incorporates data distribution information into an otherwise, pure machine learning algorithm. A common issue that both approaches attempt to address is the efficient use of information from a very limited number of samples.

Because of the setting discussed for the methods introduced in Section 10.3.3, that is, comparing several, single-sample, high-dimensional groups, no existing analytical method is applicable. The few approaches that do accommodate a comparison of more than two groups, also requires more than a single sample within each group. While some ad hoc implementation of current algorithms for microarray analysis would be required, it would also bring into question the validity of comparing results from them to those based on the approach presented in Section 10.3.3. Similarly, the method introduced in Section 10.3.2 is to address comparisons of intensity heterogeneity within and among groups of very high dimension based on a few samples per group. In order to compare results from this approach, some ad hoc approach would be required that simultaneously compares within to between heterogeneity in a nonparametric setting and thus, bring into question the validity of comparing results.

By integrating inference principles within a bioinformatics setting, we discussed two methods to facilitate the genomic characterization of phenotypic groups based upon a very limited number of samples. A separate but novel contribution from each approach is the reduction in the amount of genomic information extracted to summarize within and between group differences in expression. Unlike other available algorithms, the methods presented integrate desirable features from both LDA and CA and posit them into a formal approach for the purpose of enabling high-dimensional comparisons among groups based on very limited samples.

For the methods presented, source code and other related information on the distance-based approach of Section 10.3.2 and the gene pairing approach of Section 10.3.3 may found at www.cancerbiostats.onc.jhmi.edu/GENE_S.cfm and www.cancerbiostats.onc.jhmi.edu/HAM.cfm, respectively.

REFERENCES

Breiman L. (1984). *Classification and Regression Trees*. Belmont CA, Wadsworth International Group.

Brown M.P., Grundy W.N. et al. (2000). Knowledge-based analysis of microarray gene expression data by using support vector machines. *Proc. Natl Acad. Sci. USA* 97: 262–267.

Cortes C. and Vapnik V. (1995). Support-vector networks. *Mach. Learn.* 20: 273–297.

Efron B. and Tibshirani R. (1993). *An Introduction to the Bootstrap*. New York, Chapman & Hall.

Everitt B. (1992). *The Analysis of Contingency Tables*. London, New York, Chapman & Hall.

Fisher A., Dickson C. et al. (1922). *The Mathematical Theory of Probabilities and its Application to Frequency Curves and Statistical Methods*. New York, The Macmillan Company.

Fisher R.A. (1936). The use of multiple measurements in taxonomic problems. *Ann. Eugenics* 7: 179–188.

Furey T.S., Cristianini N. et al. (2000). Support vector machine classification and validation of cancer tissue samples using microarray expression data. *Bioinformatics* 16: 906–914.

Kowalski J. (2001). A non-parametric approach to translating gene region heterogeneity associated with phenotype into location heterogeneity. *Bioinformatics* 17: 775–790.

Kowalski J., Pagano M., and De Gmttola V. (2002) A nonparametric test of gene region heterogeneity associated with phenotype. *J. Am. Stat. Assoc.* 93: 398–408.

Kowalski J. and Powell J. (2003). Nonparametric inference for stochastic linear hypotheses: application to high-dimensional data. *Biometrika* 93: 393–408.

Kowalski J., Drake C. et al. (2004). Non-parametric, hypothesis-based analysis of microarrays for comparison of several phenotypes. *Bioinformatics* 20: 364–373.

Lin Y., Wahba G. et al. (2002). Statistical properties and adaptive tuning of support vector machines. *Mach. Learn.* 48: 115–136.

Schaner M.E., Ross D.T. et al. (2003). Gene expression patterns in ovarian carcinomas. *Mol. Biol. Cell* 14: 4376–4386.

Sollich P. (2002). Bayesian methods for support vector machines: evidence and predictive class probabilities. *Mach. Learn.* 46: 21–52.

Tusher V.G., Tibshirani R. et al. (2001). Significance analysis of microarrays applied to the ionizing radiation response. *Proc. Natl Acad. Sci. USA* 98: 5116–5121.

Vapnik V.N. (1995). *The Nature of Statistical Learning Theory*. New York, Springer-Verlag.

Vapnik V.N. (1998). *Statistical Learning Theory*. New York, Wiley-Interscience.

Weston J., Mukherjee S. et al. (2000). Feature selection for SVMs. In T.K. Leen T.G. Dietterich, and V. Tresp (eds.), *Advances in Neural Information Processing Systems, Vol. 13*. Cambridge, MA, MIT Press, pp. 668–674.

Zhang Z., Page G. et al. (2001). Applying classification separability analysis to microarray data. In S.M. Lin and K.F. Johnson (eds.), *Methods of Microarray Data Analysis: Papers from CAMDA '00*. Boston, Kluwer Academic Publishers, pp. 125–136.

Zhang Z., Page G. et al. (2002). Fishing expedition — a supervised approach to extract patterns from a compendium of expression profiles. In: S.M. Lin and K.F. Johnson (eds.), *Microarray Data Analysis II: Papers from CAMDA '01*. Boston, Kluwer Academic Publishers.

11 Modeling Affymetrix Data at the Probe Level

Tzu-Ming Chu, Shibing Deng, and Russell D. Wolfinger

CONTENTS

11.1 Introduction 197
11.2 Models 200
  11.2.1 Multiplicative Model 200
  11.2.2 Fixed Linear Model 204
  11.2.3 Mixed Linear Model 205
11.3 The Primate Example 208
11.4 Simulation Study 214
11.5 Discussion 217
References 220

11.1 INTRODUCTION

The technology of gene expression arrays, also known as microarrays, is a breakthrough biotechnology developed at the end of the last century that allows researchers to perform a single experiment on thousands of genes simultaneously (Schena et al., 1995; Lockhart et al., 1996). One of the major applications of expression array technology is finding the signature genes characterizing different samples or experimental conditions. The selected signature genes can be applied for further broad investigation, such as characterizing transcriptional activity of the cell cycle (Spellman et al., 1998), clustering cancer subtypes (Golub et al., 1999), studying of evolution among closely related species (Enard et al., 2002; Hsieh et al., 2003), and promoter investigation (Lee et al., 2002). The results of studies using expression array technology enable us to explore the function and interactions of genes on a genome-wide scale. A tremendous amount of data, containing information from thousands of genes, is generated from the expression array studies. To convert the data into statistically significant evidence and then to provide biological meaning are the primary goals of gene expression studies.

There are two major types of microarrays, cDNA microarrays (Schena et al., 1995) and oligonucleotide chips (Lockhart et al., 1996). The analysis of cDNA microarray data is discussed in Chapter 12. Although many statistical concepts can be applied to analyze oligonucleotide chip data, there are some specific characteristics of oligonucleotide chips that should be taken into account. This chapter primarily addresses three popular modeling approaches for oligonucleotide chip data at the probe level, with emphasis on statistical modeling and inferences on testing differentially expressed genes. The three modeling approaches are multiplicative modeling (Li and Wong, 2001a), mixed linear modeling (Chu et al., 2002), and fixed linear modeling (Irizarry et al., 2003). These methods are currently implemented in the software programs dChip, the SAS Microarray Solution, and Bioconductor, respectively.

The Affymetrix GeneChip® is currently the most widely used oligonucleotide expression array technology. The GeneChip® contains a probe set representing unique genes, and each probe set consists of 11–20 probe pairs. Each probe pair consists of a perfect match (PM) oligonucleotide probe, which is designed exactly complementary to a preselected 25mer of the target gene, and a mismatch (MM) probe, which is identical to PM except for one single nucleotide difference at position 13. According to Lockhart et al. (1996), the purpose of the MM probe is to serve as an internal control of hybridization specificity. More details about the design of Affymetrix GeneChips® are discussed in other chapters in this book. Although each oligo within a probe set is designed to interrogate the same gene, there are well known and very strong differences between the performances of individual probes (see Figure 11.2 and Figure 11.7). Statistical analysis and summarization of the probe-level data is therefore a critical challenge that must be addressed in order to effectively assess results from the chips. Affymetrix itself has been responsive by providing a new summary measure based on Tukey's biweight function in the software, MAS 5.0 (Affymetrix, 2001a), accompanying the chips. Although this certainly represents an improvement over their earlier methods (MAS 4.0), by its nature the Affymetrix summary prevents analysts from making their own adjustments for individual probe effects. Evidence shows that the correlations among replicate chips at the probe level are significantly higher than the correlations of signals based on MAS 5.0 (Figure 11.1). Research addressing this concern has arisen from numerous sources, including Teng et al. (1999), Schadt et al. (2000), Efron et al. (2001), Li and Wong (2001a), Lemon et al. (2002), and Irizarry et al. (2003).

A statistically optimal approach for experimental data involving many chips requires that we consider all of the PM and MM data simultaneously. This provides 40 times the data compared to traditional summary methods and gives more power for statistical inference. However, several questions arise regarding the statistical relationships of PM and MM within a probe pair, between probe pairs within a probe set, and between probe sets across arrays. Do they have a linear relationship? Are the amounts of cross-hybridization similar for PM and MM probes? To what extent does MM serve as an internal quality control? Li and Wong (2001a) investigate these and other questions in the context of a multiplicative model for the measurements, whereas Efron et al. (2001) consider scaled logarithms. Lazaridis et al. (2002) suggest using PM information only. Lemon et al. (2002) treat both PM and MM as two independent response variables. Wu et al. (2003) correct cross hybridization from

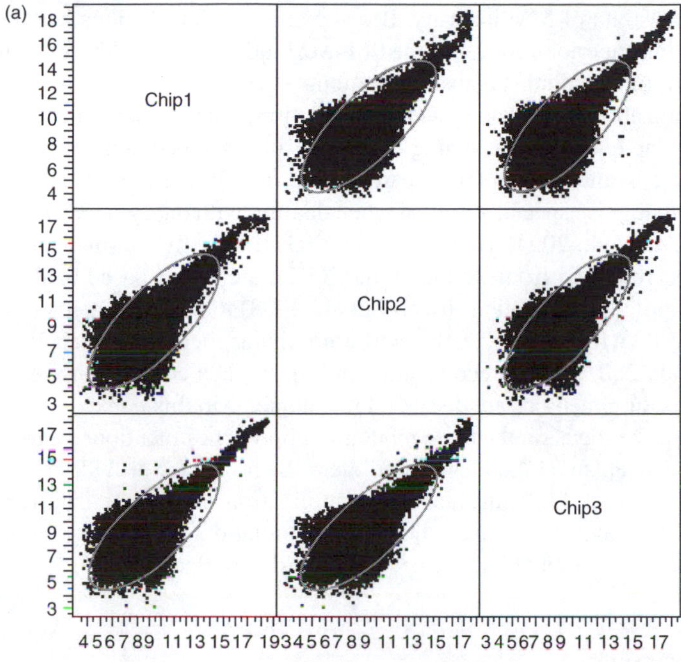

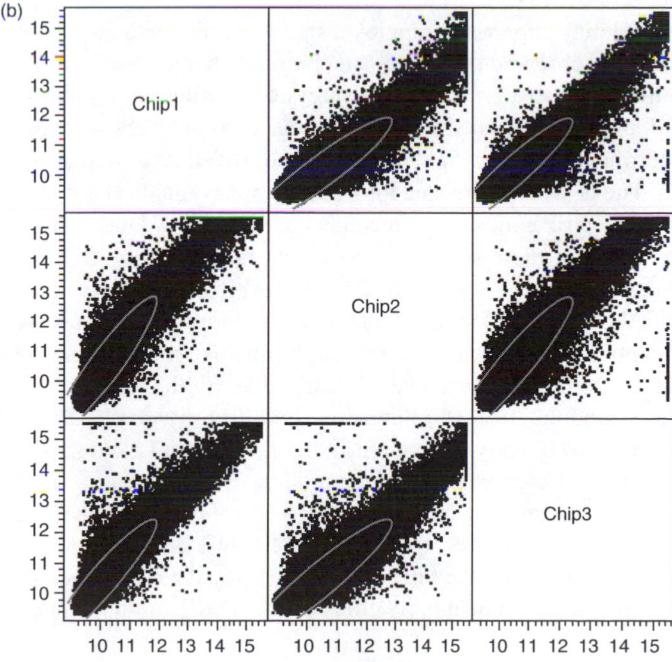

FIGURE 11.1 Scatter plots among three replicate chips (data unpublished) at log (base 2) scale. The top and bottom matrixes present the data at the signal level (from MAS 5.0) and at the probe level (PM only), respectively. The curves are 95% confidence ellipses assuming bivariate normality. The average correlations among replicate chips are 0.75 and 0.93 for signal and probe data, respectively.

GC-number strata of MM intensity. Beasley et al. (2005) demonstrate the PM-only model having superior power than the PM–MM model and the ANCOVA model with MM as covariance in most realistic situation.

Normalization to reduce systematic variation, such as batch difference in production of the chips, and scanning difference in the processing of the chips, prior to modeling is another important part of microarray analysis. This preprocessing before modeling is especially critical when dealing with data generated across different labs (Chu et al., 2004). Bolstad et al. (2003) review five normalization methods: cyclic loess modified from Dudoit et al. (2002), a contrast based method (Astrand, 2001), quantile normalization (Irizarry et al., 2003), the scaling method in Affymetrix MAS 5.0 (Affymetrix, 2001b), and a nonlinear method (Li and Wong, 2001b; Schadt et al., 2001). We forego an attempt to provide a comprehensive review here, as normalization methods are discussed in Chapter 2 in this book.

Instead, we focus on the three modeling approaches mentioned earlier, with the statistical concepts and theories behind them. Section 11.2 provides background on them, and Section 11.3 illustrates their application to primate data (Enard et al., 2002; Hsieh et al., 2003). In Section 11.4, simulation studies for comparing these three models are presented. Discussion is provided in Section 11.5.

11.2 MODELS

Parametric modeling approaches employ a statistical model to quantitatively explain the characteristics observed from data. An essential characteristic of probe level data is probe patterns that display a similar expression profiles across chips. Figure 11.2 presents probe patterns from six spiked-in genes from 14 chips (experiments A to M and experiment Q in the first batch) of the Affymetrix human (U95A) Latin Square data. The data set and related information are available at Affymetrix website, http://www.affymetrix.com/support/technical/sample_data/datasets.affx. Although the concentrations of these spiked-in probe sets range from 0 to 1024 pM and the observed intensities range from 2^6 to 2^{15}, the expression patterns are consistently repeated with the curves within each plot parallel in Figure 11.2. Li and Wong (2001a) describe this phenomenon and propose a multiplicative model to fit the probe level data. Chu et al. (2002) propose a mixed linear model for the $\log_2$ probe level specifying arrays as a random effect (Wolfinger et al., 2001). Irizarry et al. (2003) apply a fixed linear model by directly log-transforming the mean of Li–Wong's multiplicative model. The major differences between the models are the way they model stochastic error terms. The first model considers additive errors whereas the others consider multiplicative errors. Moreover, there are some subtle but important differences in model specifications among the three models. All the three models model one gene at a time, and so the subsequent derivations of the models omit an index for genes.

11.2.1 MULTIPLICATIVE MODEL

Suppose θ_i is the expression index of a gene in the ith sample. Assume that the increase of probe intensity is proportional to θ_i, but the proportions are different for different

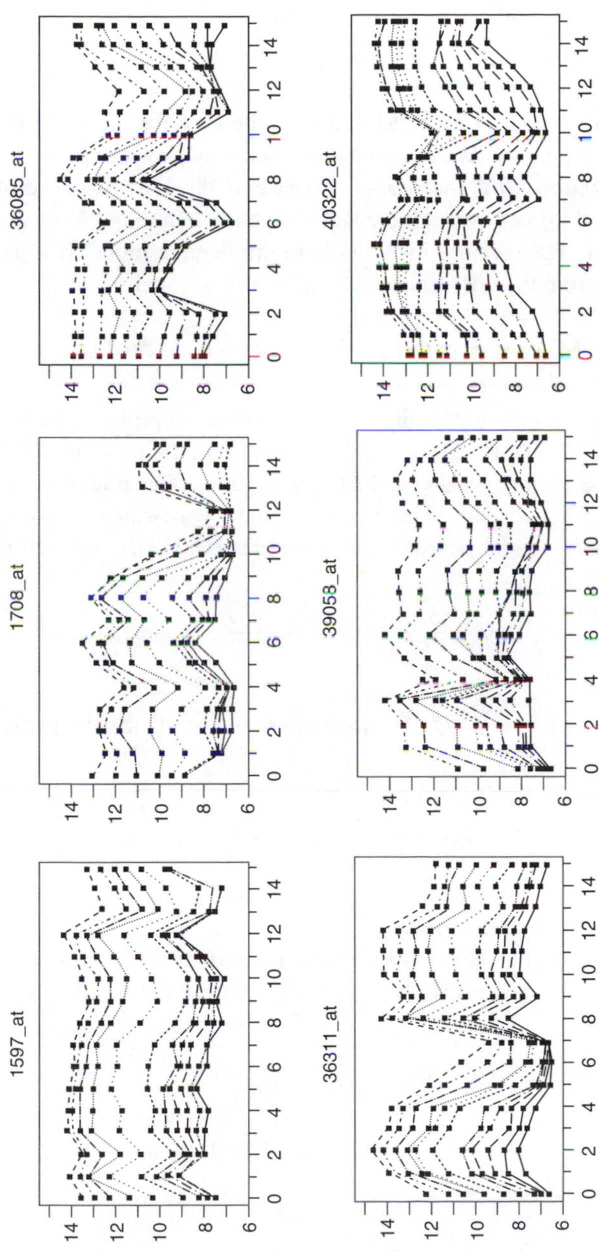

FIGURE 11.2 Probe patterns of six spiked-in probe sets from Affymetrix Human Latin Square data (U95A chips). *X*-axes indicate probe number (0 to 15). *Y*-axes indicate $\log_2$ PM intensity. The probe set identifiers are listed on top of each plot. Each plot contains 14 curves representing the 14 spiked-in concentrations (from 0 to 1024 pM) from the 14 chips.

probes. Also, assume that within a probe pair, the PM intensity increases at a higher rate than the MM intensity. Therefore, we can model PM and MM by the following:

$$\begin{aligned}\mathrm{MM}_{ij} &= \nu_j + \theta_i\alpha_j + \varepsilon_{ij}\\ \mathrm{PM}_{ij} &= \nu_j + \theta_i\alpha_j + \theta_i\phi_j + \varepsilon'_{ij}\end{aligned}$$

Here MM_{ij} and PM_{ij} represent the MM and PM intensity for the ith chip and the jth probe pair for the gene, ν_j is the baseline response of the jth probe pair due to nonspecific hybridization, α_j is the rate of increase of the MM response of the jth probe pair, ϕ_j is the additional rate of increase in the corresponding PM response, and ε random error term. The original Li–Wong's multiplicative model (Li and Wong, 2001a) was derived from the difference of the above two models:

$$Y_{ij} = \mathrm{PM}_{ij} - \mathrm{MM}_{ij} = \theta_i\phi_j + \epsilon_{ij}, \quad \Sigma\phi_j^2 = J, \quad \eta_{ij} \sim N(0, \sigma^2) \tag{11.1}$$

A constraint to make the sum of ϕ_j^2 equal to J, the number of probe pairs, is included to make the model identifiable. Assuming that the ϕ_j remain the same for the same gene in different experiments, the ϕs can be learned from a large number of chips and can be treated as known constants. Therefore, the linear least-squares estimate for θ_i given the ϕs, named the model-based expression index (MBEI), can be estimated by

$$\tilde{\theta}_i = \frac{\sum_j y_{ij}\phi_j}{\sum_j \phi_j^2} = \frac{\sum_j y_{ij}\phi_j}{J}$$

with $E(\tilde{\theta}_i) = \theta_i$ and $\mathrm{var}(\tilde{\theta}_i) = \sigma^2/J$. Then, an approximate standard error is

$$\begin{aligned}\mathrm{SE}(\tilde{\theta}_i) &= \sqrt{\hat{\sigma}_i^2/J}\\ \hat{\sigma}_i^2 &= \sum_j R_{ij}^2/(J-1)\end{aligned}$$

where R_{ij} is the residual from the model.

Li and Wong (2001b) introduce a Q statistic for constructing confidence interval for fold change between two arrays based on the conditional means and standard errors above. Suppose

$$\tilde{\theta}_1 \sim N(\theta_1, \delta_1^2), \quad \tilde{\theta}_2 \sim N(\theta_2, \delta_2^2)$$

where θ_1 and θ_2 are the real expression levels. Let $r = \theta_1/\theta_2$ and

$$Q = \frac{(\tilde{\theta}_1 - r\tilde{\theta}_2)^2}{(\delta_1^2 + \delta_2^2 r^2)}$$

Q can be shown to be a pivotal quantity involving r and has a χ^2 distribution with 1 degree of freedom (Wallace, 1988). Hence a confidence interval can be constructed using Q. By substituting standard error estimates and inverting the confidence interval on Q, a confidence interval of r, the true fold change, can be derived. The same concept can be extended for constructing average fold change of expression between two groups. Furthermore, the significant differentially expressed gene between two groups at a given significance level can be statistically tested based on a fixed-level confidence interval of fold change.

A simulation study in Chu et al. (2003) shows that this Q-test is too liberal. Alternatively, a two-group t-test can be performed by grouping the MBEIs and standard errors of MBEIs based on their associations. Assuming that $\tilde{\theta}_{1i} \sim N(\theta_1, \tilde{\delta}_{1i}^2)$, $i = 1, \ldots, n_1$, are MBEIs for group 1 and $\tilde{\theta}_{2i'} \sim N(\theta_2, \tilde{\delta}_{2i'}^2)$, $i' = 1, \ldots, n_2$, are MBEIs for group 2,

$$\hat{\theta_1} = \frac{\sum_i \tilde{\theta}_{1i}/\tilde{\delta}_{1i}^2}{\sum_i 1/\tilde{\delta}_{1i}^2}$$

and

$$\hat{\theta_2} = \frac{\sum_{i'} \tilde{\theta}_{2i'}/\tilde{\delta}_{2i'}^2}{\sum_{i'} 1/\tilde{\delta}_{2i'}^2}$$

are linear unbiased minimum variance estimates of θ_1 with variance $(\sum_i 1/\tilde{\delta}_{1i}^2)^{-1}$ and θ_2 with variance $\left(\sum_{i'} 1/\tilde{\delta}_{2i'}^2\right)^{-1}$ (Exercises 7.40, Casella and Berger, 1990), respectively. Hence, a two-group t-test with unequal variances (refer to Section 2.2.3 in this book for details) can be performed for testing

$$H_0\colon \theta_1 = \theta_2 \quad \text{vs.} \quad H_\text{A}\colon \theta_1 \neq \theta_2 \tag{11.2}$$

As usual, a Satterthwaite approximation (Rosner, 1995) for degrees of freedom is preferable here.

Li and Wong (2003) propose a test considering measurement accuracy that is similar in concept to include a random array effect in a linear mixed modeling approach (Chu et al., 2002). Assume that array-specific expression levels of a gene in replicated arrays are independently observed and that $\tilde{\theta}_i \sim N(\theta_i, \tilde{\delta}_i^2)$. Instead of a common mean for these $\tilde{\theta}_i$s, array-specific expression is considered as a random variable, $\theta_i \sim N(\mu, \tau^2)$. Considering the total mRNA is split for replication, μ is the expression level in the total mRNA and τ is the variation introduced after replication. In other words, a small random error is accumulated for each individual chip. Therefore, $\tilde{\theta}_i \sim N(\mu, \tau^2 + \tilde{\delta}_i^2)$. The two-group t-test discussed above can be applied here if there is an estimate for the unknown parameter τ^2. Li and Wong (2003) suggest a resampling approach to perturb $\tilde{\theta}_i$ to estimate τ^2. Specifically, resampled MBEIs $\tilde{\beta}_i$s are drawn from $N(\tilde{\theta}_i, \tilde{\delta}_i^2)$ and the sample variance of resampled $\tilde{\beta}_i$s is a "resampled estimate of τ^2," denoted $\tilde{\tau}^2$. After resampling a certain number of times, they apply the average of resampled estimates, $\tilde{\tau}^2$s, as an estimate of τ^2, denoted $\hat{\tau}^2$. This approach was implemented in dChip v1.3 with resampling 20 times.

Instead of this resampling approach, τ^2 can be estimated as follows:

$$\begin{aligned}
\hat{\tau}^2 &= E\left\{\frac{1}{I-1}\sum_i\left(\tilde{\beta}_i - \bar{\tilde{\beta}}_.\right)^2\right\} \\
&= \frac{1}{I-1}\left\{E\left(\sum_i \tilde{\beta}_i^2\right) - E\left(I\bar{\tilde{\beta}}_.^2\right)\right\} \\
&= \frac{1}{I-1}\left\{\sum_i\left(\tilde{\theta}_i^2 + \tilde{\delta}_i^2\right) - \frac{1}{I}E\left(\sum_i \tilde{\beta}_i\right)^2\right\} \\
&= \frac{1}{I-1}\left\{\sum_i\left(\tilde{\theta}_i^2 + \tilde{\delta}_i^2\right) - \frac{1}{I}E\left(\sum_i \tilde{\beta}_i^2 + 2\sum_{i\neq i'}\tilde{\beta}_i\tilde{\beta}_{i'}\right)\right\} \\
&= \frac{1}{I-1}\left\{\sum_i\left(\tilde{\theta}_i^2 + \tilde{\delta}_i^2\right) - \frac{1}{I}\left(\sum_i \tilde{\theta}_i^2 + \sum_i \tilde{\delta}_i^2 + 2\sum_{i\neq i'}\tilde{\theta}_i\tilde{\theta}_{i'}\right)\right\} \\
&= \frac{1}{I-1}\left\{\sum_i\left(\tilde{\theta}_i^2 + \tilde{\delta}_i^2\right) - \left(I\bar{\tilde{\theta}}_.^2 + \frac{1}{I}\sum_i \tilde{\delta}_i^2\right)\right\} \\
&= \frac{1}{I-1}\sum_i\left(\tilde{\theta}_i - \bar{\tilde{\theta}}_.\right)^2 + \frac{1}{I}\sum_i \tilde{\delta}_i^2
\end{aligned}$$

This estimate is the variance of MBEIs (between-chip sample variance) plus their mean of standard errors (average within-chip variance).

11.2.2 Fixed Linear Model

Considering the measurements of intensities to be distributed lognormally with stochastic errors being additive in the log-scale (multiplicative in original scale), leads to the following linear model:

$$Y_{ij} = \mu_i + \alpha_j + \varepsilon_{ij}, \quad \varepsilon_{ij} \sim N(0, \sigma^2) \tag{11.3}$$

where indexes i and j indicate chip number and probe number, respectively, Y_{ij} represents log-transformed response, μ_i represents log scale expression index of chip i, α_j represents an affinity effect of probe j, and ε_{ij} is stochastic error. The μ_i will be called FMEI (fixed model expression index) in this chapter for convenient purpose. For identifiability of parameters, the constraint $\sum_j \alpha_j = 0$ is applied. This model is a classical two-factor ANOVA model and can be presented in matrix notation as follows:

$$\mathbf{Y} = \mathbf{X}\boldsymbol{\beta} + \boldsymbol{\varepsilon}, \quad \boldsymbol{\varepsilon} \sim N(\mathbf{0}, \sigma^2\mathbf{I})$$

$\mathbf{Y}$ is a vector of log-transformed responses, $\mathbf{X}$ is a design matrix, $\boldsymbol{\beta}$ is a parameter matrix, $\boldsymbol{\varepsilon}$ is a vector of random error, $\mathbf{I}$ is an identity matrix with the size equal to the number of all responses applied in the model. The least-squares estimate of $\boldsymbol{\beta}$ is

$$\hat{\boldsymbol{\beta}} = (\mathbf{X}'\mathbf{X})^{-1}\mathbf{X}'\mathbf{Y}$$

with variance

$$\text{var}(\hat{\boldsymbol{\beta}}) = \sigma^2(\mathbf{X}'\mathbf{X})^{-1}$$

To test for significantly different changes in gene expression, the following hypotheses are compared:

$$H_0\colon\ \mu_{(1)} = \mu_{(2)} \quad \text{vs.} \quad H_\text{A}\colon\ \mu_{(1)} \neq \mu_{(2)} \tag{11.4}$$

Under the ANOVA model above, the BLUE (best linear unbiased estimate) of $\mu_{(k)}$ is

$$\hat{\mu}_{(k)} = \frac{1}{\#(A_k)} \sum_{i \in A_k} \hat{\mu}_i = \boldsymbol{\eta}_k' \hat{\boldsymbol{\beta}}$$

where A_k is the subset of the chips in group k, $\#(A_k)$ represents the cardinality of A_k, $\hat{\mu}_i$ is the least-squares estimate of μ_i, $\boldsymbol{\eta}_k$ is a indicator vector of corresponding parameters in $\hat{\boldsymbol{\beta}}$. Therefore, the following statistic can be applied for testing Equation 11.4:

$$\begin{aligned}
&\frac{\hat{\mu}_{(1)} - \hat{\mu}_{(2)}}{\text{SE}(\hat{\mu}_{(1)} - \hat{\mu}_{(2)})} \sim t_d \\
&\hat{\mu}_{(1)} - \hat{\mu}_{(2)} = (\boldsymbol{\eta}_1 - \boldsymbol{\eta}_2)'\hat{\boldsymbol{\beta}} \\
&\text{SE}(\hat{\mu}_{(1)} - \hat{\mu}_{(2)}) = \sigma[(\boldsymbol{\eta}_1 - \boldsymbol{\eta}_2)'(\mathbf{X}'\mathbf{X})^{-1}(\boldsymbol{\eta}_1 - \boldsymbol{\eta}_2)]^{1/2} \\
&d = \#\text{ of responses} - \#\text{ of parameters}
\end{aligned}$$

11.2.3 Mixed Linear Model

In ANOVA-type of analyses, additive effects are specified to partition the total variability of observations. The fixed model previously described considers only effects modeling the mean, whereas the mixed model also allows random effects, which provide a simple method for modeling correlations between observations. Fixed effects are those effects with a well defined, finite number of levels and only these finite levels are of interest in the experiment. For fixed effects, we estimate each level and do testing among all levels or comparisons between levels to see if they are significant. Random effects are those effects considered to be drawn from an infinite population having some probability distribution, usually normal. For random effects, we estimate the parameters of this probability distribution (variance components in the normal case) and possibly also individual effect estimates properly shrunken towards zero. Inclusion of random effects also allows inferences about the fixed effects to be made to broader populations.

For oligonucleotide chip experiments, we typically consider genotype (such as strain and species if applicable), treatment, and probe effects to be fixed, and because of potentially complex experimental sources of variation such as alternative splicing and SNP, it is typically sensible to include two-way interactions of these effects as well.

Effects impacting chips can be considered random, because they are the accumulation of small experimental sources of noise during the preparation of the samples. Putting these all together, the following is a typical mixed model for probe-level data analysis (Chu et al., 2002):

$$Y_{ijkl} = \mu + \mathrm{S}_i + \mathrm{T}_j + \mathrm{ST}_{ij} + \mathrm{P}_k + \mathrm{SP}_{ik} + \mathrm{TP}_{jk} + \mathrm{A}_{l(ij)} + \varepsilon_{ijkl} \tag{11.5}$$

with normality assumptions,

$$\mathrm{A}_{l(ij)} + \varepsilon_{ijkl} \sim N(0, \sigma_a^2 + \sigma^2)$$

$$\mathrm{cov}(\mathrm{A}_{l(ij)} + \varepsilon_{ijkl}, \mathrm{A}_{l'(i'j')} + \varepsilon_{i'j'k'l'}) = \begin{cases} \sigma_a^2 + \sigma^2 & \text{if } (i,j,k,l) = (i',j',k',l') \\ \sigma_a^2 & \text{if } (i,j,l) = (i',j',l') \text{ but } k \neq k' \\ 0 & \text{otherwise} \end{cases}$$

Here, Y_{ijkl} is the log-transformed response of the ith strain with the jth treatment at the kth probe in the lth replicate. The symbols μ, S, T, ST, P, SP, TP and A represent grand mean, strain, treatment, strain by treatment interaction, probe, strain by probe interaction, treatment by probe interaction, and array effects, respectively. The $\mathrm{A}_{l(ij)}$s are assumed to be independent and identically distributed normal random variables with mean 0 and variance σ_a^2. Unlike the fixed linear model, which assumes that all intensities within a probe set are independent, the mixed model considers the probe intensities from the same chip to be correlated. This is reasonable because they interrogate the same gene. The ε_{ijkl}s are assumed to be independent identically distributed normal random variables with mean 0 and variance σ^2, and are independent of the $\mathrm{A}_{l(ij)}$s.

The fixed model in the previous section is a special case of the mixed model here with all interactions involved and probe and array random effect omitted. The mixed model can be easily extended to include more known effects, higher-order interactions, and additional random effects to model different levels of correlation like the one often induced by biological replicates.

The above model can be written in matrix form as follows:

$$Y = X\boldsymbol{\beta} + Z\boldsymbol{\gamma} + \boldsymbol{\varepsilon}$$

$$\begin{bmatrix} \boldsymbol{\gamma} \\ \boldsymbol{\varepsilon} \end{bmatrix} \sim N\left(\begin{bmatrix} \mathbf{0} \\ \mathbf{0} \end{bmatrix}, \begin{bmatrix} \mathbf{G} & \mathbf{0} \\ \mathbf{0} & \mathbf{R} \end{bmatrix} \right)$$

$$\mathbf{V} = \mathbf{ZGZ}' + \mathbf{R}$$

Here, $\mathbf{V}$ is the covariance matrix of $\mathbf{Y}$. The unknown parameters $\boldsymbol{\beta}$, $\boldsymbol{\gamma}$, $\boldsymbol{G}$, and $\boldsymbol{R}$, need to be estimated. When $\boldsymbol{V}$ is known, $\boldsymbol{\beta}$ can be obtained by minimizing the generalized least squares (GLS) criterion,

$$(\mathbf{Y} - \mathbf{X}\beta)'\mathbf{V}^{-1}(\mathbf{Y} - \mathbf{X}\beta)$$

When $\mathbf{V}$ is unknown, one can minimize the estimated GLS, in which some reasonable estimate for $\mathbf{V}$ is inserted. Therefore, finding a reasonable estimate of $\mathbf{V}$ ($\mathbf{G}$ and $\mathbf{R}$)

becomes the first goal here. A rich history of solutions to this problem is available in the statistics literature (Hartley and Rao, 1967; Patterson and Thompson, 1971; Harville, 1977; Laird and Ware, 1982; Jennrich and Schluchter, 1986), and most concludes that maximum likelihood (ML)-based methods are the best in many situations. ML and restricted/residual maximum likelihood (REML) are two typical likelihood-based methods (Wu et al., 2001). These two methods maximize the following two corresponding log likelihood functions:

$$\text{ML:}\ l(\mathbf{G},\mathbf{R}) = -\frac{1}{2}\log|\mathbf{V}| - \frac{1}{2}\mathbf{r}'\mathbf{V}^{-1}\mathbf{r} - \frac{n}{2}\log(2\pi)$$

$$\text{REML:}\ l_R(\mathbf{G},\mathbf{R}) = -\frac{1}{2}\log|\mathbf{V}| - \frac{1}{2}\log|\mathbf{X}'\mathbf{V}^{-1}\mathbf{X}| - \frac{1}{2}\mathbf{r}'\mathbf{V}^{-1}\mathbf{r} - \frac{n-p}{2}\log(2\pi)$$

Here $\mathbf{r} = \mathbf{Y} - \mathbf{X}(\mathbf{X}'\mathbf{V}^{-1}\mathbf{X})^{-}\mathbf{X}'\mathbf{V}^{-1}\mathbf{Y}$, n is the size of Y and p is the rank of X. Newton–Raphson can be applied to maximize these log likelihood functions. Once estimates of $\mathbf{G}$ and $\mathbf{R}$, denoted $\hat{G}$ and $\hat{R}$, are provided, solving the mixed model equations (Henderson, 1984) provides estimates of $\boldsymbol{\beta}$ and $\boldsymbol{\gamma}$:

$$\begin{bmatrix} \mathbf{X}'\hat{\mathbf{R}}^{-1}\mathbf{X} & \mathbf{X}'\hat{\mathbf{R}}^{-1}\mathbf{Z} \\ \mathbf{Z}'\hat{\mathbf{R}}^{-1}\mathbf{X} & \mathbf{Z}'\hat{\mathbf{R}}^{-1}\mathbf{Z} + \hat{\mathbf{G}}^{-1} \end{bmatrix} \begin{bmatrix} \hat{\beta} \\ \hat{\gamma} \end{bmatrix} = \begin{bmatrix} \mathbf{X}'\hat{\mathbf{R}}^{-1}\mathbf{Y} \\ \mathbf{Z}'\hat{\mathbf{R}}^{-1}\mathbf{Y} \end{bmatrix}$$

The solutions can also be written as follows:

$$\hat{\boldsymbol{\beta}} = (\mathbf{X}'\hat{\mathbf{V}}^{-1}\mathbf{X})^{-}\mathbf{X}'\hat{\mathbf{V}}^{-1}\mathbf{Y}$$

$$\hat{\boldsymbol{\gamma}} = \hat{\mathbf{G}}\mathbf{Z}'\hat{\mathbf{V}}^{-1}(\mathbf{Y} - \mathbf{X}\hat{\boldsymbol{\beta}})$$

Here the minus sign ($^{-}$) denotes a generalized inverse (Searle, 1971). If $\mathbf{G}$ and $\mathbf{R}$ are known, $\hat{\boldsymbol{\beta}}$ is the BLUE of $\boldsymbol{\beta}$, and $\hat{\boldsymbol{\gamma}}$ is the best linear unbiased predictor (BLUP) of $\boldsymbol{\gamma}$ (Searle, 1971; Harville, 1988, 1990; McLean et al., 1991; Robinson, 1991). The estimated covariance matrix of $(\hat{\boldsymbol{\beta}}, \hat{\boldsymbol{\gamma}})$ is

$$\mathbf{C} = \begin{bmatrix} \mathbf{X}'\mathbf{R}^{-1}\mathbf{X} & \mathbf{X}'\mathbf{R}^{-1}\mathbf{Z} \\ \mathbf{Z}'\mathbf{R}^{-1}\mathbf{X} & \mathbf{Z}'\mathbf{R}^{-1}\mathbf{Z} + \mathbf{G}^{-1} \end{bmatrix}^{-}$$

Since $\mathbf{R}$ and $\mathbf{G}$ are usually unknown, an estimate, $\hat{\mathbf{C}}$ of $\mathbf{C}$ is obtained by applying $\hat{\mathbf{R}}$ and $\hat{\mathbf{G}}$ for $\mathbf{R}$ and $\mathbf{G}$. Under this situation, the appropriate acronyms become EBLUE and EBLUP. Here "E" represents "empirical."

Let $\boldsymbol{\lambda} = (\boldsymbol{\beta}, \boldsymbol{\gamma})$. For testing differential expression, a wide variety of hypotheses can be tested by setting up an appropriate $\mathbf{L}$ matrix.

$$H_0\text{: } \mathbf{L}\boldsymbol{\lambda} = \mathbf{0} \quad \text{vs.} \quad H_{\text{A}}\text{: } \mathbf{L}\boldsymbol{\lambda} \neq \mathbf{0} \tag{11.6}$$

Typically, when comparing expressions between two groups, the $\mathbf{L}$ will be a single row with 1s and -1s for the corresponding elements of group 1 and group 2, respectively, and 0s for the rest. For example, to compare strain group1 with strain group 2, the

L can be set equal to a horizontal vector with the second element 1, the third element -1, and 0 for the rest, $(0 \ 1 \ -1 \ 0 \cdots 0)$, assuming the second and third parameters in $\boldsymbol{\lambda}$ are S_1 and S_2.

The following F statistic can be used for testing Equation 11.6:

$$F = \boldsymbol{\lambda}\mathbf{L}'(\mathbf{L}'\hat{\mathbf{C}}\mathbf{L})^{-1}\mathbf{L}\boldsymbol{\lambda}/\text{rank}(\mathbf{L})$$

F has an approximate F-distribution, with rank(**L**) numerator degree of freedom and $\hat{\nu}$ denominator degree of freedom. $\hat{\nu}$ can be estimated by different approaches (McLean and Sanders, 1988; Stroup, 1989). While it is possible to use advanced repeated measures adjustments for certain designs (Catellier and Muller, 2000) or Satterthwaite-type methods (McLean and Sanders, 1988; Kenward and Roger, 1997), we use a straightforward containment method (SAS/STAT Proc Mixed documentation, 1999) for the analyses below. When **L** is a single row, the square root of F is a t-statistic:

$$t = \frac{\mathbf{L}\boldsymbol{\lambda}}{(\mathbf{L}\hat{\mathbf{C}}\mathbf{L}')^{1/2}} \sim t_{\hat{\nu}}$$

In the multiplicative and linear fixed modeling approaches, the expression indexes, $\tilde{\theta}_i$ and $\hat{\mu}_i$ in the two corresponding models in the previous two sections, can be estimated directly. Similarly, the mixed model expression index (MMEI) can be derived by summing up the corresponding estimates (e.g., $\hat{S}_i + \hat{T}_j + S\hat{T}_{ij} + \hat{A}_{l(ij)}$) from the model.

11.3 THE PRIMATE EXAMPLE

In genetic evolution, chimpanzees and orangutans are two sibling relatives of humans. Among their DNA sequences, the average sequence divergences are about 1.24%, 3.08%, and 3.12% for human–chimpanzee, human–orangutan, and chimpanzee–orangutan (Chen and Li, 2001), respectively. However, there are obvious differences in morphology, physiology, and cognitive abilities between these species. Many of these differences may be due to quantitative differences in transcription (King and Wilson, 1975). One of the most interesting tasks in evolutionary genetics is applying gene expression to identify genes associated with evolutionary selection.

Data from http://email.eva.mpg.de/~khaitovi/supl1/affymetrix.html contain probe data (stored in CEL files) from 28 chips. The 7 experimental subjects include 3 adult male chimpanzees, 3 adult male humans, and 1 adult male orangutan. Duplicate samples of both brain and liver tissues were taken from each subject. In total, 28 tissue samples were utilized in 28 chips. Refer to Enard et al. (2002) for details about the samples. The data were generated for comparing the rates of divergence between brain and liver among the three species (especially interested in human–chimpanzee comparison) based on an evolutionary tree constructed by neighbor-joining method on a distance matrix based on gene expression data at the gene level (signal indexes, average differences of PM–MM, generated from Affymetrix MAS 4.0). The major finding of Enard et al. (2002) is that the branch from the central node to the joining node of the three humans is almost twice as long as the relative branch on the liver tree. This leads to their conclusion that gene expression had diverged more rapidly in human brain.

It can be expected that the more genes that are significantly differentially expressed, the greater the divergence. Based on this, we investigate the number of significantly different genes between brain and liver among these species to see which tissue makes these species more divergent. We apply the three probe-level modeling approaches for finding genes significantly differentially expressed in brain and liver between human and chimpanzee. Specifically, the following comparable hypotheses will be tested.

- Hypotheses for the multiplicative model:
  1. H_0: $\theta_{\mathrm{HB}} = \theta_{\mathrm{CB}}$ vs. H_A: $\theta_{\mathrm{HB}} \neq \theta_{\mathrm{CB}}$
  2. H_0: $\theta_{\mathrm{HL}} = \theta_{\mathrm{CL}}$ vs. H_A: $\theta_{\mathrm{HL}} \neq \theta_{\mathrm{CL}}$

  Here, θ_{HB}, θ_{CB}, θ_{HL}, and θ_{CL} represent the mean expression indexes of a gene from human brain, chimpanzee brain, human liver, and chimpanzee liver, respectively.
- Hypotheses for the fixed linear model:
  1. H_0: $\mu_{\mathrm{HB}} = \mu_{\mathrm{CB}}$ vs. H_A: $\mu_{\mathrm{HB}} \neq \mu_{\mathrm{CB}}$
  2. H_0: $\mu_{\mathrm{HL}} = \mu_{\mathrm{CL}}$ vs. H_A: $\mu_{\mathrm{HL}} \neq \mu_{\mathrm{CL}}$

  Here, μ_{HB}, μ_{CB}, μ_{HL}, and μ_{CL} represent the mean of $\log_2$ expression indexes of a gene from human brain, chimpanzee brain, human liver, and chimpanzee liver, respectively.
- Hypotheses for the mixed linear model:
  1. H_0: $S_{\mathrm{H}} + \mathrm{ST}_{\mathrm{HB}} = S_{\mathrm{C}} + \mathrm{ST}_{\mathrm{CB}}$ vs. H_A: $S_{\mathrm{H}} + \mathrm{ST}_{\mathrm{HB}} \neq S_{\mathrm{C}} + \mathrm{ST}_{\mathrm{CB}}$
  2. H_0: $S_{\mathrm{H}} + \mathrm{ST}_{\mathrm{HL}} = S_{\mathrm{C}} + \mathrm{ST}_{\mathrm{CL}}$ vs. H_A: $S_{\mathrm{H}} + \mathrm{ST}_{\mathrm{HL}} \neq S_{\mathrm{C}} + \mathrm{ST}_{\mathrm{CL}}$

  Here, S_{H} and S_{C} represent the species main effects of human and chimpanzee, respectively. $\mathrm{ST}_{\mathrm{HB}}$, $\mathrm{ST}_{\mathrm{BC}}$, $\mathrm{ST}_{\mathrm{LH}}$, and $\mathrm{ST}_{\mathrm{LC}}$ represent the four levels of the species by tissue interactions corresponding to the subscript letters.

The 28 chips consisted of 14 U95A chips for the brain tissues and 14 U95Av2 chip for the liver tissues. Most of the probes in U95A and U95Av2 chips are the same, except for a few probe sets (26 probe sets in U95A and 25 probe sets in U95Av2), which were excluded from the analyses here. In total, 12,600 probe sets were analyzed. Prior to modeling, the average intensities across the 28 chips, excluding the lowest and highest 1% of the data, are applied as baseline for loess normalization. The three models in Section 11.2 then fit the normalized data and the above hypotheses are tested.

After fitting the models, the associated signal indexes are constructed. A scatter plot matrix in Figure 11.3 compares $\log_2$ transformed MBEI (log2_MBEI), FMEI, and MMEI. The boxes indicate the correlation coefficients based on 352,800 (12600×28) expression indexes. The three expression indexes are highly consistent with each other with correlation coefficients 0.908, 0.906, and 0.997 for pairs (log2_MBEI, FMEI),

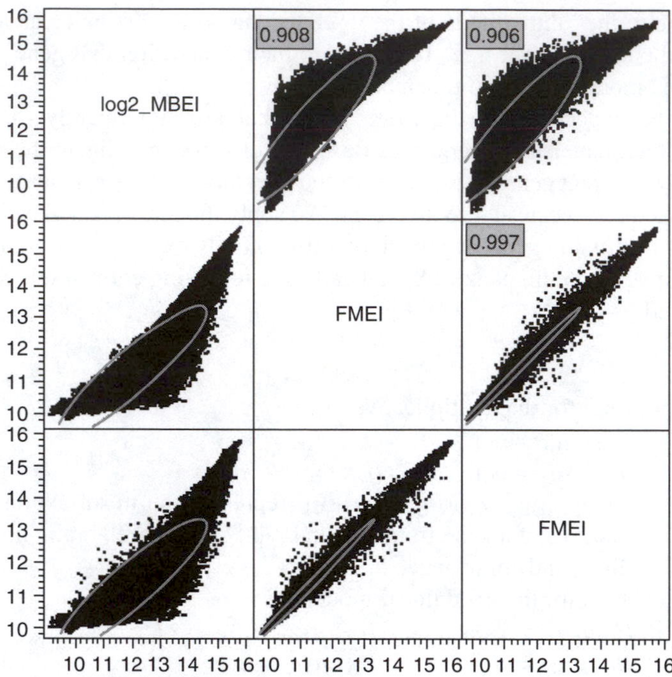

FIGURE 11.3 Scatter plots of signal indexes. Log2_MBEI represents the $\log_2$ transformed MBEI. The boxes indicate the correlation coefficients. The curves are 95% confidence ellipses assuming bivariate normality.

(log2_MBEI, MMEI), and (FMEI, MMEI), respectively. Moreover, log2_MBEI tends to be larger than FMEI and MMEI.

With such high correlation coefficients, it may be expected that the testing significances (*p*-values) for the hypotheses are also consistent among these three modeling approaches. However, the significances are actually much less consistent as shown by the volcano plot and scatter plot matrix comparisons in Figure 11.4 and Figure 11.5. Volcano plots compare estimates (i.e., the numerator of the test statistic) vs. the associated negative $\log_{10}$ *p*-values for all genes. Since larger estimates tend to have higher significance, the plot usually has a "V" shape. The horizontal red line in the plots indicates the Bonferroni cutoff (5.792) computed over all comparisons. As presented in volcanoes, the fixed linear model selects many more significant genes than the other two models do.

In Figure 11.6, the correlation coefficients among the negative $\log_{10}$ *p*-values are 0.537, 0.636, and 0.688 for comparing human to chimpanzee brain tissues and 0.468, 0.565, and 0.699 for comparing human to chimpanzee liver tissues. Even with the correlation of FMEI and MMEI being as high as 0.997, the test significances based on the two associated models only have correlation slightly less than 0.7. The multiplicative model has less agreement of significances to the other two models.

In testing for differentially expressed genes based on the three models, three sets with various numbers of genes are selected for both brain and liver tissues.

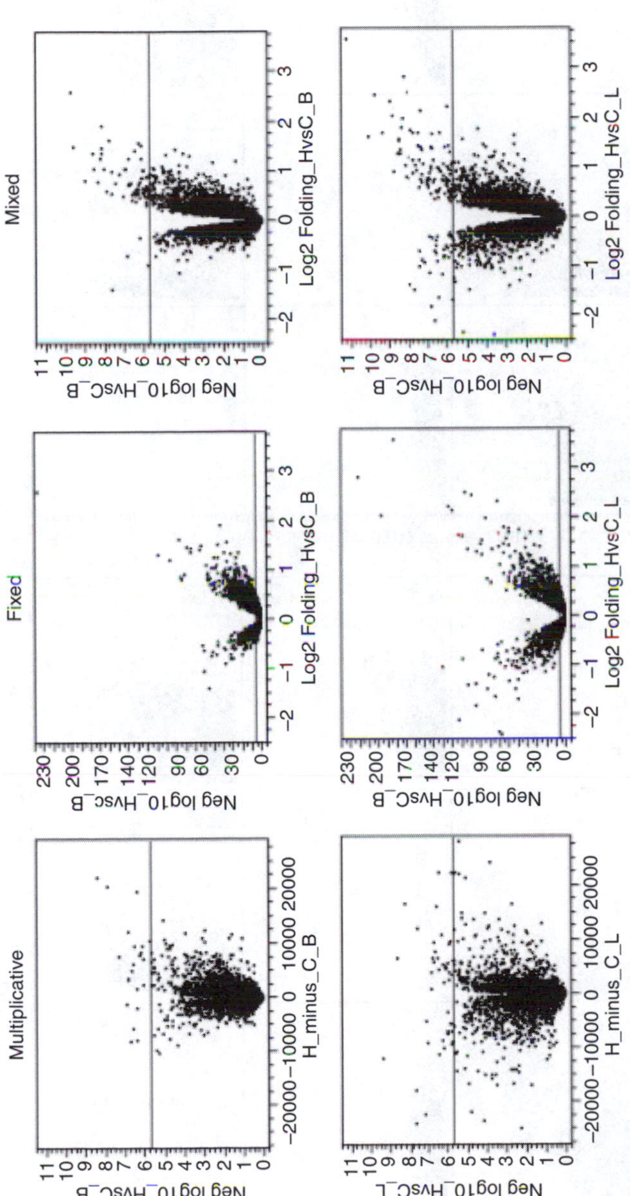

FIGURE 11.4 Volcano plots. The top and bottom plots present the results from testing the first and second hypotheses, respectively, under each model. *X*-axes indicate the estimated mean difference between human and chimpanzee under the multiplicative model and the estimated $\log_2$ fold change between human and chimpanzee under the other two models. *Y*-axes indicate $-\log_{10}$ *p*-value. The horizontal lines indicate the Bonferroni's cutoff (5.702).

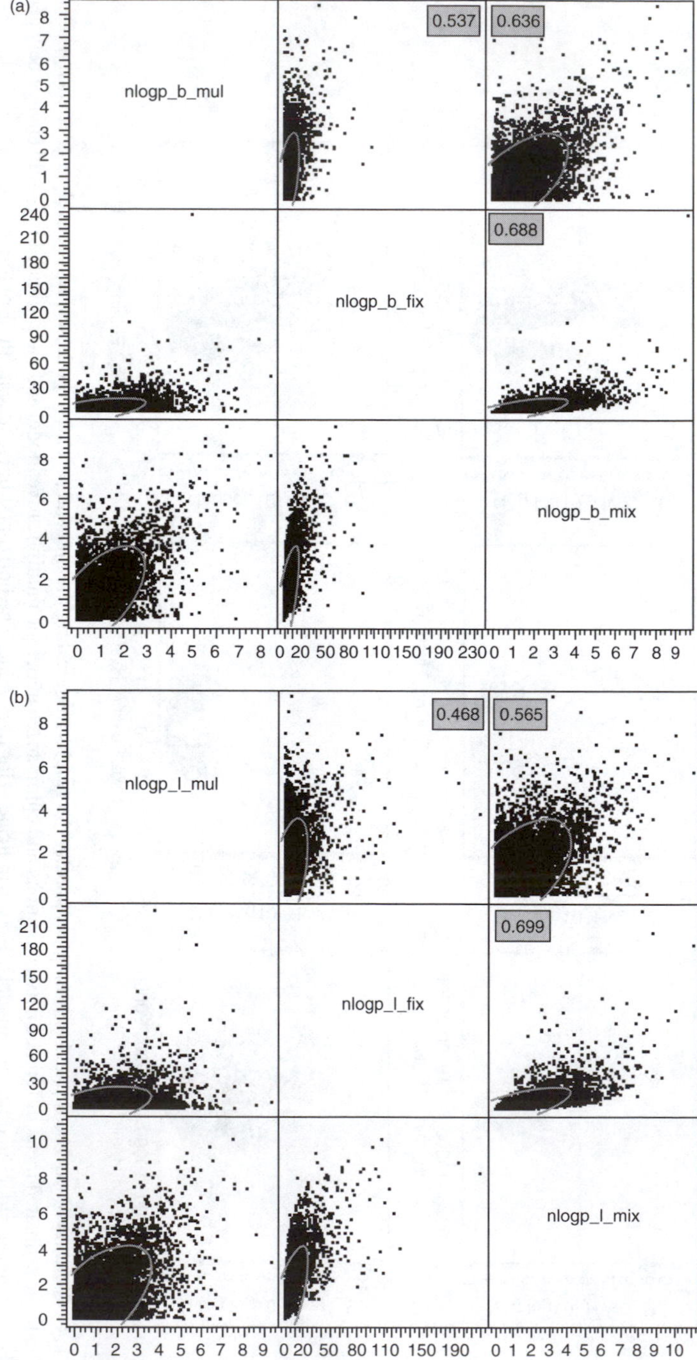

FIGURE 11.5 Scatter plots among $-\log_{10}$ *p*-values. (a) *p*-values generated from comparing human brain tissue to chimpanzee brain tissue. (b) *p*-values generated from comparing human liver tissue to chimpanzee liver tissue.

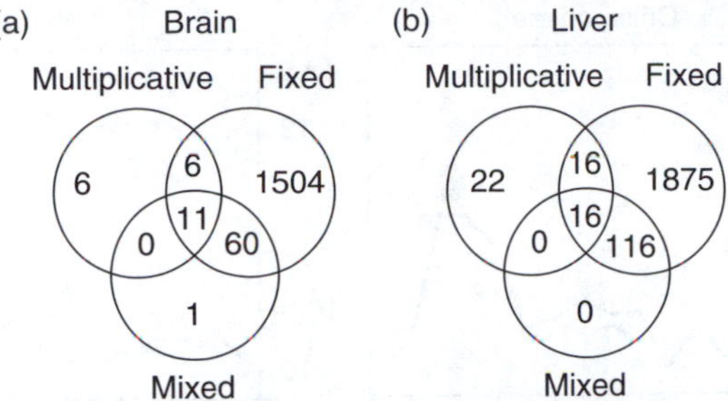

FIGURE 11.6 Number of significant genes from comparing human and chimpanzee in brain (a) and liver (b) tissues among the three models. The numbers listed within the circles represent the corresponding number of significant genes based on 5.702 (the Bonferroni cutoff for $-\log_{10}(p\text{-values})$).

Figure 11.6 presents Venn diagrams of the number of significant genes selected based on Bonferroni's adjustment for comparing human and chimpanzee in brain (a) and liver (b) tissues among the three models. The multiplicative model selects the least number of significant genes whereas the fixed linear model selects many more significant genes. The genes selected by the mixed linear model are almost selected by the fixed linear model, but they are not all in the top rank from the fixed linear model. However, it is agreed by the three models that the number of significantly different genes in liver is larger than the number of significantly different genes in brain. Therefore, the three approaches tend to favor that the gene expression divergence between human liver and chimpanzee liver is greater than that between human brain and chimpanzee brain. This is a substantially different conclusion from Enard et al. (2002). Refer to Hsieh et al. (2003) for more discussion.

A key reason for the fixed linear model selecting many more significant genes is the degrees of freedom of the test statistic. When the mixed linear model is used for balanced split plot designs (Chu et al., 2002) with each probe set within a chip as the whole-plot unit and each individual probe as the subplot unit, the degrees of freedom equal to the number of chips minus the total degrees of freedom of whole plot effects in the model. Refer to Littell et al. (1996) and Steel et al. (1997) for the details of split plot designs. The degrees of freedom of whole plot effects in the mixed model include 1 for the grand mean, 2 for the species effect, 1 for the tissue effect, 2 for the species by tissue interaction, and 14 (rank of $\mathbf{Z}$ matrix) for the array random effect. Therefore, the degrees of freedom of the t-test applied for the mixed linear model is 8. The Satterthwaite degrees of freedom for the multiplicative model are mostly between 7 and 10. In contrast, the fixed linear model considers all intensities to be independent and the degrees of freedom are the number of observations applied minus the total degrees of freedom in model. This is typically 28 (arrays) $\times$ 16 (probes) $-$ 1 (grand mean) $-$ 27 (μ, FMEI) $-$ 15 (α, affinity effect of probe) $=$ 405. With such large degrees of freedom, the t-distribution is almost the same as a normal distribution and the t-test is very similar to a z-test. This leads the fixed linear model to having smaller p-values in general.

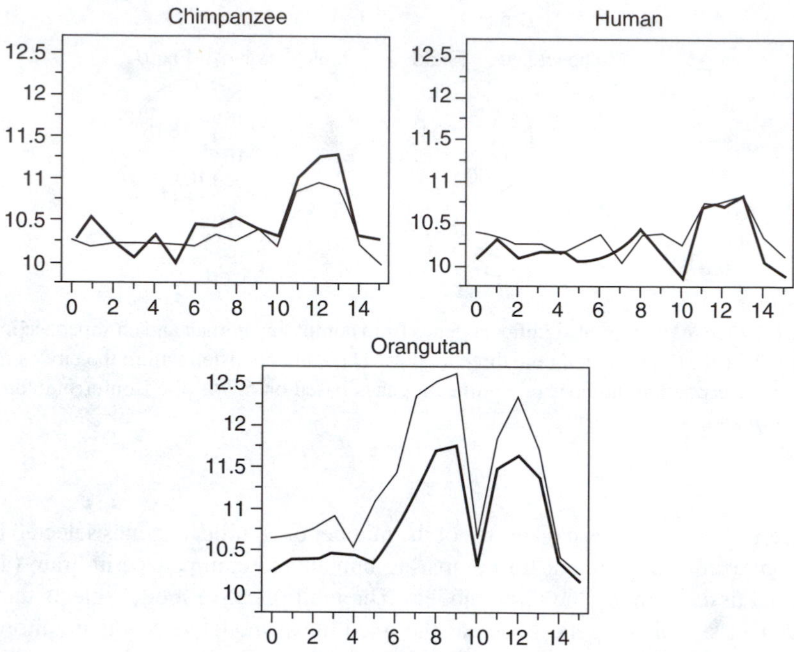

FIGURE 11.7 Expression profile of probe set 36622_at (AI989422) from six brain chips of three subjects. X-axes indicate probe identifier (from 0 to 15). Y-axes indicate $\log_2$ intensities of PM probes.

Among the three modeling approaches, only the mixed linear model incorporates interactions involving probes, such as species by probe and tissue by probe interaction for the primate data. When diversity of DNA sequence among experimental subjects is a concern, including interactions involving probes in the model specification can be useful. In the primate data, 57% of the probe sets (7,209 out of 12,600) have significant species by probe interaction, even under the strict Bonferroni cutoff of 5.7 for $-\log_{10} p$-values. Figure 11.7 presents an example probe set 36622_at (AI989422) in brain tissues among the three experimental subjects (with two replicates in each of chimpanzee, human, and orangutan). The expression pattern of PM probes from brain tissue of orangutan is different from the ones of chimpanzee and human. The pattern of MM probes and paired differences also show similar results (data not shown). A previous study (Chu et al., 2002) reported that 9% of probe sets exhibit a significant strain by probe interaction in the ionizing radiation response data from Tusher et al. (2001).

11.4 SIMULATION STUDY

Even when MBEI, FMEI, and MMEI highly agree with each other with paired correlations larger than 0.9, the discrepancies between the sets of significant genes

TABLE 11.1
Design of Simulated Data

	Control					
Array	1	2	3	4	5	6
Strain	**A**	**A**	**A**	**B**	**B**	**B**
Replicate	1	2	3	1	2	3
	Treated					
Array	7	8	9	10	11	12
Strain	**A**	**A**	**A**	**B**	**B**	**B**
Replicate	1	2	3	1	2	3

selected by the three modeling approaches are large. To further investigate this phenomenon, simulation studies were conducted by simulating data under each model and performing cross validation from fitting all models. Probe set 200772_x_at (BF686442) of 18 U133A chips from a previous study (unpublished) was randomly selected and applied to fit the three models for extracting the parameters to be used in the simulations. The 18 U133A chips were applied for 3 treatments in 2 strains with 3 replicate, which is a two-factor factorial design as the primate data. The template mixed linear model is applicable as symbols S and T represent strain and treatment effect, respectively. With all the parameters in the three models being estimated, they are used to create three simulated data sets with the design setting in Table 11.1.

More specifically, the real data of probe set 200772_x_at was used to fit model 11.1 and obtain estimates of (θ_i's, ϕ_j's, σ^2, τ^2), to fit model 11.3 and obtain estimates of (μ_i's, α_j's, and σ^2), and to fit model 11.5 and obtain estimates of (μ, S_i, T_j, ST_{ij}, P_k, SP_{ik}, TP_{jk}, σ^2, σ_a^2). We then used the following formulas to create simulated data sets of the three simulated scenarios:

1. Assuming the multiplicative model is the true model:

$$\mathrm{PM}_{ij}(t) = \begin{cases} \mathrm{Normal}_i(\bar{\theta}, \hat{\tau}^2) \times \hat{\phi}_j + \mathrm{Normal}_{ijt}(0, \hat{\sigma}^2), & \text{if } j \leq 6 \\ \mathrm{Normal}_i(\delta \times \bar{\theta}, \hat{\tau}^2) \times \hat{\phi}_j + \mathrm{Normal}_{ijt}(0, \hat{\sigma}^2), & \text{if } j > 6 \end{cases}$$

$$\bar{\theta} = \frac{1}{12} \sum \tilde{\theta}_{ijl}$$

$$\delta = 1, 1.05, 1.1, 1.2, 1.3, 1.4, 1.5, 1.6, 1.8, 2$$

$$t = 1, \ldots, 1000$$

2. Assuming the fixed linear model is the true model:

$$\log_2(\mathrm{PM}_{ij}(t)) = \bar{\mu} + \log_2(\delta) \times I_{\{j>6\}} + \hat{\alpha}_k + \mathrm{Normal}_{ijt}(0, \hat{\sigma}^2)$$

$$\bar{\mu} = \frac{1}{12} \sum_i \hat{\mu}_i$$

$$\delta = 1, 1.05, 1.1, 1.2, 1.3, 1.4, 1.5, 1.6, 1.8, 2$$

$$t = 1, \ldots, 1000$$

3. Assuming the mixed linear model is the true model:

$$\log_2(\mathrm{PM}_{ijkl}(t)) = \hat{\mu} + \log_2(\delta) \times I_{\{j \in A_2\}} + \hat{P}_k + \mathrm{Normal}_{l(ij)t}(0, \hat{\sigma}_a^2) + \mathrm{Normal}_{ijklt}(0, \hat{\sigma}^2)$$

$$\delta = 1, 1.05, 1.1, 1.2, 1.3, 1.4, 1.5, 1.6, 1.8, 2$$

$$t = 1, \ldots, 1000$$

A_2 is the subset of treated group.

Here δ indicates the assumed fold change for the treated group and was allowed to range from 1 to 2 to create 10 simulated cases under each scenario. The corresponding parameters are listed in Table 11.2. Under each simulated case, 1000 arrays are created with the given embedded fold change between treated and control groups and no change between strain groups. Therefore, testing the strain effects allows us examine how well the tests control Type 1 error (i.e., the false positive rate). Testing the treatment effects allows us to simultaneously examine power. The nominal significance level for the tests is set to 0.05. Under each simulated scenario, the simulated data were fitted by each of the three models and the results are presented in Figure 11.8.

The dramatic result from this simulation study is that the t-test from the fixed linear model is radically too liberal when wrongly specified. This represents a clear danger in failing to model appropriate interactions and correlations and provides strong evidence that the much larger number of significant genes for the fixed linear model from Figure 11.6 is in fact due to an excessive number of false positives. It also invalids the power comparison for this model in the lower left and lower right graphs. In contrast, the t-test (adjusted for array random effect) from the multiplicative model is too conservative under all the three scenarios. This could be caused by an overestimate of array variability, $\hat{\tau}^2$. The t-test from the mixed linear model controls false positive rate well under all the three scenarios and is as powerful as the t-test from the fixed linear model under Scenario II (without array random effect embedded in simulated data). Based on these simulation results, when selecting differentially expressed genes, the multiplicative model tends to be conservative, the fixed linear model tends to be liberal, and the mixed model is on target with power as-good-as or better than the other two methods, even under incorrect specification.

TABLE 11.2
Parameters Applied in Simulation Study

	Equation		
	11.1	**11.3**	**11.5**
Parameters			
$\bar{\theta}$	3765.83		
$\bar{\mu}$		11.7706	
$\hat{\mu}$			11.7706
Probe parameters (ϕ, α, P)			
$\hat{\phi}_1$	0.5217	−0.6536	−0.6536
$\hat{\phi}_2$	0.5198	−0.8276	−0.8276
$\hat{\phi}_3$	1.1640	0.5164	0.5164
$\hat{\phi}_4$	1.0396	0.3584	0.3584
$\hat{\phi}_5$	1.8001	1.1544	1.1544
$\hat{\phi}_6$	0.4962	−0.7166	−0.7166
$\hat{\phi}_7$	0.8386	0.0004	0.0004
$\hat{\phi}_8$	0.9731	0.2804	0.2804
$\hat{\phi}_9$	0.5999	−0.4356	−0.4356
$\hat{\phi}_{10}$	0.5721	−0.5346	−0.5346
$\hat{\phi}_{11}$	1.4825	0.8584	0.8584
Random effects			
$\hat{\tau}$	881.1552		
$\hat{\sigma}_a$			0.3607
$\hat{\sigma}$	604.5715	0.4745	0.3083

11.5 DISCUSSION

High quality data is essential for reliable analytical results from Affymetrix data. The variation among probes within a probe set is significantly higher than the variation among chips. The expression pattern of probes is reliable and repeatable and at a much finer level of granularity than the summarized signal data from MAS 5.0. We consider three statistical models for probe-level data here. The multiplicative model (Li and Wong, 2001) fits the data in its original scale with additive stochastic error term whereas the fixed (Irizarry et al., 2003) and mixed linear models (Chu et al., 2002) fit the data on the log scale with additive stochastic error term (i.e., multiplicative in the original scale).

The mixed linear model adds complexity to the fixed linear model by including probe-involved interactions and array random effects to account for the correlation among probes within a probe set. Putting probe by strain (or species) interaction can be very useful to help adjust for sequence diversity among experimental subjects. The primate data gives a good example to this point. The multiplicative model does not

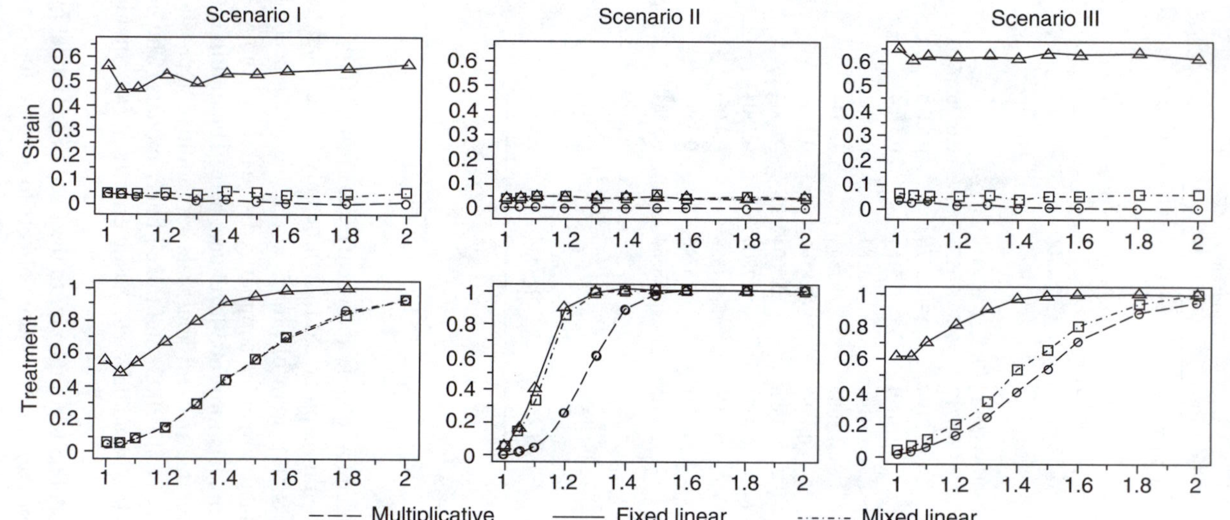

FIGURE 11.8 Comparison of simulation results for the three scenarios. The top and bottom plots represent the results of testing strain (simulated false positive rate) and treatment (simulated testing power) effects, respectively. The simulated data from the three scenarios are generated based on the multiplicative (Scenario I), fixed linear (Scenario II), and mixed linear (Scenario III) models. The dashed, solid, and dashed dots curves represent the results from fitting the multiplicative (11.1), fixed linear (11.3), and mixed linear (11.5) models, respectively.

have array random effects in its model specification but it adjusts for this effect during hypothesis testing. The fixed linear model neither includes array random effect in the model nor adjusts it in hypothesis testing.

The three models produce highly correlated expression indexes (MBEI, FMEI, and MMEI). More comparison of expression indexes can be found on Affycomp, a benchmark for Affymetrix GeneChip expression measures (http://affycomp.biostat.jhsph.edu/). The comparisons in Affycomp are totally based on expression indexes themselves without the associated standard errors calculated at the probe level. According to the concept of sufficient statistics, standard errors contain important information under the proposed model and should be taken into account for statistical inferences. By doing so, the three models behave strikingly different in selecting significant genes from hypothesis testing.

The primate data are used for the example here to investigate the diversities of brain and liver between human and chimpanzee. The fixed linear model selects many more significant genes than the other two models do; however, the simulation study reveals that this is likely due to an excessive number of false positives. The multiplicative model selects the fewest significant genes with certain proportion of them different from the other two selected. The Venn diagrams (Figure 11.6) indicate the significant genes selected by the mixed linear model are almost selected by the fixed linear model; however, they are not all in the top rank of the significant genes set from the fixed linear model.

Considering arrays to be a random effect and adjusting the test statistics appropriately, the degrees of freedom in the adjusted t-test from the multiplicative model are close to the degrees of freedom in the t-test from the mixed linear model in the example of the primate data. The t-test in the fixed linear model has very large degrees of freedom and is virtually a z-test. Despite the difference of the significant genes selected, the three models all select more significant gene from comparing human and chimpanzee liver tissues than from comparing human and chimpanzee brain tissues. This suggests that liver genes may have diverged more than brain genes between human and chimpanzee; probably due to environmental factors such as diet.

A simulation study investigates the discrepancies between hypothesis tests among the three models. The simulated scenarios are created assuming each model is the true model in turn. A hypothetical experimental design is applied with real treatment effects and no strain effect. With these settings, control of false positive rate can be inspected by testing the strain effect and power can be simultaneously examined by testing the treatment effect. All the three models in each scenario fit the simulated data to examine the effects under incorrect specification. The simulation results reveal that the multiplicative model is generally conservative, the fixed linear model is much too liberal under its two wrongly specified scenarios, and the mixed model is valid in all the cases. Specifying arrays as a random effect induces a correlation across probes from the same probe set, which may be desirable from a biological point of view. Also, the modeling-based approaches can be used to test for the existence of array random effects. The mixed model provides robust statistical inference regardless whether array random effects exist or not.

REFERENCES

Affymetrix (2001a). New statistical algorithms for monitoring gene expression on GeneChip® probe arrays. Technical Report, Affymetrix. http://www.affymetrix.com/support/technical/technotes/statistical_algorithms_technote.pdf

Affymetrix (2001b). Statistical algorithms reference guide. Technical Report, Affymetrix.

Astrand, M. (2001). Normalizing oligonucleotide arrays. Unpublished Manuscript. http://www.math.chalmers.se/~magnusaa/maffy.pdf

Beasley, T.M., Holt, J.K., and Allison, D.B. (2005). Comparison of linear weighting schemes for perfect match and mismatch gene expression levels from microarray data. *American Journal of Pharmacogenomics*, 5: 197–205.

Bolstad, B.M., Irizarry, R.A., Astrand, M., and Speed, T.P. (2003). A comparison of normalization methods for high density oligonucleotide array data based on bias and variance. *Bioinformatics* 19: 185–193.

Casella, G. and Berger, R.L. (1990). *Statistical Inference*. Duxbury Press, Belmont, CA.

Catellier, D.J. and Muller, K.E. (2000). Tests for Gaussian repeated measures with missing data in small samples. *Statistics in Medicine* 19: 1101–1114.

Chen, F. and Li, W. (2001). Genomic divergence between human and other hominoids and the effective population size of the common ancestor of humans and chimpanzees. *American Journal of Human Genetics* 68: 444–456.

Chu, T., Weir, B., and Wolfinger, R.D. (2002). A systematic statistical linear modeling approach to oligonucleotide array experiments. *Mathematical Biosciences* 176: 35–51.

Chu, T., Weir, B.S., and Wolfinger, R.D. (2003). Comparison of Li–Wong and loglinear mixed models for the statistical analysis of oligonucleotide arrays. *Bioinformatics* 20: 500–506.

Chu, T., Deng, S., Wolfinger, R., Pauls, R., and Hamadeh, H.K. (2004). Cross-site comparison of gene expression data reveals high similarity. *Environmental Health Perspectives* 112: 449–455.

Dudoit, S., Yang, Y.H., Callow, M.J., and Speed, T.P. (2002). Statistical methods for identifying genes with differential expression in replicated cDNA microarray experiments. *Statistica Sinca* 12: 111–139.

Enard, W., Khaitovich, P., Klose, J., Zoellner, S., Heissig, F., Giavalisco, P., Nieselt-Struwe, K., Muchmore, E., Varki, A., Ravid, R., Doxiadis, G.M., Bontrop, R.E., and Pääbo, S. (2002). Intra- and interspecific variation in primate gene expression patterns. *Science* 296: 340–343.

Efron, B., Tibshirani, R., Storey, J.D., and Tusher, V. (2001). Emprical Bayes of a microarray experiment. *Journal of the American Statistical Association* 96: 1151–1160.

Golub, T.R., Slonim, D.K., Tamauo, P., Huard, C., Hassenbeek, M., Mesirov, J.P., Coller, H., Loh, M.L., Downing, J.R., Caligiuri, M.A., Bloomfield, C.D., and Lander, E.S. (1999). Molecular classification of cancer: discovery of class prediction by gene expression monitoring. *Science* 286: 531–537.

Hartley, H.O. and Rao, J.N.K. (1967). Maximum-likelihood estimation for the mixed analysis of variance model. *Biometrika* 54: 93–108.

Harville, D.A. (1977). Maximum likelihood approaches to variance component estimation and to related problems. *Journal of the American Statistical Association* 72: 320–338.

Harville, D.A. (1988). Mixed-model methodology: theoretical justifications and future directions. In *Proceedings of the Statistical Computing Section*. American Statistical Association, New Orleans, pp. 41–49.

Harville, D.A. (1990). BLUP (best linear unbiased prediction) and beyond. In *Advances in Statistical Methods for Genetic Improvement of Livestock*. Springer-Verlag, Berlin, pp. 239–276.

Henderson, C.R. (1984). *Applications of Linear Models in Animal Breeding*. University of Guelph, Ontario.

Hsieh, W., Chu, T., Wolfinger, R., and Gibson, G. (2003). Mixed model reanalysis of primate data suggests tissue and species biases in oligonucleotide-based gene expression profiles. *Genetics* 165: 747–757.

Irizarry, R.A., Hobbs, B., Collin, F., Beazer-Barclay, Y.D., Antonellis, K.J., Scherf, U., and Speed, T.P. (2003). Exploration, normalization, and summaries of high density oligonucleotide array probe level data. *Biostatistics* 4: 249–264.

Jennrich, R.I. and Schluchter, M.D. (1986). Unbalanced repeated-measures models with structured covariance matrices. *Biometrics* 42: 805–820.

Kenward, M.G. and Roger, J.H. (1997). Small sample inference for fixed effects from restricted maximum likelihood. *Biometrics* 53: 983–997.

King, M.C. and Wilson, A.C. (1975). Evolution at two levels in humans and chimpanzees. *Science* 188: 107–116.

Laird, N.M. and Ware, J.H. (1982). Random-effects models for longitudinal data. *Biometrics* 38: 963–974.

Lazardis, E.N., Sinibaldi, R., Bloom, G., Mane, S., and Jove, R. (2002). A simple method to improve probe set estimates from oligonucleotide arrays. *Mathematical Biosciences* 176: 53–58.

Lee, T.I., Rinaldi, N.J., Robert, F., Odom, D.T., Bar-Joseph, Z., Gerber, G.K., Hannett, N.M., Harbison, C.T., Thompson, C.M., Simon, I., Zeitlinger, J., Jennings, E.G., Murray, H.L., Gordon, D.B., Ren, B., Wyrick, J.J., Tagne, J., Volkert, T.L., Fraenkel, E., Gifford, D.K., and Young, R.A. (2002). Transcriptional regulatory networks in *Saccharomyces cerevisiae*. *Science* 298: 799–804.

Lemon, W.J., Palatini, J.J.T., Krahe, R., and Wright, F.A. (2002). Theoretical and experimental comparisons of gene expression indexes for oligonucleotide arrays. *Bioinformatics* 18: 1470–1476.

Li, C. and Wong, W.H. (2001a). Model-based analysis of oligonucleotide arrays: expression index computation and outlier detection. *Proceedings of the National Academy of Sciences, USA* 98: 31–36.

Li, C. and Wong, W.H. (2001b). Model-based analysis of oligonucleotide arrays: Model validation, design issues and standard error application. *Genome Biology* 2: 0032.1–0032.11.

Li, C. and Wong, W.H. (2003). DNA-Chip Analyzer (dChip). *Statistics for Biology and Health — Methods and Software*. Springer-Verlag, Berlin, pp. 120–141.

Littell, R.C., Milliken, G.A., Stroup, W.W., and Wolfinger, R.D. (1996). *SAS System for Mixed Models*. SAS Institute Inc., Cary, NC.

Lockhart, D., Dong, H., Byrne, M., Follettie, M., Gallo, M., Chee, M., Mittmann, M., Wang, C., Kobayashi, M., Horton, H., and Brown, E.L. (1996). Expression monitoring by hybridization to high-density oligonucleotide arrays. *Nature Biotechnology* 14: 1675–1680.

McLean, R.A. and Sanders, W.L. (1988). Approximating degrees of freedom for standard errors in mixed linear models. In *Proceedings of the Statistical Computing Section*. American Statistical Association, New Orleans, pp. 50–59.

McLean, R.A., Sanders, W.L., and Stroup, W.W. (1991). A unified approach to mixed linear models. *The American Statistician* 45: 54–64.

Patterson, H.D. and Thompson, R. (1971). Recovery of inter-block information when block sizes are unequal. *Biometrika* 58: 545–554.

Robinson, G.K. (1991). That BLUP is a good thing: the estimation of random effects. *Statistical Science* 6: 15–51.

Rosner, B. (1995). *Fundamentals of Biostatistics*, 4th ed. Duxbury Press, Belmont, CA.

SAS Institute Inc. (1999). *SAS/STAT Software Version 8*. SAS Institute Inc. Cary, NC.

Schadt, E., Li, C., Su, C., and Wong, W.H. (2000). Analyzing high-density oligonucleotide gene expression array data. *Journal of Cellular Biochemistry* 80: 192–202.

Schadt, E., Li, C., Eliss, B., and Wong W.H. (2001). Feature extraction and normalization algorithms for high-density oligonucleotide gene expression array data. *Journal of Cellular Biochemistry* **37** (Suppl): 120–125.

Schena, M., Shalon, D., Davis, R.W., and Brown, P.O. (1995). Quantitative monitoring of gene expression patterns with a complementary DNA microarray. *Science* 270: 467–470.

Searle, S.R. (1971). *Linear Models*. John Wiley & Sons, New York.

Spellman, P.T., Sherlock, G., Zhang, M.Q., Iyer, V.R., Anders, K., Eisen, M.B., Brown, P.O., Botstein, D., and Futcher, B. (1998). Comprehensive identification of cell cycle-regulated genes of the yeast *Saccharomyces cerevisiae* by microarray hybridization. *Molecular Biology of the Cell* 9: 3273–3297.

Steel, R.G.D, Torrie, J.H., and Dickey, D.A. (1997). *Principles and Procedures of Statistics: A Biometrical Approach*, 3rd ed. McGraw-Hill Inc., New York.

Stroup, W.W. (1989). Predictable functions and prediction space in the mixed model procedure. Applications of mixed models in Agriculture and Related Disciplines. Southern Cooperative Series Bulletin Vol. 343, Baton Rouge, Louisiana Agricultural Experiment Station, pp. 39–48.

Teng, Chi-Hse, Nestorowicz, A., and Reifel-Miller, A. (1999). Experimental designs using Affymetrix geneChips. *Nature Genetics* 23: 78, Poster Abstracts.

Tusher, V., Tibshirani, R., and Chu, G. (2001). Significance analysis of microarrays applied to the ionizing radiation response. *Proceedings of National Academy of Sciences, USA* 98: 5116–5121.

Wallace, D. (1988). The Behrens–Fisher and Fieller–Creasy problems. In *Lecture Notes in Statistics 1, R.A. Fisher: An Appreciation* S.E. Fienberg and D.V. Hinkley (eds.). Springer-Verlag, Berlin, pp. 119–147.

Wolfinger, R.D., Gibson, G., Wolfinger, E.D., Bennett, L., Hamadeh, H., Bushel, P., Afshari, C., and Paules, R.S. (2001). Assessing gene significance from cDNA microarray expression data via mixed model. *Journal of Computational Biology*, 8: 625–637.

Wu, C., Gumpertz, M.L., and Boos, D.D. (2001). Comparison of GEE, MINQUE, ML, and REML estimating equations for normally distributed data. *The American Statistician* 55: 125–130.

Wu, Z., Irizarry, R.A., Gentleman, R., Martinez Murillo, F., and Spencer, F. (2003). A model based background adjustment for oligonucleotide expression arrays. Submitted. http://biosun01.biostat.jhsph.edu/~ririzarr/papers/

Zhang, L., Miles, M., and Aldape, K. (2003). A model of molecular interactions on short oligonucleotide microarrays: implications for probe design and data analysis. *Nature Biotechnology* 21: 818–821.

12 Parametric Linear Models

Christopher S. Coffey and Stacey S. Cofield

CONTENTS

12.1 Introduction 223
12.2 Existing Methods for Two-Group Comparisons 225
12.2.1 Fold-Change Analysis 225
12.2.2 Two-Group *t*-Test with Equal Variances 227
12.2.3 Two-Group *t*-Test with Unequal Variances 229
12.2.4 Statistical Analysis of Microarray 231
12.3 Existing Methods for Linear Models 231
12.3.1 Fixed-Effects Models 232
12.3.2 Mixed-Effects Models 236
12.4 A Comparison of the Methods 239
12.5 Summary 241
References 242

12.1 INTRODUCTION

Microarray experiments allow researchers to simultaneously examine the expression levels of multiple genes under different conditions. Regardless of whether the experiment utilizes cDNA or oligonucleotide (Affymetrix) chips, the primary goal of many microarray studies is to determine which genes differ with regards to expression levels across conditions. Although most current microarray studies compare expression levels across two conditions, comparisons among multiple conditions are possible. This chapter will summarize the use of parametric linear models for such comparisons.

For the remainder of this chapter, we assume that the appropriate parametric assumptions have been met. Most notably, these methods assume that the gene expression levels arise from a normal distribution (bell-shaped curve), the expression levels have been appropriately transformed prior to analysis, or the sample size is large enough to apply asymptotic (large sample) normality. The widespread use of the parametric procedures described here should be tempered by the fact that these assumptions may not hold for many microarray experiments since the sample size for each gene is small (often less than 10). Furthermore, depending on the application, it may be necessary to perform additional normalization to adjust for background and across array differences before testing for differences in expression levels. Such

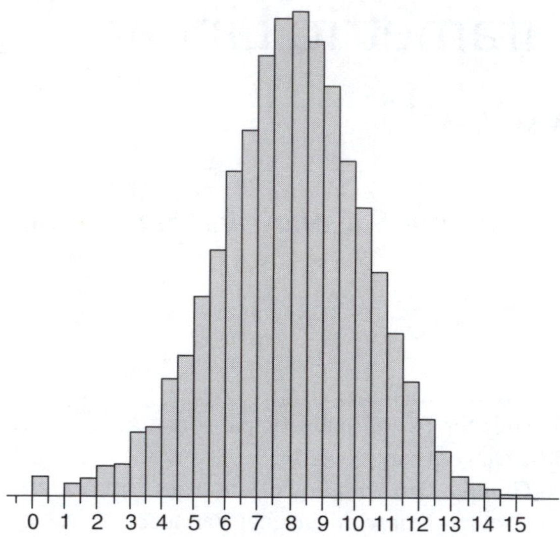

FIGURE 12.1 Distribution of $\log_2$ responses for yeast subset (1412 genes, 34,085 observations).

normalization procedures are discussed in Chapter 2 and we assume that, if needed, any such procedure has already been applied to the observed data.

Parametric linear models have been well studied in statistics over the past century and have been applied in a wide variety of fields (Neter et al., 1996). In fact, nearly every introductory statistics course includes discussions regarding statistical techniques such as the Student's t-test, linear regression, and analysis of variance (ANOVA). All of these can be thought of as parametric linear models and tied together under the general linear model framework (Muller and Fetterman, 2003). Hence, most statistical software packages include procedures for fitting these models and most researchers have a basic understanding of how to interpret their results. Not surprisingly, as the interest in microarray experiments has increased, several variations of parametric linear models have been proposed for identifying or characterizing differentially expressed genes (Golub et al., 1999; Efron et al., 2001; Jin et al., 2001; Kerr and Churchill, 2001a; Wolfinger et al., 2001; Chu et al., 2002; Dudoit et al., 2002). From simplest to most complex, these models include fold-change analysis, two-group comparisons (including the t-test), fixed-effects linear models, and linear models including both fixed and random effects (i.e., mixed models). Each of these techniques has been used to analyze microarray data, to varying degrees of success. In the following sections, each method will be defined and briefly discussed.

We will illustrate each of these techniques using the *Saccharomyces cerevisiae* swi/snf (yeast) mutation study data set, available at http://genome-www.stanford.edu/Saccharomyces/ (Sudarsanam et al., 2000). This experiment consists of four test conditions (mutant) and a wild-type strain, used to investigate mutants deleted for a gene encoding one conserved (snf2) or unconserved (swi1) component in rich or minimal media (type of preparation). Minimal media is a growth solution and rich media is the same solution with added sugars, growth factors, and hormones. The experimental strains are arrayed in triplicate and the wild-type strain is used as a reference sample on each array, resulting in 12 arrays (Eisen et al., 1998). The wild-type strain is always labeled with Cy5 in channel 2 and the experimental strains are always labeled with Cy3 in channel 1. Each log base-2 ($\log_2$) transformed intensity measurement was used as an individual observation; the response value is in $\log_2$ units and not ratio units.

We will use a subset of 1412 genes (34,085 observations) from the swi1mini and wild-type strains. The distribution of responses is shown in Figure 12.1 and appears to approximate a normal distribution.

12.2 EXISTING METHODS FOR TWO-GROUP COMPARISONS

Let n_1 and n_2 represent the sample sizes in the two conditions. Note that, by design, $n_1 = n_2 = n$ for many microarray experiments. However, due to missing data or other problems, certain instances may lead to analyses involving comparisons of groups with unequal sample sizes. Hence, we consider the more general situation. Let Y_{1ig} and Y_{2ig} represent the intensity of the $\log_2$ response for the ith observation for gene g in groups 1 and 2, respectively ($i = 1, \ldots, n_i; g = 1, \ldots, G$). Some researchers still prefer to use the ratio of two channels as the response. However, the major advantage of using the log-scale measurements as the outcome is that it allows one to adjust for a dye effect. For example, if the red dye hybridizes differently than the green dye, the expected ratio of the responses is not equal to one and, if only the ratio is utilized as the outcome, there is no way to separate out the dye effect from the true group differences (Dudoit et al., 2002).

12.2.1 FOLD-CHANGE ANALYSIS

The most basic method for comparing expression levels in two conditions is the "fold-change." Fold-change analysis for a gene is determined by computing the average $\log_2$ ratio of the expression levels under the two conditions, or equivalently, computing the average difference in $\log_2$ transformed expression values between the two conditions. For gene g, define the sample means (average intensity) for the $\log_2$ (response) values in the two conditions as

$$\bar{y}_{1g} = \frac{\sum_{i=1}^{n_1} y_{1ig}}{n_1} \quad \text{and} \quad \bar{y}_{2g} = \frac{\sum_{i=1}^{n_2} y_{2ig}}{n_2}$$

Then the fold-change can be defined as

$$F_g = \bar{y}_{2g} - \bar{y}_{1g}, \quad g = 1, \ldots, G$$

For a given gene, a positive fold-change means that the swi1mini strain was more expressed than the wild-type sample and conversely, a negative fold-change means the swi1mini strain was less expressed than the wild-type sample. For comparison purposes, one typically assumes that all genes with absolute fold-changes greater than some arbitrarily chosen cutoff level are differentially expressed. For example, we could conclude that any genes with a fold-change greater than one, that is, those with a twofold increase or decrease in expression for a particular gene, are considered to be differentially expressed. However, fold-change analysis is not a statistical method in the strict sense. Hence, there are no defined statistical properties for hypothesis testing and one cannot compute confidence intervals or p-values for the comparison.

Using the two-array subset of the yeast data described in Section 12.1, the comparison of the swi1mini mutant strain to the wild-type strain resulted in a range of observed absolute fold-change differences of 0.0002 to 6.35. A total of 123 genes produced absolute fold-changes ≥ 1 and, based on the criteria described above, would be deemed differentially expressed.

Figure 12.2 plots the standard deviation vs. the fold-change for each gene.

Fold-change is a measure of absolute change and will only find significant genes that lie beyond the arbitrary threshold, without any consideration for the variability in the values used to estimate the differential expression. This can lead to two types of problems. First, the fold-change criteria will fail to detect genes having small, but important, changes in expression levels that occur with little variation. Among biological systems, such numerically small changes in expression can be critical for some genes. Second, all genes with large fold-changes will be considered differentially expressed regardless of the amount of variation around the observed expression levels. Thus, the fold-change may incorrectly determine that genes with large fold-changes and large variation are differentially expressed. For example, Figure 12.2 illustrates that the 123 genes with absolute fold-change differences greater than or equal to one have wide ranges of variability. It is unlikely that all of these genes would

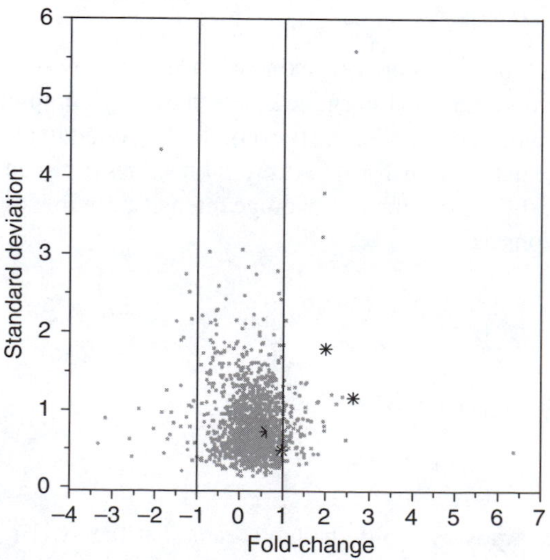

FIGURE 12.2 Fold-change by standard deviation.

remain significantly differentially expressed if the gene variance was considered in addition to the fold-change value.

The four genes marked by an $*$ will be followed throughout this chapter to illustrate the differences in conclusions that can be obtained using the various methods. The two rightmost points marked by an $*$ will be used to illustrate differences for genes with a similar fold-change but different variability. The fold-change values corresponding to these genes are approximately 2.0 and 2.5, respectively. A fold-change analysis suggests that both of these genes have similar evidence of differential expression. However, the fact that the values observed for the gene with a fold-change near 2.5 are much more stable (i.e., have a lower variance) would seem to provide us with more confidence that this gene is truly differentially expressed. On the other hand, the large variability associated with the first gene suggests that the "true" fold-change could be substantially greater or less than two.

12.2.2 Two-Group *t*-Test with Equal Variances

The *t*-statistic allows for simultaneous consideration of differential expression and variability (Dudoit et al., 2002). A large absolute *t*-statistic suggests that the expression level is significantly altered in one group compared to the other. Detailed discussions of the *t*-test may be found in most basic statistical textbooks, such as, Rosner (1995).

Let μ_{1g} and μ_{2g} represent the true (unknown) expression levels for gene g in the two conditions, respectively. The equal-variance two-group *t*-test approach considers the two-sided hypothesis of equal mean expression levels for a gene:

$$H_0\colon\ \mu_{1g} = \mu_{2g} \quad \text{vs.} \quad H_A\colon\ \mu_{1g} \neq \mu_{2g}$$

As before, let $\bar{y}_{1g}$ and $\bar{y}_{2g}$ represent the sample means for the $\log_2$ (response) values in the two conditions for gene g $(g = 1, \ldots, G)$. Furthermore, define the overall estimated variance(s) of gene gs expression level in $\log_2$ units as

$$s_g^2 = \frac{\sum_{i=1}^{2} \sum_{j=1}^{n_i} (y_{ijg} - \bar{y}_{ig})^2}{n_1 + n_2 - 2}$$

It then follows that the equal variance *t*-statistic comparing gene expression in the two groups is

$$t_g = \frac{\bar{y}_{2g} - \bar{y}_{1g}}{\sqrt{s_g^2(1/n_1 + 1/n_2)}}$$

Note that the numerator of the *t*-statistic is similar to the fold-change. As a consequence, the hypothesis test carried out by the *t*-statistic is identical to the fold-change comparison, but accounts for gene-specific variability in computing the test statistic.

The process of dividing each difference by its estimated variability serves to "standardize" the fold-change. Since the t-test examines the relative change in expression, it will identify genes with small differential expression if this difference occurs with little variation. Such differences cannot be found using the fold-change method.

By accounting for the observed error in our measurements, we can utilize the distributional properties of the t-statistic under the null hypothesis to determine a cutoff value to assist researchers in determining whether the observed differences in expression level are likely to be explained by chance or appear to be real differences. In choosing a cutoff, researchers can choose a significance level at which they choose to work. This significance level is chosen in such a way as to control the probability of making a type 1 error, or incorrectly concluding that there is a differential gene expression among the conditions when no difference truly exists. After choosing a significance level, the distributional properties of the t-statistic under the null distribution can be used to choose the critical value or cutoff point. We then conclude that all genes with a t-statistic (standardized fold-change) above this critical value are "statistically significant," and hence differentially expressed. For example, if the desired significance level is 0.05, we can conclude that there is less than a 5% probability that we would observe a difference in gene expression levels greater than or equal to the observed difference by chance alone.

In addition, the null distribution of the test statistic can be utilized to calculate a p-value in order to quantify the likelihood that the observed result is due to chance. This allows different researchers to make their own conclusions based on their own notions of the negative impact introduced by a type 1 error. Hence, if the p-value associated with the comparison for any particular gene equals 0.03, a researcher who chooses to work at the 0.05 significance level would conclude that there is sufficient evidence for differential expression whereas a researcher who prefers to work at the more stringent 0.01 significance level would not conclude that there is enough evidence. The choice of significance level only controls the probability of making a type 1 error for that single hypothesis test. We have not addressed the multiple testing issues introduced by simultaneous comparison of thousands of genes, as this issue is addressed in Chapter 15. This problem is not unique to the t-test, but rather is a concern with all of the parametric methods described in this chapter.

Using the two-array subset of the yeast data described in Chapter 15, the comparison of the swi1mini mutant strain to the wild-type strain, using an equal-variance t-test, resulted in 48 genes with a p-value less than 0.05. Note that the effect size for the t-test for each gene is the same as the fold-change value calculated in Chapter 15. The p-values ranged from 3.1×10^{-8} to 0.9998.

For the hypothesis test associated with the t-statistic to be valid, the intensity responses must be independent of one another. This assumption may not hold for many microarray experiments, as it is likely that the intensity responses on the same slide are likely to be related.

The form of the t-statistic that we have described utilizes gene-specific variance estimates in the denominator. Due to the small sample sizes common with microarray studies, there are few observations to estimate each of these gene-specific variances. This can lead to biased estimates, particularly due to the fact that the distribution of a sample variance tends to be skewed toward the low side. One solution to this problem would be to assume that there is equal variance across all of the genes and pool all of the gene-specific variance estimates together into a single variance estimate. The advantage of this approach is that we are able to use more observations to estimate the variance and, hence, our estimator should have much better properties. However, since it seems unlikely that one can reasonably assume equal variance for all genes, this approach should be avoided.

12.2.3 Two-Group *t*-Test with Unequal Variances

The t-test described above is valid under the assumption that gene expression levels have an equal variance under the two conditions. This is reflected in the manner with which we obtain the variance estimate that appears in the denominator of the test statistic. When the variance in the two conditions is not equal, an alternative version of the t-test is produced while accounting for this difference.

For gene g ($g = 1, \ldots, G$), define the sample means for the $\log_2$ (response) values in the two conditions, $\bar{y}_{1g}$ and $\bar{y}_{2g}$, as before and define the overall estimated variance of gene gs expression level in $\log_2$ units for the two groups as

$$s_{1g}^2 = \frac{\sum_{i=1}^{n_1}(y_{1ig} - \bar{y}_{1g})^2}{n_1 - 1} \quad \text{and} \quad s_{2g}^2 = \frac{\sum_{i=1}^{n_2}(y_{2ig} - \bar{y}_{2g})^2}{n_2 - 1}$$

It then follows that the t-statistic comparing gene expression in the two cell lines accounting for unequal variances is

$$t_g = \frac{\bar{y}_{2g} - \bar{y}_{1g}}{\sqrt{s_{1g}^2/n_1 + s_{2g}^2/n_2}}$$

Unlike the test statistic with equal variances, the distribution of this test statistic has only an approximate t-distribution. Several methods exist for estimating the appropriate degrees of freedom needed to apply the approximation. We will use the commonly applied Satterthwaite approximation (Rosner, 1995).

Using the two-array subset of the yeast data described in Section 12.1, the comparison of the swi1mini mutant strain to the wild-type strain, using an unequal-variance t-test with a Satterthwaite approximation, resulted in 60 genes with a p-value less than 0.05. Figure 12.3 shows the plot of the p-value associated with the t-test by the effect size for each gene. Note that the effect size for the t-test for each gene is the same as the fold-change value calculated in Section 12.1. The p-values ranged from 3.0×10^{-8} to 0.9998. Note that many genes found "significant" by the fold change method are no longer deemed significant when the variability for expression levels is accounted for (these genes are found in the upper right and left corners of Figure12.3).

Of the two rightmost genes marked with an $*$, the strength of evidence supporting differential expression is greater for the gene with smaller variance. In fact, of the four genes being tracked throughout this chapter, this was the only gene found to be significant by the t-test method. This gene has the largest fold-change of the four genes but also had the second largest gene variation (Figure 12.2). The t-test method

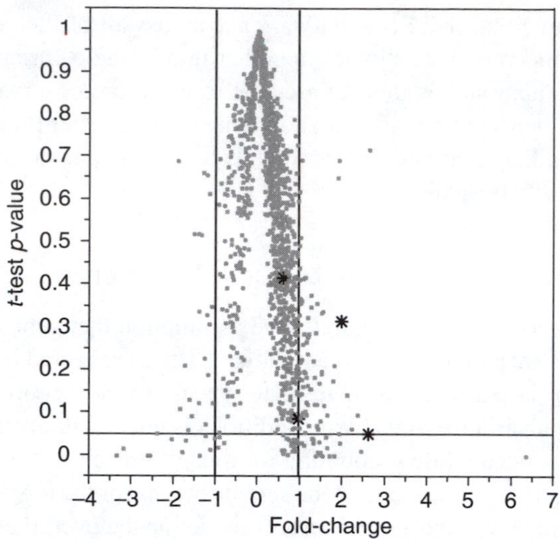

FIGURE 12.3 Fold-change by t-test p-value for all genes.

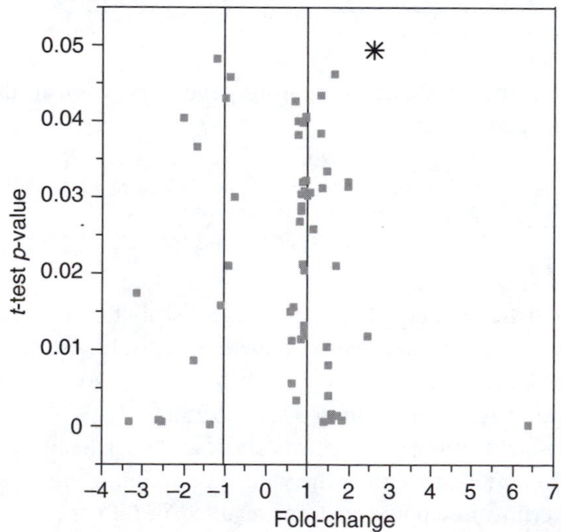

FIGURE 12.4 Fold-change by t-test p-value for genes with a significant p-value.

accounts for enough gene variation for this gene to be found significant, as opposed to the gene that has a similar fold-change but a larger variation.

Figure 12.4 shows the fold-change values plotted against the p-values for the 60 significant genes. The genes with fold-change values between -1 and 1 would not have been found significant using the fold-change method in Section 12.1. Using the t-test, a measure of relative change, finds genes with smaller differential expression to be significant than by using fold-change alone.

12.2.4 Statistical Analysis of Microarray

Even though the t-statistics are a vast improvement over simple fold-change approaches, they are not without concerns of their own. Specifically, some genes included in the study are likely to have small differences in expression level that occur with little or no variability. As a consequence, the absolute t-statistic for these genes will be large. Although one can make the argument that these are in fact statistically significant, it is not at all clear what type of scientific relevance can be given to such small differences.

Tusher et al. (2001) have proposed a modification known as the statistical analysis of microarrays (SAM) to address this problem. SAM computes a statistic d_g for each gene g, analogous to the equal variance t-statistic except that a small positive constant is added to the denominator:

$$d_g = \frac{\bar{y}_{2g} - \bar{y}_{1g}}{\sqrt{s_g^2(1/n_1 + 1/n_2)} + s_0}$$

where s_0 is a small positive constant or "fudge factor," based on the median standard deviation for all genes (Tusher et al., 2001). The "fudge factor" is used as a correction to place a boundary on the size of the denominator and prevent those genes with extremely small fold-changes and variation from being selected as differentially expressed. An implementation of the SAM algorithm is available for download from the internet at http://www-stat.stanford.edu/%7Etibs/SAM.

Even though computationally similar to parametric methods, SAM utilizes a permutation-based, rank order method for testing significant differences in expression levels in a microarray experiment. A false discovery rate (FDR; Hochberg and Benjamini, 1990) can be calculated by dividing the median number of false positive genes across all permutations by the number of observed significant genes.

We mention SAM here because it is so closely related to the parametric t-test. However, by utilizing the FDR, SAM does not depend on the assumption of normality required for the validity of the hypothesis tests obtained with the t-tests. Hence, the SAM technique may be thought of as nonparametric in nature and the results of applying the SAM technique to our example data set will not be presented here.

12.3 EXISTING METHODS FOR LINEAR MODELS

Both SAM and the t-tests consider within gene variation and are a substantial improvement over fold-change approaches. However, neither method accounts for other sources of variability, such as the possible variation in responses due to dye or array. In any experiment, known sources of variation should be accounted for when analyzing the data. Linear model-based procedures, such as ANOVA and subsequent extensions to mixed-effects models, are commonly used to examine for significant differences between groups while accounting for multiple sources of variation (Muller and Fetterman, 2003). Such model-based procedures are equally applicable for controlling multiple sources of variation in microarray studies (Jin et al., 2001; Kerr and Churchill, 2001a).

Even though the connection is not immediately clear, many of the issues involved in analyzing complex microarray data are similar to issues in agricultural studies, which led to substantial developments in experimental design during the last century (Fisher, 1949; Kirk, 1983; Kerr and Churchill, 2001b). Although design issues are addressed in Chapter 7, we note that involving a statistician during the design stage of a microarray study can pay huge dividends during data analysis. The old adage of "garbage in, garbage out" applies to microarray studies just as with any other type of data analysis.

The type of linear modeling that is appropriate for any given situation depends on the type of effects included in the model. The term fixed-effects applies to those effects that are the only effects of interest to the investigator and would be repeated exactly if the experiment was repeated. In the analysis stage, we typically want to compare the average outcomes among the various levels of a fixed-effect. On the contrary, the term random-effect applies to those effects that are assumed to represent a random sample from some population. In experiments involving random-effects, we are often interested in making inferences to the population as a whole rather than the specific values observed in our experiment. Parallel to these definitions, fixed-effects linear models describe linear models that involve only fixed-effects and random-effects linear models describe linear models that involve only random-effects. Models involving both fixed and random effects are referred to as mixed models (Laird and Ware, 1982; Littell et al. 1996). In the following sections, we discuss the use of both fixed- and mixed-effects linear models for the analysis of microarray data.

12.3.1 Fixed-Effects Models

A fixed-effects model that is assumed linear in the parameters falls into a general linear model framework. A general linear model is defined as:

$$\mathbf{y} = \mathbf{X}\boldsymbol{\beta} + \boldsymbol{\varepsilon}$$

If n represents the total sample size and p represents the number of unknown parameters required to specify the fixed-effects of interest, then $\mathbf{y}$ is an $n \times 1$ vector of observed $\log_2$ (responses), $\mathbf{X}$ is an $n \times p$ design matrix of known constants for the fixed-effects, $\boldsymbol{\beta}$ is a $p \times 1$ vector of unknown parameters, and $\boldsymbol{\varepsilon}$ is an $n \times 1$ vector of random errors. Furthermore, it is generally assumed that the elements of $\boldsymbol{\varepsilon}$ are independent with mean 0 and variance σ^2, that is, $\varepsilon_{ij} \sim N(0, \sigma^2)$. In matrix terminology, this implies that $\boldsymbol{\varepsilon} \sim N(\mathbf{0}, \sigma^2\mathbf{I}_n)$ (Neter et al., 1996).

Note that the two-group t-test can be thought of as a special case of the linear model by defining $\mathbf{X}$ to be an $n \times 2$ matrix with the first column as all 1s to represent the intercept and the second column representing an indicator variable, which is equal to 1 if the observation is from group 2 and equal to 0 if the observation is from group 1. Corresponding to the definition of the design matrix, the $\boldsymbol{\beta}$ vector contains two elements: β_0 representing the intercept and β_1, the difference between the mean outcome value for group 2 and the mean outcome value for group 1. In scalar form, this two-group comparison can be written as:

$$y_{ij} = \beta_0 + \beta_1 x_1 + \varepsilon_{ij}, \quad j = 1, 2$$

where

$$x_1 = \begin{cases} 0, & \text{if } j = 1 \\ 1, & \text{if } j = 2 \end{cases}$$

and $\varepsilon_{ij} \sim N(0, \sigma^2)$. It follows that $\mu_1 = \beta_0$ and $\mu_2 = \beta_0 + \beta_1$ represent the mean expression levels in groups 1 and 2, respectively. It then follows that the test of differential gene expression, which is equivalent to the two-group t-test assuming equal variances among the two groups, is provided by testing $H_0: \beta_1 = 0$ in this linear model. Note that this parameterization of the two group t-test corresponds to reference-cell coding using the first group as the reference-cell. Such coding schemes are commonly used by default for categorical variables in many statistical software packages, although we note that alternative coding schemes exist and may prove more useful to researchers in certain instances (Muller and Fetterman, 2003).

Unlike the two-group t-test, the fixed-effect linear model allows for the comparison of more than two groups or the control of other fixed-effects by including additional terms in the model. When these additional terms are associated with the outcome, the error (residual) variance can be substantially decreased. Since the denominator of the test statistic used for comparing the two groups involves this error variance, this allows for greater power in the comparison of the two groups. Hence, in microarray experiments, accounting for known sources of variation increases the power of the experiment to observe significant differences in expression levels for a given gene.

Kerr and Churchill (2001b) have identified four basic sources of variation in microarray data: groups (varieties), genes, dyes, and arrays. Note that the dye effect appears only in cDNA microarray studies. For oligonucleotide data, the models must be modified by removing any dye effects or interactions. The four main effects measure the overall variation in expression levels associated with differences due to groups, arrays, dyes, and genes, respectively. For example, a dye main effect would appear in the data if there was a trend for expression levels to be consistently higher for the red vs. the green dye as a result of the red dye hybridizing better. The "full" model containing the four factor main effects and all possible two-way, three-way, and four-way interactions would involve 16 parameters. However, Kerr and Churchill have argued against using many of the higher order interactions due to issues regarding confounding or difficulty in interpretation. The variety by gene (VG) interactions measure the differences in expression levels between groups for a particular gene and are the effects of primary interest. Hence, the test of whether this interaction is nonzero is equivalent to the test of whether the expression levels for a given gene differ between the two groups. Other interactions that Kerr and Churchill have proposed for inclusion in the model include (1) dye by variety (DV) interactions to account for run differences, (2) array by gene (AG) interactions to account for spot-to-spot variability, and (3) dye by gene (DG) interactions to account for gene-specific dye differences.

Define Y_{ijkg} to be the expression level for array i, dye j, variety k, and gene g. Depending on which sources of variation are thought to be important, different models may be of more use in any particular study. For example, one choice of model would include parameters to account for the four-factor main effects and interaction terms to account for the VG interaction (the effects of interest), spot-to-spot variation, and interactions between dyes and genes:

$$y_{ijkg} = \mu + A_i + D_j + V_k + G_g + (\mathrm{VG})_{kg} + (\mathrm{AG})_{ig} + (\mathrm{DG})_{jg} + \varepsilon_{ijkg}$$

For estimation, we assume the ε_{ijkg}s are independent, identically distributed random variables with mean 0 and variance σ_g^2. Note that this assumes homogeneity of variance for observations observed on the same gene, but allows for heterogeneity of variance for expression levels measured in different genes. For hypothesis testing, the additional assumption that the errors follow a normal distribution is required. Kerr et al. (2002) also propose an alternate model, which replaces the single degree of freedom group effect with the array $\times$ dye (AD) interaction term:

$$y_{ijkg} = \mu + A_i + D_j + (\mathrm{AD})_{ij} + G_g + (\mathrm{VG})_{kg} + (\mathrm{AG})_{ig} + (\mathrm{DG})_{jg} + \varepsilon_{ijkg}$$

Typically, since the random errors are assumed to differ on a gene-by-gene basis, Wolfinger et al. (2001) suggested separating the single model described above into two separate models. The first model contains the "global" components $[A_i, D_j, (\mathrm{AD})_{ij}]$ and can be thought of as the normalization model

$$y_{ijkg} = \mu + A_i + D_j + (\mathrm{AD})_{ij} + r_{ijkg}$$

The normalized residuals from the first model are then utilized as the outcomes in separate models for the gene-specific effects for every gene of interest

$$r_{ijkg} = G_g + (\mathrm{VG})_{kg} + (\mathrm{DG})_{jg} + (\mathrm{AG})_{ig} + \varepsilon_{ijkg}$$

As before, the VG interaction term captures the effects of interest.

One of the biggest advantages of this ANOVA model is that it is quite simple to extend to more complex microarray designs. For example, if we wish to compare expression levels for four groups, this simply increases the degrees of freedom associated with the VG interaction term. However, all of the previous discussion remains pertinent. This may not be true for many of the current advances for analyzing microarray data, which are based only on the assumption of a two-group comparison.

For the yeast data, fitting an individual model for each gene, the fixed-effects would be strain and array, with strain being the effect of interest. Note that, in this example, there is no dye effect since the wild-type strain were always labeled with Cy5 in channel 2 and the experimental strains were always labeled with Cy3 in channel 1. As such, any dye effect cannot be separated from any observable group

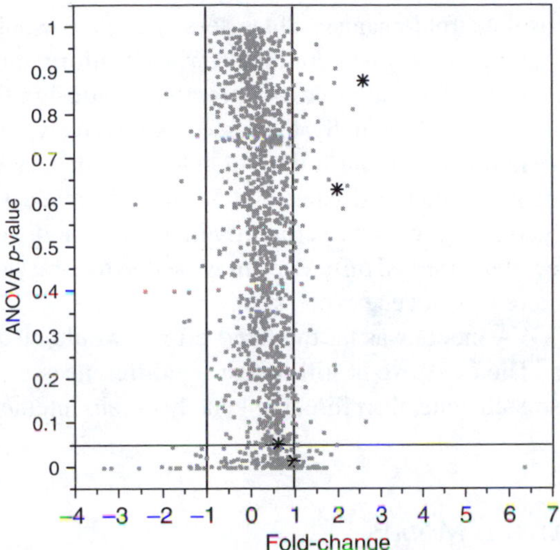

FIGURE 12.5 Fold-change by ANOVA *p*-value for all genes.

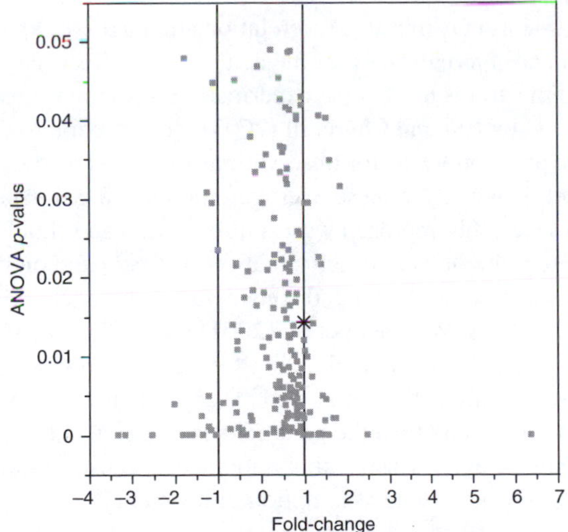

FIGURE 12.6 Fold-change by ANOVA *p*-value for significant genes.

effect. There were 193 genes with a significant strain comparison ($p < 0.05$, range $1.0 \times 10^{-9} - 0.9993$). Figure 12.5 shows the fold-change by the ANOVA *p*-value for all genes, while Figure 12.6 shows the same for the 193 significant genes. One of the genes being tracked by all methods is called significant by the ANOVA method and one is marginally significant, with a *p*-value less than 0.06. These two genes are the genes with the smaller fold-change and smaller gene variation (Figure 12.2).

Genes with absolute fold-change values less than one would not have been detected by fold-change alone. Incorporating variance information for each gene, as well as for the array effect, allowed for previously nonsignificant genes to be classified as significantly differentially expressed. Conversely, some genes previously classified as significant are no longer of interest. For example, consider the marked gene with an observed fold-change of 2.5. This difference is significant by the *t*-test method but accounting for array variability in a linear model rendered this gene insignificant. Hence, the observed difference may be due to array-to-array variability rather than true differences in gene expression.

If a single ANOVA model was fit, the fixed-effects would be gene, strain, array and gene by strain. The objective is still to test for differences in the swi1mini and wild-type stains for each gene, therefore, the gene by strain interaction would be the effect of interest.

12.3.2 Mixed-Effects Models

Fixed-effects models are an effective way to analyze microarray data since these models allow for the inclusion of factor effects. However, these models are based on the assumption of independent observations and homogenous variation among all expression levels obtained for a single gene. The presence of random-effects in many microarray experiments may introduce correlation among these expression levels. For example, in a typical microarray experiment, the array effect may be considered a random-effect, as the arrays used in the experiment represent a subset of all possible arrays. As discussed by Cui and Churchill (2003), one solution to this problem is to use the average expression levels for observations observed on the same level of the random-effect and then analyze these average outcome values using a fixed-effects linear model. However, this approach wastes information and should be avoided if at all possible. A better alternative in such situations is to expand on the general linear models that we have discussed so far and consider mixed models, which include both random- and fixed-effects. Wolfinger et al. (2001) utilize the two-stage specification of the model described above while allowing for random array and spot effects that follow a normal distribution. Kerr et al. (2002) argue that a reasonable case can be made to go a step further and treat the gene effects and gene interactions as random. However, they caution against using this approach because of concerns about the appropriateness of the underlying assumption of normality.

In matrix notation, the mixed-effects model is defined as:

$$\mathbf{y} = \mathbf{X}\boldsymbol{\beta} + \mathbf{Z}\mathbf{u} + \boldsymbol{\varepsilon}$$

If n represents the total sample size, p represents the number of unknown fixed-effect parameters, and q represents the number of unknown random-effect covariance parameters, then $\mathbf{y}$ is an $n \times 1$ vector of observed $\log_2$ (responses), $\mathbf{X}$ is an n by p design matrix of known constants for the fixed-effects, $\boldsymbol{\beta}$ is a $p \times 1$ vector of unknown fixed-effect parameters, $\mathbf{Z}$ is an $n \times q$ design matrix of known constants for the random-effects, $\mathbf{u}$ is a $q \times 1$ vector of unknown random-effect parameters, and $\boldsymbol{\varepsilon}$ is an $n \times 1$ vector of random errors. The assumptions on $\mathbf{u}$ and $\boldsymbol{\varepsilon}$ are: $E[\mathbf{u}] = \mathbf{0}$, $E[\boldsymbol{\varepsilon}] = \mathbf{0}$, $\text{var}[\mathbf{u}] = \mathbf{G}$ and $\text{var}[\boldsymbol{\varepsilon}] = \mathbf{R}$, where $\mathbf{u}$ and $\boldsymbol{\varepsilon}$ are uncorrelated. In matrix form, the assumptions are defined as:

$$E\begin{bmatrix}\mathbf{u}\\ \boldsymbol{\varepsilon}\end{bmatrix} = \begin{bmatrix}\mathbf{0}\\ \mathbf{0}\end{bmatrix} \quad \text{and} \quad \text{var}\begin{bmatrix}\mathbf{u}\\ \boldsymbol{\varepsilon}\end{bmatrix} = \begin{bmatrix}\mathbf{G} & \mathbf{0}\\ \mathbf{0} & R\end{bmatrix}$$

These assumptions yield the variance of $\mathbf{y}$:

$$\begin{aligned}\text{var}(\mathbf{y}) &= \text{var}(\mathbf{X}\boldsymbol{\beta} + \mathbf{Zu} + \boldsymbol{\varepsilon})\\ &= \text{var}(\mathbf{X}\boldsymbol{\beta}) + \text{var}(\mathbf{Zu}) + \text{var}(\boldsymbol{\varepsilon})\\ &= \mathbf{Z}\text{var}(\mathbf{u})\mathbf{Z}' + \text{var}(\boldsymbol{\varepsilon})\\ &= \mathbf{ZGZ}' + \mathbf{R}\end{aligned}$$

Note that the fixed effect linear model can be thought of as a special case of the mixed-effects model where $\mathbf{Z} = \mathbf{0}$ and $\mathbf{R} = \sigma^2\mathbf{I}_n$.

To illustrate the use of a mixed model for microarray data, consider the gene-specific model fit during the second stage of the Wolfinger et al. (2001) approach. This model assumes that there are two sources of random error (1) pure error (σ^2) and (2) error due to array variability (σ^2_{A}). If there are n total observations and k arrays, then the $\mathbf{X}$ matrix for this model is identical to the $\mathbf{X}$ matrix from the fixed-effects models with the exception that the columns used to capture the array effects are removed, since these effects are now considered random rather than fixed. The random array effects are captured by the $\mathbf{Z}$ matrix, which consists of an $n \times k$ matrix with a 1 in column k if an observation appears on array k and a 0 in column k, otherwise. This assumed variance structure corresponds to the assumption that

$$\mathbf{R} = \sigma^2\mathbf{I}_n \quad \text{and} \quad \mathbf{G} = \sigma^2_A\mathbf{I}_k$$

As a consequence, this covariance structure induces correlation among multiple expression levels for a given gene observed on the same array, although genes on separate arrays remain independent. From this assumption, it follows that

$$\text{var}(Y_{ijgk}) = \sigma^2 + \sigma^2_A$$

and

$$\text{cov}(Y_{ijgk}, Y_{i'j'gk'}) = \begin{cases}\sigma^2_A, & \text{if } k = k'\\ 0, & \text{otherwise}\end{cases}$$

This mixed-effects model allows for estimation of differences between factor levels, estimation of gene variability, and estimation of random array variability. The hypothesis tests of interest can be obtained from this model as well.

With complex mixed models, one must take great care in choosing the correct denominator degrees of freedom for use in any given test. Littell et al. (1996) provide a detailed discussion regarding the choice of test statistics in the mixed model. Although researchers would be wise to involve a statistician in any of the analyses described in this chapter, it is imperative to seek a statistician's input when conducting analyses using complex mixed models since incorrect specification of the test statistic can potentially lead to substantially inflated type 1 errors.

For the yeast data, fitting an individual mixed-effects model for each gene, the fixed-effect is strain and the random-effect is array. The objective is to test for differences in the swi1mini and wild-type strains for each gene. There were 202 genes with a significant strain comparison ($p < .05$, range $1.0 \times 10^{-9} - 0.9992$). Figure 12.7 shows the fold-change by the p-value for all genes, while Figure 12.8 shows the same

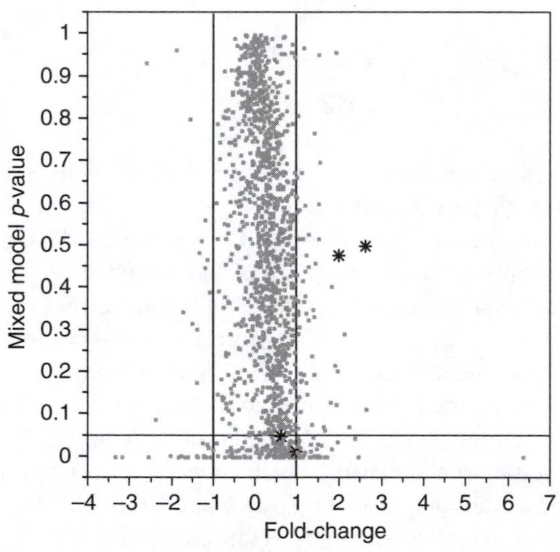

FIGURE 12.7 Fold-change by mixed model p-value for all genes.

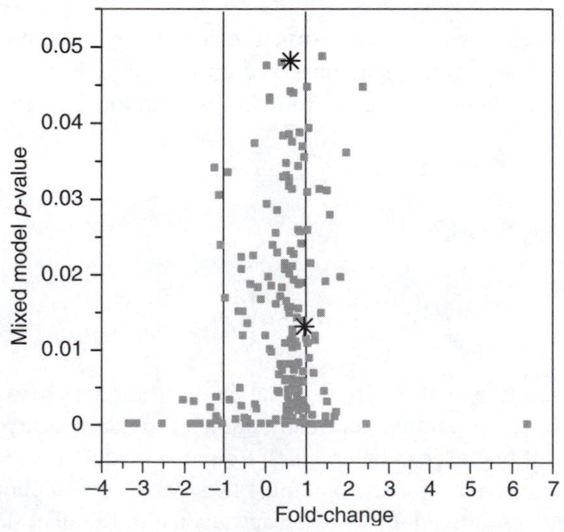

FIGURE 12.8 Fold-change by mixed model p-value for significant genes.

for the 202 significant genes. Again, the two genes with the smallest fold-change, of the four being tracked, are found significant by the mixed model method.

If a single mixed-effects model was fit, the fixed-effects would be gene, strain, and gene by strain and the random-effect would be array. The objective is still to test for differences in the swi1mini and wild-type stains for each gene, therefore, the gene by strain interaction would be the effect of interest.

Even though the mixed model approach suggested by Wolfinger et al. (2001) accounts for multiple sources of variation, the models are fit on a gene-by-gene basis and, hence, assume that the effects of genes are independent of each other. This assumption seems very unlikely, as one would expect that the expression levels for several genes involved in the same biological process are likely to be highly correlated. In theory, the mixed models approach can easily be modified to account for correlation across genes. However, in practice, the implementation of such models would require larger sample sizes than those currently utilized with microarray studies.

12.4 A COMPARISON OF THE METHODS

In total, 282 genes were found significant by at least one method (fold-change, unequal-variance t-test, ANOVA, and mixed-effects models). Of the 282 genes, 81 genes were found significant by only one method, 132 by only two methods, 43 by three methods, and 26 were significant by all four methods. Of the 81 genes significant by only one method, 65 were by fold-change, 6 by t-test, 5 by ANOVA, and 5 using a mixed model (Table 12.1). Of the 132 significant by two methods, 4 were significant by fold-change and t-test, 5 by fold-change and mixed model, 2 by t-test and mixed model, and 121 by ANOVA and mixed model. Of the 43 significant by three methods, 2 were significant by fold-change, t-test, and mixed model; 21 were significant by fold-change, ANOVA, and mixed model; and 20 were significant by t-test, ANOVA, and mixed model.

Grouping the 217 genes found significant by any method, other than fold-change alone, into 5 significance groups results in 10 genes that are significant by the t-test method only, 5 that are significant by ANOVA only, 14 that are significant by mixed model or mixed model and t-test, 162 that are significant by both linear model methods (ANOVA and mixed model) with or without significance by the t-test method, and 26 genes that were found significant by all methods (Figure 12.9).

In analyzing the data, with methods that incorporated an estimate for gene variance, we found that large fold-changes in expression were often associated with p-values not indicative of statistical change, and conversely subtle fold-changes were often highly significant. For example, the 65 genes that were significant by fold-change alone had a mean variation of 1.8 with a range of absolute fold changes from 1 to 2.6. These genes were not found significant by any other method when gene variability was considered.

Considering the four genes that were tracked by all methods, each method incorporated gene variability in a manner that changed the gene significance across the methods. The gene with a fold-change of 2.01 and a standard deviation of 2.55 was found significant only by the fold-change method. The observed difference may be

TABLE 12.1
Number of Significant Genes by Method or Combination of Methods (N = 1412)

# of methods (*n*)	Methods	# significant
Zero (1130)	None	1130
One (81)	Fold-change only	65
	t-test only	6
	ANOVA only	5
	Mixed model only	5
Two (132)	Fold-change and *t*-test	4
	Fold-change and mixed model	5
	t-test and mixed model	2
	ANOVA and mixed model	121
Three (43)	Fold-change, *t*-test, and mixed model	2
	Fold-change, ANOVA, and mixed model	21
	t-test, ANOVA, and mixed model	20
Four (26)	All methods	26

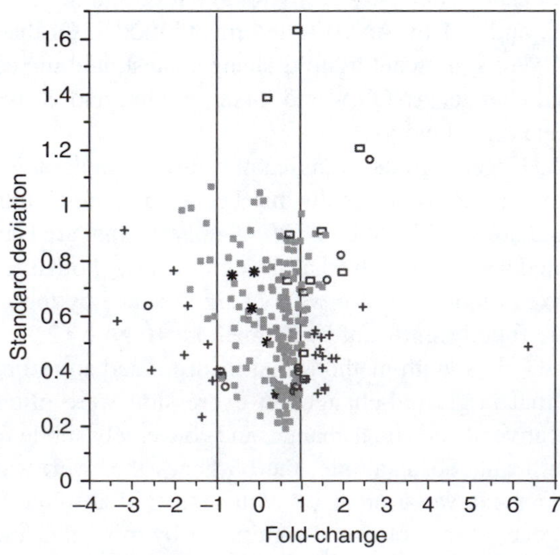

FIGURE 12.9 Fold-change by standard deviation, by significant methods: ○: *t*-test only; ∗: ANOVA only; □: mixed model or mixed model plus *t*-test; ■: linear model (ANOVA and mixed model) with or without significant *t*-test; +: all methods.

due only to gene variability. The gene with a fold-change of 2.64 and a standard deviation of 2.17 is found significant by the fold-change and *t*-test methods. The observed difference is significant by the *t*-test method but accounting for array variability in a linear model rendered this gene insignificant — this observed difference may be due to array variability. The gene with a fold-change of 0.98 and standard deviation of 0.69, is barely nonsignificant by the fold-change method, is marginally significant when incorporating gene variation in the *t*-test ($p = 0.08$), but becomes significant when array variation is considered in the ANOVA and mixed-effect models. The gene variation is estimated at 0.38 and 0.26 when array is considered a fixed- or random-effect, respectively. The gene with the smallest fold-change of 0.64 and a standard deviation of 0.73 was found to be significant only when array was considered a random-effect in the mixed model method. When array was considered a fixed-effect, the estimate for gene variation was reduced to 0.50, resulting in a marginally significant difference ($p = 0.052$). However, the estimate for gene variation was reduced to 0.21 in the mixed model method, resulting in a significant observed difference.

12.5 SUMMARY

We have discussed five existing methodologies for the analysis of microarray experiments: fold-change analysis, *t*-tests, SAM, fixed-effects linear models, and mixed models. Fold-change measures should be avoided since these do not account for the degree of variability in the data. The *t*-test, either assuming equal or unequal variances across groups, can be used to standardize the observed differences and provide a measure of how likely it is that the observed differences can be explained by chance. The SAM approach modifies the *t*-statistic by adding a small constant to the denominator and may be used to prevent finding "significant" differences for genes with extremely small differences in expression levels that occur with little variation. When researchers either want to control for multiple sources of variation or there are more than two groups to compare, general parametric model-based procedures may be applied. When all sources of variation are considered to be fixed-effects, a fixed-effects ANOVA model can be used to obtain the relevant hypothesis tests. Alternately, if some of the sources of variation are considered to be random-effects, a mixed model can be utilized for the same purpose.

All of the methods described in this chapter assume that the appropriate assumptions have been met. Because the purpose of this chapter is to demonstrate applications of parametric linear models to microarray data, we have not addressed the issues associated with using residual diagnostics to assess the adequacy of the normality assumption since this issue is not unique to microarray analysis. However, we stress the importance of such diagnostic techniques and suggest that users who employ the methods described in this chapter consult discussions of regression diagnostics in an introductory statistics textbook (e.g., Rosner, 1995; Neter et al., 1996). When researchers are confident that these assumptions are valid, we feel that mixed-effects models are an appropriate method of analysis. When the assumption of normality does not hold, the hypothesis tests reported from the model-based procedures described here are not valid. Since microarray experiments typically involve small sample sizes for each gene (typically less than 10), large sample arguments do not apply and researchers need to ensure that the assumption of normality is valid before applying these parametric techniques. When one has reason to doubt that the assumption of normality is met, the data can either be normalized by an appropriate transformation before applying these methods or researchers should consider alternate methods that do not rely on the parametric assumptions associated with methods described in this chapter. Such nonparametric methods are described in Chapter 13. An important property of the procedures described in this chapter is that the estimators of average differences in expression levels are valid even without the assumption of normality. Hence, when one is concerned about the validity of the normality assumption, these procedures can be used to obtain the parameter estimates and then alternate procedures such as bootstrapping (Efron and Tibshirani, 1998) can be used to conduct hypothesis tests that do not depend on the assumption of normality.

Most microarray experiments can be thought of as exploratory analyses for the purpose of determining which subsets of genes should be explored in greater detail (i.e., determine the best allocation of limited future resources). As such, any significant findings will need to be validated by replication in future studies. A major advantage of the use of parametric linear models for analyzing microarray data lies in the fact that well documented, easy-to-use procedures for such models are available in most standard statistical software packages. Furthermore, the t-test does not require special statistical software since summary measures and the t-statistic can be computed using standard spreadsheet software. Hence, the parametric techniques described here can be thought of as the quick and dirty approach for analyzing microarray data. As with any analysis, researchers are encouraged to analyze their data using both parametric techniques and techniques that do not rely the assumption of normality. Clearly, if consistent results are obtained using the alternate methods, more credibility can be given to the results obtained. Finally, researchers should be encouraged to utilize greater replication for microarray studies in order to account for all possible sources of variation, obtain valid measures for differences in expression levels among the groups, and provide sufficient diagnostic measures to evaluate the adequacy of the assumptions.

REFERENCES

Chu, T., Weir, B., and Wolfinger, R. (2002). A systematic statistical linear modeling approach to oligonucleotide array experiments. *Mathematical Biosciences* 176: 35–51.

Cui, X., and Churchill, G.A. (2003). Statistical tests for differential expression in cDNA microarray experiments. *Genome Biology* 4: 210 (epub).

Dudoit, S., Yang, Y., Callow, M., and Speed, T. (2002). Statistical methods for identifying differentially expressed genes in replicated cDNA microarray experiments. *Statistica Sinica* 12: 111–139.

Efron, B., and Tibshirani, R.J. (1998). *An Introduction to the Bootstrap*. Chapman & Hall/CRC Press, London/Boca Raton, FL.

Efron, B., Tibshirani, R., Storey, J., and Tusher, V. (2001). Empirical Bayes analysis of a microarray experiment. *Journal of the American Statistical Association* 96: 1151–1160.

Eisen, M., Spellman, P., Brown, P., and Botstein, D. (1998). Cluster analysis and display of genome-wide expression patterns. *Proceedings of the National Academy of Sciences, USA* 95: 14863–14868.

Fisher, R.A. (1949). *The Design of Experiments*. Oliver and Boyd Ltd., Edinburgh.

Golub, T., Slonim, D., Tamayo, P., Huard, C., Gaasenbeek, M., Mesirov, J., Coller, H., Loh, M., Downing, J., Caligiuri, M., Bloomfield, C., and Lander, E. (1999). Molecular classification of cancer: class discovery and class prediction by gene expression monitoring. *Science* 286: 531–537.

Hochberg, Y., and Benjamini, Y. (1990). More powerful procedures for multiple significance testing. *Statistics in Medicine* 9: 811–818.

Jin, W., Riley, R.M., Wolfinger, R.D., White, K.P., Passador-Gurgel, G., and Gibson, G. (2001). The contribution of sex, genotype and age to transcriptional variance in *Drosophila melanogaster*. *Nature Genetics* 29: 389–395.

Kerr, M.K., and Churchill, G.A. (2001a). Statistical design and the analysis of gene expression microarray data. *Genetical Research* 77: 123–128.

Kerr, M.K., and Churchill, G.A. (2001b). Experimental design for gene expression microarrays. *Biostatistics* 2: 183–201.

Kerr, M.K., Afshari, C.A., Bennett, L., Bushel, P., Martinez, J., Walker, N. and Churchill, G.A. (2002). Statistical analysis of a gene expression microarray experiment with replication. *Statistica Sinica* 12: 203–218.

Kirk, R.E. (1982). *Experimental Design*. Brooks Cole Publishing Co., Pacific Grove, CA.

Laird, N.M., and Ware, J.H. (1982). Random-effects models for longitudinal data. *Biometrics* 38: 963–974.

Littell, R., Milliken, G., Stroup, W., and Wolfinger, R. (1996). *SAS (c) Systems for Mixed Models*. SAS Institute Inc., Cary, NC.

Muller, K.E., and Fetterman, B.A. (2003). *Regression and ANOVA: A Unified Perspective*. SAS Institute Inc., Cary, NC.

Neter, J., Kutner, M.H., Nachtsheim, C.J., and Wasserman, W. (1996). *Applied Linear Statistical Models*, 4th ed. R.D. Irwin, Chicago, IL.

Rosner, B. (1995). *Fundamentals of Biostatistics*, 4th ed. Duxbury Press, Belmont, CA.

Sudarsanam, P., Iyer, V., Brown, P., and Winston, F. (2000). Whole-genome expression analysis of snf/swi mutants of *Saccharomyces cerevisiae*. *Proceedings of the National Academy of Sciences, USA* 97: 3364–3369.

Tusher, V., Tibshirani, R., and Chu, G. (2001). Significance analysis of microarrays applied to the ionizing radiation response. *Proceedings of the National Academy of Sciences, USA* 98: 5116–5121.

Wolfinger, R., Gibson, G., Wolfinger, E., Bennett, L., Hamadeh, H., Bushle, P., Afshari, C., and Paules, R. (2001). Assessing gene significance from cDNA microarray expression data via mixed models. *Journal of Computational Biology* 8: 625–637.

13 The Use of Nonparametric Procedures in the Statistical Analysis of Microarray Data

T. Mark Beasley, Jacob P.L. Brand, and Jeffrey D. Long

CONTENTS

13.1 Introduction 246
13.2 Motivating Example 248
13.3 Nonparametric Bootstrap 249
13.3.1 Nonparametric Bootstrap Methodology 249
13.3.2 Statistics for Bootstrap Methods 250
13.3.2.1 Parametric Equal Variance t-Test 250
13.3.2.2 Parametric Unequal Variance t-Test 251
13.4 Permutation-Based Nonparametric Methods 251
13.4.1 Permutation Test Based on Mean Differences 252
13.4.2 Permutation Tests Based on Information from All Genes 252
13.4.3 Rank-Based Approach 254
13.4.4 Null Hypothesis for Rank-Based Statistics 255
13.4.5 Cliff's δ Statistic 257
13.4.6 Kolmogorov–Smirnov Statistic 259
13.5 Chebby Checker Methods 259
13.5.1 Chebby Checker Method 1: $CC_{(1)}$ 260
13.5.2 Chebby Checker Method 2: $CC_{(2)}$ 260
13.5.3 Chebby Checker Method 3: $CC_{(3)}$ 260
13.6 Discussion 260
References 262

13.1 INTRODUCTION

One of the most widely used experimental designs for microarray studies involves obtaining gene expression data on samples of cases from two or more ($J \geq 2$) populations that differ with respect to some characteristic. Thus, analysis of variance (ANOVA) models are typically used to test whether gene expression levels differ across the populations for each gene. It can be expressed with the following linear model:

$$Y_{ij(k)} = \mu_{*(k)} + \beta_{j(k)} + \varepsilon_{ij(k)} \tag{13.1}$$

where j refers to the J conditions, groups, or populations (i.e., between-subjects factor), i refers to the n_j samples nested within the jth group, k refers to the K number of genes, $\varepsilon_{ij(k)}$ is a random error vector for the kth gene, $\beta_{j(k)}$ is the differential expression effect for the kth gene, and $N = \Sigma n_j$ is the total number of subjects. $Y_{ij(k)}$ is the expression level. A log transformation is often used in order for $\beta_{j(k)}$ to reflect differential expression ratio. An F-ratio, $F_{(k)}$, can be used to test whether there are statistically significant differences in expression for the kth gene:

$$F_{(k)} = \frac{\sum_{j=1}^{J} n_j(\bar{Y}_{j(k)} - \bar{Y}_{*(k)})^2/(J-1)}{\sum_{j=1}^{J} \sum_{i=1}^{n_j} (Y_{ij(k)} - \bar{Y}_{j(k)})^2/(N-J)} \tag{13.2}$$

The F-ratio is distributed as $F_{[(J-1),(N-J)]}$ under the null hypothesis:

$$H_{0(k)}: \beta_{j(k)} = 0, \quad \text{for all } j \text{ and for the } k\text{th gene.} \tag{13.3}$$

In using the parametric F-ratio, the random error components ($\varepsilon_{ij(k)}$) for the kth gene are assumed to be independent and identically distributed with a mean of zero, a homoscedastic (constant) variance, ($\sigma^2_{\varepsilon(k)}$), and normal shape for each group (i.e., NID[0, $\sigma^2_{\varepsilon(k)}$]). By requiring identical error distributions, it can be assured that a rejection of the null hypothesis in Equation 13.2 is due to *shifts* (differences) among *location parameters*. Furthermore, assuming normal error distributions *means* as estimates of location and the parametric F-ratio will yield the maximum statistical power for rejecting Equation 13.3.

There are three major issues that complicate the application of the parametric F-ratio to data from microarray experiments. First, this large number of variables (K) to small number of total cases (N) ratio renders multivariate solutions difficult and in some cases intractable. Thus, investigators often conduct multiple univariate tests to evaluate whether expression levels differ across the populations (groups) for each gene, thus yielding thousands of statistical tests and presenting challenging multiple testing issues. As Gibson (2002) stated, "... it is not clear how to assess the appropriate level of significance for microarrays in which thousands of comparisons are performed..." (p. 20). However, if investigators test each gene separately without any correction for multiple testing, then the experiment-wise Type I error rate will be unacceptably high (Tusher et al., 2001). Therefore, with thousands of variables to test, microarray researchers are encouraged to apply a correction for multiple testing (Allison and Coffey, 2002; Wu, 2001).

Second, because of the relatively high costs of microarray of research, the sample sizes tend to be quite small. For example, Kayo et al. (2001) used only $n_j = 3$ monkeys per group when studying aging and caloric restriction. Corominola et al. (2001) used only $n_j = 4$ humans per group when studying obesity and diabetes in humans. Currently, sample sizes in microarray studies rarely exceed $N = 20$.

Finally, gene expression data is often noisy and frequently demonstrates departures from normality and homoscedasticity. At times these departures can be extreme (Hunter et al., 2001) with no known distributional form (Troyanskaya et al., 2002). Moreover, it is extremely unlikely that the normality and homoscedasticity assumptions will hold for every one of the thousands of genes tested.

This problem of a large K/N ratio is pernicious because it presents several other complications. First, the parametric F-ratio in Equation 13.2 is known to be robust to most violations of the normality assumption; however, this robustness relies on the central limit theorem and therefore depends on the degree of nonnormality (i.e., skew and kurtosis) and sample sizes. Due to small sample sizes, parametric statistical tests of the differences between the mean levels of gene expression for each of the genes will be more sensitive to the assumed distributional forms of the expression data (i.e., normality); therefore, resulting p-values may not be accurate when there are departures from normality. Moreover, by applying an alpha adjustment due to the large number of tests, the resulting alpha will be extremely small. It is well documented that even when parametric tests are robust at, for example, the $\alpha = .05$ level under violations of normality (Boneau, 1960; Box and Watson, 1962), they are often far less robust when very small alpha levels are examined (Bradley, 1968; Hotelling, 1961). Thus, microarray researchers may not be able to rely on the asymptotic properties and robustness of parametric tests with such small sample sizes and small alphas. Unfortunately, it is believed that the normality assumption is violated frequently in microarray research; however, estimates of skew and kurtosis are also unstable with small samples. Therefore, applied researchers cannot test whether data are normally distributed with any reasonable degree of precision or power (Esteban et al., 2001). In this context, nonparametric methods that do not assume a specific distribution of data are desirable.

One form of nonparametric inference provides an approach to testing whether two or more samples differ in their *distribution* of gene expression without assuming normality. This often referred to as "fully nonparametric" inference. Another form of nonparametric inference provides a *robust alternative* to testing whether two or more samples differ in their mean gene expression level without assuming the test statistic approximates, a known statistical distribution under the null hypothesis.

The basic idea behind nonparametric methods is to estimate a referent (e.g., null) distribution (T_0) of a test statistic, $\hat{t}$, by constructing a corresponding "null statistic" t^* under the *complete* null hypothesis that *all* the genes have no differential expression:

$$H_0\colon \beta_{j(k)} = 0, \quad \text{for all } j \text{ and all } k \text{ genes} \tag{13.4}$$

Under the complete null hypothesis, it is reasoned that a large number of samples drawn from the distribution of the null statistic (t^*) provides a good approximation to T_0, the null distribution of $\hat{t}$. Thus, the null statistic, t^*, is computed from multiple resamples of the actual data and the distribution of the t^* values ($\hat{T}_0$) is used to estimate the referent distribution such that statistical inference can be accomplished by using $\hat{t}$ and its estimated referent distribution, $\hat{T}_0$, without relying on distributional assumptions or asymptotic theory. The performance of such methods depends on how well the distribution of t^* approximates the null distribution of T_0. There are two major approaches to estimating a null referent distribution using resampling: Permutation and Bootstrap. Currently, the most popular way of constructing t^* is to take it as a permutational version of T_0, which has been used in microarray applications (Xu et al., 2002). Bootstrap methods offer a potential by more powerful alternative to permutation-based tests. A third nonresampling method based on Chebyshev's Inequality will be discussed briefly.

13.2 MOTIVATING EXAMPLE

To motivate these issues, consider a hypothetical example in which a microarray experiment examining the changes in gene expression in response to the application of TNF-α in normal rheumatoid arthritis synovial fibroblasts (RASF) cells compared with RASF cell where Ad-IκB-DN has blocked the action of TNF-α. RASF are abnormal cells found in the synovium around joints in individuals with rheumatoid arthritis (RA) and are associated with inflammation of the joints (Mountz and Zhang, 2001). In addition, consider that these $J = 2$ groups are compared for differential expression across 12,625 genes. In this two-sample case, the null hypothesis in Equation 13.2 reduces to:

$$H_{0(k)}: \mu_{1(k)} = \mu_{2(k)} \tag{13.5}$$

where $\mu_{j(k)}$ is the population mean for the jth group on the kth gene. In this case, the F-ratio (13.2) reduces to the independent samples t-test (see Section 13.3.2.1).

Using standard adjustments for multiple tests yields extremely small alphas, which reduces statistical power. For example, with $\alpha = .05$ and $k = 12{,}625$ variables (genes) the Bonferroni adjustment results in $\alpha_{\text{BON}} = \alpha/k = 0.00000396$; whereas, the Dunn–Šidák (Dunn, 1961; Šidák, 1967) correction yields a slightly larger value, $\alpha_{\text{DS}} = (1 - (1 - \alpha)^{1/k}) = 0.00000406$. These adjustments are based on conducting multiple independent tests. However, it is extremely likely that there is some correlation (i.e., dependence) among several tests conducted on data from the same study, and thus, the Bonferroni and Dunn–Šidák adjustments tend to be overcorrections for the probability of at least one Type I error. Allison and Beasley (1998) proposed a Monte Carlo-based procedure for adjusting alpha when multiple dependent tests are conducted. However, with the small sample sizes available to microarray researchers, it may not be possible to obtain stable and accurate correlation coefficients necessary for this procedure. Other approaches use gene-specific scatter (i.e., Tusher et al., 2001) or empirical Bayes methodology (Efron and Tibshirani, 2002) and false discovery rate methods (Benjamini and Hochberg, 2000) to incorporate information from other genes and potentially reduce the number of tests conducted (see Section 13.2.4). Allison et al. (2002) proposed a method wherein the p-values from the multiple univariate tests are used in context with false discovery rate methods to reduce the number of tests in second level analyses. Suppose that after using Allison et al. (2002) procedure, the number of suspected genes has been determined to be $k = 800$. With $\alpha = .05$ and $k = 800$ tests (suspected genes), the Bonferroni adjustment results in $\alpha_{\text{BON}} = \alpha/k = 0.0000625$. Therefore, even with such methods for reducing test size, a large number of tests still results, which calls for a very small per-test alpha level, regardless of the adjustment procedure.

Eventhough we will focus on the two sample tests based on the example presented, the methods herein do generalize to multiple groups and more complicated designs (see Kerr and Churchill, 2000, Chapter 7). Since we restrict our following discussion to the two-sample experiment, let $x_1, \ldots, x_m$ and $y_1, \ldots, y_n$ be two independent samples with sample sizes m and n, with $N = n + m$, for notational convenience.

13.3 NONPARAMETRIC BOOTSTRAP

For nonparametric bootstrap tests, resamples are generated by drawing *with replacement* from the data (Efron and Tibshirani, 1993). The bootstrap is a method to "nonparametrically" produce the referent distribution and compute p-values. It is specified by a test-statistic $\hat{t}$ computed from the original two samples $x_1, \ldots, x_m$ and $y_1, \ldots, y_n$ and a bootstrap test-statistic t^* computed from any bootstrap resample $x_1^*, \ldots, x_m^*, y_1^*, \ldots, y_n^*$ where $x_1^*, \ldots, x_m^*$ is randomly drawn from $x_1, \ldots, x_m$ with replacement and $y_1^*, \ldots, y_n^*$ is randomly drawn from $y_1, \ldots, y_n$ with replacement and independent from $x_1^*, \ldots, x_m^*$. Before we further describe the details in the nonparametric bootstrap method, we first describe the general framework of the bootstrap in the two-sample hypotheses situation.

13.3.1 Nonparametric Bootstrap Methodology

Let T_0 be the reference distribution of $\hat{t}$ under the null-hypothesis H_0 (13.4) and assume that $\hat{t} > 0$, which can be realized by taking the absolute value of $\hat{t}$. The unknown p-value with respect to H_0 based on $\hat{t}$ and T_0 can then be expressed as $P_{T_0}(t > \hat{t})$, where t is a random variable with probability distribution T_0 and $\hat{t}$ is treated as a fixed value. Let the bootstrap test-statistic t^* be defined in such a way that the probability distribution of t^* under randomly drawing the bootstrap resamples with replacement from the original samples $\hat{T}_0$ can be regarded as an approximation of the unknown reference distribution T_0 of $\hat{t}$ under H_0. With the bootstrap method, the unknown p-value $P_{T_0}(t > \hat{t})$ is approximated by the probability $P_{\hat{T}_0}(t^* > \hat{t})$ where t^* is a random variable with probability distribution $\hat{T}_0$ and $\hat{t}$ is treated as a fixed value.

Let $x_1^*, \ldots, x_m^*, y_1^*, \ldots, y_n^*$ be a particular bootstrap resample generated by random sampling with replacement. For the two-group design, the probability that this particular resample occurs equals $p = p_x p_y$ with $p_x = b_x/m^m, p_y = b_y/n^n, b_x = m!/(k_1^x! \cdots k_m^x!)$ and $b_y = n!/(k_1^y! \cdots k_n^y!)$, and k_i^x and k_j^y are the numbers of times the observation x_i occurs in $x_1^*, \ldots, x_m^*$ and the observation y_j occurs in $y_1^*, \ldots, y_n^*$ (Fisher and Hall, 1991). Therefore, b_x and b_y are the numbers of permutations of $x_1^*, \ldots, x_m^*$ and $y_1^*, \ldots, y_n^*$.

For small sample sizes ($m \leq n \leq 7$) commonly used in microarray studies, the bootstrap resamples are generated by means of exhaustive enumeration rather than by random sampling with replacement via simulation, which is the standard approach in the nonparametric bootstrap (Fisher and Hall, 1991).

Let $x_{1(\nu)}^*, \ldots, x_{m(\nu)}^*, y_{1(\nu)}^*, \ldots, y_{n(\nu)}^*$, $\nu = 1, \ldots, W$

$$W = \binom{2m-1}{m}\binom{2n-1}{n} \tag{13.6}$$

be all possible unordered bootstrap resamples, which can be obtained by generating $x_{1(\nu)}^*, \ldots, x_{m(\nu)}^*$ as a random sample with replacement from $x_1, \ldots, x_m$ and generating $y_{1(\nu)}^*, \ldots, y_{n(\nu)}^*$ as a random sample with replacement from $y_1, \ldots, y_n$. Here an unordered sample means that the order of its values is not taken into account while with ordered samples, this order is as well considered. For instance, the two samples {1.7, 2.3, 3.8, 1.7, 2.3} and {1.7, 1.7, 2.3, 2.3, 3.8} represent the same unordered samples but are different ordered samples.

Let $t_1^*, \ldots, t_W^*$ be the bootstrap test statistics corresponding to $x_{1(\nu)}^*, \ldots, x_{m(\nu)}^*$, $y_{1(\nu)}^*, \ldots, y_{n(\nu)}^*$, and $\nu = 1, \ldots, W, t_{(1)}^* \leq t_{(2)}^* \leq \cdots \leq t_{(W)}^*$ be their ordered values and $p_{(1)}, \ldots, p_{(W)}$ the corresponding probabilities of occurrence of these ordered test-statistics $t_{(1)}^*, \ldots, t_{(W)}^*$ with $p_{(i)} = b_{(i)}/m^m n^n$ using Fisher's formula. It can be shown that $b_{(i)} = b_{(i)}^x b_{(i)}^y$ with $b_{(i)}^x$ and $b_{(i)}^y$ the number of permutations of the first and second group of the corresponding bootstrap resample. Also let s be a nonnegative integer such that $t_{(s)}^* \leq \hat{t} \leq t_{(s+1)}^*$, where $t_{(0)}^* = t_{(0)}^* - 1$ and $t_{(W+1)}^* = t_{(W)}^* + 1$, the bootstrap p-value $P_{\hat{T}_0}(t^* > \hat{t})$ is given by:

$$P_{\hat{T}_0}(t^* > \hat{t}) = P_{\hat{T}_0}(t^* \geq t^*{}_{(s+1)}) = \sum_{i=s+1}^{W} P_{\hat{T}_0}(t^* = t^*{}_{(i)}) = \sum_{i=s+1}^{W} b_{(i)}/m^m n^n = r/B \tag{13.7}$$

with $r = \sum_{i=s+1}^{W} b_{(i)}$ and $B = m^m n^n$. To avoid p-values equal to zero the bootstrap p-value is given by the Monte Carlo p-value $(r+1)/(B+1)$ as proposed in Davison and Hinkley (1997). Notice that r equals the sum of corresponding permutations of bootstrap resamples for which $t^* \geq \hat{t}$. We describe the definitions of $\hat{t}$ and t^* in the following section.

13.3.2 Statistics for Bootstrap Methods

13.3.2.1 Parametric Equal Variance *t*-Test

The original sample test statistic $\hat{t}$ can be defined as the parametric t-test statistic:

$$\hat{t} = |\bar{q}/\mathrm{SE}(\bar{q})| = \sqrt{F} \quad \text{in Equation 13.2} \tag{13.8}$$

where $\bar{q} = (\bar{x} - \bar{y})$, $\text{SE}(\bar{q})$ is the standard error of $\bar{q}$ given by $\text{SE}(\bar{q}) = s\sqrt{1/m + 1/n}$, s is the pooled standard deviation given by $s = \sqrt{((m-1)s_x^2 + (n-1)s_y^2)/(m+n-2)}$, s_x^2 the sample variance of $x_1, \ldots, x_m$, and s_y^2 the sample variance of $y_1, \ldots, y_n$. The bootstrap test statistic $t^*{}_\nu$ is given by $t^*{}_\nu = |\bar{q}_\nu^* - \bar{q}|/\text{SE}_\nu^*(\bar{q})$, where the bootstrap standard error $\text{SE}_\nu^*(\bar{q})$ is computed from the bootstrap resample in exactly the same way as $\text{SE}(\bar{q})$ is computed from the original sample and is given by $\text{SE}_\nu^*(\bar{q}) = s_\nu^*\sqrt{1/m + 1/n}$, with the pooled standard deviation $s^*{}_\nu$ given by $s^*{}_\nu = \sqrt{((m-1)s^{*2}_{x(\nu)} + (n-1)s^{*2}_{y(\nu)})/(m+n-2)}$, where $s^{*2}_{x(\nu)}$ is the sample variance of $x^{*(\nu)}_1, \ldots, x^{*(\nu)}_m$ and $s^{*2}_{y(\nu)}$ is the sample variance of $y^{*(\nu)}_1, \ldots, y^{*(\nu)}_n$. Because $t^*{}_\nu$ is undefined when $\text{SE}^*{}_\nu(\bar{q}) = 0$, the p-value is given by $(r+1)/(B+1)$, where $B = m^m n^n$, $r = \sum_{\nu=1}^{W} b_x^{(\nu)} b_y^{(\nu)} I_\nu$ with $b_x^{(\nu)}$ and $b_y^{(\nu)}$ the number of permutations of $x^{*(\nu)}_1, \ldots, x^{*(\nu)}_m$ and $y^{*(\nu)}_1, \ldots, y^{*(\nu)}_n$ and I_ν the indicator variable defined by:

$$I_\nu = \begin{cases} 1, & \text{if } |\bar{q}_\nu^* - \bar{q}| \geq \dfrac{\text{SE}_\nu^*(\bar{q})}{\text{SE}(\bar{q})}|\bar{q}| \\ 0, & \text{if } |\bar{q}_\nu^* - \bar{q}| < \dfrac{\text{SE}_\nu^*(\bar{q})}{\text{SE}(\bar{q})}|\bar{q}| \end{cases} \tag{13.9}$$

13.3.2.2 Parametric Unequal Variance *t*-Test

It is well known that if the two samples have unequal variances, then using the pooled sample variance leads to a biased estimator of the standard error of the numerator. This can be fixed simply by using the two-sample unequal-variance t-statistic. This variant of the nonparametric bootstrap with pivoting is similar to the previous one except that $\text{SE}(\bar{q})$ is given by $\text{SE}(\bar{q}) = s\sqrt{s_x^2/m + s_y^2/n}$ and $\text{SE}^*{}_\nu(\bar{q})$ is given by $\text{SE}^*{}_\nu(\bar{q}) = \sqrt{s^{*2}_{x(\nu)}/m + s^{*2}_{y(\nu)}/n}$. The resulting test statistic $\hat{t}$ here is the unequal variance t-test statistic. Because $\text{SE}^*{}_\nu(\bar{q}) = 0$ for mn of the bootstrap resamples, the minimal possible p-value is equal to $(mn+1)/(m^m n^n + 1)$.

13.4 PERMUTATION-BASED NONPARAMETRIC METHODS

Permutation-based methods are similar to bootstrap approaches, except that they resample *without replacement* to create a referent distribution. With small sample sizes, it is not difficult to exhaust the number of unique permutations in order to create an exact test. For the two-sample case, the number of possible permutations is

$$r = \binom{m+n}{n} \tag{13.10}$$

The minimal possible p-value for two-sample permutation-based tests is

$$p = \begin{cases} 2/r, & \text{for equal sample size} \\ 1/r, & \text{for unequal sample size} \end{cases} \tag{13.11}$$

TABLE 13.1
Minimum Two-Tailed *p*-values for the Kolmogorov–Smirnov (KS), Wilcoxon–Mann–Whitney (WRS), and Bootstrap Procedures

n	KS	WRS	Bootstrap
3	0.20	0.10	0.02
4	0.0572	0.0285714	0.0016327
5	0.0158	0.0079365	0.0001260
6	0.0044	0.0021645	0.0000094
7	0.0012	0.0005828	0.0000007

This difference is because the empirical distribution of test statistic is symmetric for equal sample sizes and may be asymmetric for unequal sample sizes. Table 13.1 shows the minimum two-tailed p-values of the permutation-based Mann–Whitney U test for several two-group balanced designs with small sample sizes. When sample sizes are larger, these methods often rely on asymptotic theory and use a known statistical distribution (e.g., chi-square) as a referent distribution. However, for more complicated test statistics, the approximation of the estimated referent distribution ($\hat{T}_0$) to a known statistical distribution is often questionable, even with larger sample sizes.

13.4.1 Permutation Test Based on Mean Differences

Gadbury et al. (2003) proposed a method in which the absolute mean difference $\hat{q}$ computed from the two original samples is compared to all absolute mean differences $q_1, \ldots, q_r$ computed from all r (13.10) possible permutations of these two samples. Under the null-hypothesis $H_{0(k)}$ in Equation 13.3 that both samples are generated from identical probability distributions, the empirical distribution of $q_1, \ldots, q_r$ reflects the probability distribution of $\hat{q}$ under $H_{0(k)}$. The p-value for $\hat{q}$ is then given by the proportion of absolute mean differences $q_1, \ldots, q_r$ computed from all possible permutations of the two samples, which tie or exceed the original sample absolute mean difference $\hat{q}$. Bootstrap methods could be applied to this test based on mean difference; however, Brand et al. (2003) have shown that pivotal statistics (e.g., t-test) must be used for bootstrap resampling to yield valid test statistics. Dudoit et al. (2002) suggested the parametric t-test instead of absolute mean difference as the test statistic for permutation. Since the t-test is a pivotal statistic, it can be used in bootstrap application (see Section 13.3).

13.4.2 Permutation Tests Based on Information from All Genes

Several methods that utilize resampling to construct a referent distribution based on null statistics from all genes have been proposed (Tusher, et al., 2001; Efron and Tibshirani, 2002). Under the *complete* null hypothesis 13.4, if *all* the genes have no differential expression, then the permutational distribution of $\hat{t}$ (i.e., the distribution of the null statistic t^*) provides a good approximation to the null distribution of $\hat{t}$. However, with these methods for gene expression data, unlike in other traditional permutation tests such as rank-based tests and a permutation-based t-test (e.g., Dudoit et al., 2002), multiple $H_{0(k)}$ are tested for many genes and the permutational distribution of a statistic is *pooled over all genes*. This results in a less discrete (more continuous) permutation referent distribution and thus, may provide more statistical power, especially with smaller sample sizes.

For the two-sample experiment, an independent-samples t-test is usually in order. However, the unequal variance t-test (Section 13.3.2.2) is often suggested (Tusher et al., 2000). Furthermore, adding a small constant (s_0) to the denominator to stabilize the t-statistic avoids a near-zero denominator for $\hat{t}$ or t^*, which helps control for an inflated value of the test statistic and a reduction in statistical power. The Empirical Bayes (EB) method (Efron and Tibshirani, 2002) defines the constant as the 90th percentile of the distribution of denominators for the $k = 1$ to K genes. The significance analysis of microarrays (SAM) approach (Tusher et al., 2001) uses a more sophisticated approach by choosing the constant as value from the distribution of $\hat{t}$ denominators (see Section 13.3.2.1) that will minimize the coefficient of variation, which is calculated as the square root of the denominator for the kth gene divided by the mean expression level for the kth gene.

These approaches seem to be reasonable for practical purposes and can have better performance than the usual t-statistic (Lönnstedt and Speed, 2002), but unfortunately they can hinder a tractable analysis. For example, even under the normality assumption, the statistic no longer simply has a t distribution. In EB and SAM, it is assumed that under $\mathbf{H}_0$ (Equation 13.4), the $\hat{t}$ statistics of all K genes have the same distribution, leading to the use of the t^* statistics of all the genes to estimate a reference distribution for $\hat{t}$. As usual, this assumption may or may not hold in practice. Nevertheless, at least, this assumption is weaker than that of the t-test (Equation 13.8). Even if the expression levels of different genes are from different distributions, as long as these distributions belong to the same member of a location-scale family, which includes the normal distribution required by the t-test, then the assumption holds (Lehmann, 1998, p. 21).

These methods involve testing multiple $H_{0(k)}$ (Equation 13.3) for many genes and the permutational distribution of a statistic is pooled over all genes. In practice, because real data usually has genes with differential expression, then the distribution of t^* is a mixture of the (estimated) null distribution of $\hat{t}$ (for genes with no differential expression) and a permutation distribution of $\hat{t}$ for genes with differential expression, with the latter introducing extra variation. Hence, using the null statistic t^* may not estimate the null distribution of $\hat{t}$ very well, especially when there are a relatively large proportion of genes with differential expression and the magnitude of their expression change is large. It is possible that in SAM, statistical inference can be accomplished using the permutational distribution of $\hat{t}$, as in other permutation-based tests. However, because the permutational distribution of $\hat{t}$ can be more dispersed than the null distribution of $\hat{t}$, use of the former can lead to reduced statistical power.

Furthermore, strictly speaking, the EB approach requires the estimation (or use) of the null distribution of $\hat{t}$. Therefore, it seems reasonable to expect that the performance of EB and SAM can be improved if the null distribution of the test statistic can be better estimated.

An alternative method of constructing a test statistic and a null statistic was proposed in Pan et al. (2001). In this method, the null distribution is estimated using finite mixture models. Zhao and Pan (2003) pointed out its weakness and proposed two modifications, which may have low efficiency because of their reduced degrees of freedom in estimating the variances. Allison et al. (2001) have proposed a similar method that is detailed in Chapter 4. Pan (2003) also suggested partitioning each of the two samples into two (almost) equally-sized sub-samples. This allows for the centering of the data so that the mean of the numerator of the null statistic is always zero, regardless of whether the complete null hypothesis 13.4 holds or not.

13.4.3 Rank-Based Approach

A commonly used nonparametric approach involves ranking the data form 1 to N and computing the appropriate test statistics, which is the basis for the Wilcoxon (1945) Rank Sum (WRS) test as well as many other rank-based methods. Ranks may be considered preferable to the actual expression level data (Tsodikov et al., 2002). First, if the numbers assigned to the observations have no meaning by themselves but attain meaning only in an ordinal comparison with the other observations, the actual data contain no additional information than the ranks contain (Conover, 1980). Whether this argument applies to gene expression data from microarray studies is debatable. From the extremely large number of genes tested, gene expression data frequently demonstrates departures from extreme departures normality (Hunter et al., 2001) with no known distributional form. Tsodikov et al. (2002) have noted that rank-transformed data will provide a robust adjustment for normalizing expression level data in that ranks are likely to be less sensitive to outliers and model misspecifications than other normalization procedures. Moreover, even if the numbers have meaning but the distributions function is not Gaussian, the probability theory for the test statistics cannot be derived if the distribution function is unknown. Yet, the probability theory of statistics based on ranks is relatively straightforward and does not depend on the distribution in many cases (Bradley, 1968).

The referent distribution for the WRS is based on all unique combinations of ranks for one of the two groups and hence has a bounded distribution, especially with smaller samples. WRS generalizes to the $J > 2$ groups situation and therefore can also be performed by an ANOVA approach where the between-groups rank sum or squares are computed and divided by the known total rank sum of squares to create a test statistic KW that approximates χ^2 with d$f = J - 1$ asymptotically (Kruskal and Wallis, 1952):

$$\mathrm{KW}_{(k)} = \frac{\sum_{j=1}^{J} n_j(\bar{R}_{j(k)} - [(N+1)/2])^2}{(N[N+1])/12} \tag{13.12}$$

KW is statistically equivalent to the two-sample WRS and Mann and Whitney (1946) *U* tests (Hollander and Wolfe, 1973).

13.4.4 Null Hypothesis for Rank-Based Statistics

With the assumption that the error components are independent, have a constant variance, and are sampled from a normal distributed (NID[0, $\sigma^2_{\varepsilon(k)}$]), the parametric test has "assumed away" everything but differences in location, which allows the parametric test to focus on a single parameter. The constant variance (homoscedasticity) assumption leads to the expectation that the groups will have identical variances. The normality assumption excludes differences in skew and kurtosis. All that can be left are difference in location. Assuming normality, the mean is the maximum likelihood estimate of location and a parametric test (2 or 7) is the most powerful, statistically. The NID[0, $\sigma^2_{\varepsilon(k)}$] assumption simply allows the researcher to focus on a single parameter. Thus, the null hypothesis is that the groups have identical means, and by assuming NID[0, $\sigma^2_{\varepsilon(k)}$], the null hypothesis actually involves sampling from identical normal distributions.

For a two-sample design, to test a hypothesis similar to Equation 13.5, rank-based approaches can be used in order to relax the normality assumptions by assuming that the error components for each group have constant variance and are random variables from identical distributions (i.e., IID[0, $\sigma^2_{\varepsilon(k)}$]), not necessarily the normal. All that can be left are differences (shifts) in location; however without assuming normality, the mean may not be the most powerful measure of location. If IID[0, $\sigma^2_{\varepsilon(k)}$] can be assumed, then rank-based tests evaluate a shift model. Furthermore, rank-based tests (e.g., KW) are more powerful than parametric procedures (e.g., *t*-test), especially with identically skewed error distributions. This provides a *robust alternative* to parametric procedures for testing hypotheses such as Equation 13.3 or Equation 13.4, allowing direct inference concerning location parameters (Akritas et al., 1997).

The asymptotic relative efficiency (ARE) of the rank-based WRS test is comparable to the parametric *t*-test (Equation 13.7); however, the contrary does not hold. If the errors are NID[0, $\sigma^2_{\varepsilon(k)}$], the ARE of the WRS compared to the *t*-test is 0.955. If the two populations differ only in their location parameters (i.e., IID[0, $\sigma^2_{\varepsilon(k)}$]), the ARE is never lower than 0.864 but may be as high as infinity (Hodges and Lehmann, 1956). Thus, the rank-based WRS test is often much more powerful that the parametric independent samples *t*-test, especially when error distributions are identically distributed with extreme skew. Therefore, rank-based tests can be "safer to use" in many situations (Conover, 1980).

However, rank-based approaches may not provide a robust alternative if simply applied because of *any* violation of model assumptions. For example, Zimmerman (1996) demonstrated that rank transformed scores "inherit" the heterogeneity of variance in the original data. This leads to inflated Type I error rates when using KW as a robust alternative for testing differences in location parameters when the groups differ in their sample sizes, variance, and shape (Wilcox, 1993; Zimmerman, 1996; Brunner and Munzel, 2000). Based on the Welch (1947) correction, Brunner and Munzel (2000) developed a test for location differences when groups differ in variance; however, it does not provide control of Type I error rates for sample sizes less than 50.

Strictly, the most commonly used parametric procedures test differences in location (e.g., means) because other distributional differences are assumed not to exist (i.e., NID[0, $\sigma^2_{\varepsilon(k)}$]). Thus, in order to test hypotheses concerning shifts in location parameters the assumptions of independence, homogeneity of variance, and identical shape must still preside (Serlin and Harwell, 2001). Specifically, credible inferences about means require the assumption that the population distributions are symmetric (Koch, 1969; Serlin and Harwell, 2001); whereas, credible inferences concerning location parameters in general, require the assumption that the population distributions are of identical shape, not necessarily symmetric (i.e., IID[0, $\sigma^2_{\varepsilon(k)}$]). This frequently overlooked detail is a major reason that so much attention has been given recently to rank-based procedures as tests of "stochastic homogeneity" (Vargha and Delaney, 1998), "distributional equivalence" (Agresti and Pendergast, 1986; Beasley, 2000), or "fully nonparametric hypotheses" (Akritas and Arnold, 1994). Therefore, without assuming IID[0, $\sigma^2_{\varepsilon(k)}$], rank-based methods are "*fully nonparametric*" tests and are *not* viewed as *robust alternatives* to normal theory methods for testing differences in means (Equation 13.3). Rather, statistically significant fully nonparametric tests may be attributed to differences among *any* distributional characteristic (e.g., location, dispersion, and shape).

As a departure from parametric models that test differences means, general "fully nonparametric" models, which specify that only observations in different groups are governed by different distribution functions, have been developed for a variety of experimental designs (Akritas and Arnold, 1994; Brunner and Langer, 2000). Specifically, fully nonparametric means that these methods make no assumptions about the distribution of the error term (e.g., normality; constant variance). Thus, a sufficiently large test statistic indicates that the two groups significantly differ in their *distribution* of expression levels. We conclude that if a test statistic becomes large enough to become a "significant result" when the normality or homoscedasticity assumptions are not met, even though population means are identical, then it is still a valuable result to microarray researchers (see Cliff, 1996 for a similar argument in behavioral sciences).

Strictly, WRS and KW tests the null-hypothesis that both samples are generated from identical probability distributions:

$$H_{0(k)}: G(x)_{(k)} = H(y)_{(k)} \tag{13.13}$$

where G and H are distribution functions for the two independent samples. As previously mentioned, assuming NID[0, $\sigma^2_{\varepsilon(k)}$] or IID[0, $\sigma^2_{\varepsilon(k)}$] differences in probability distributions can only be due to shifts in location parameters. Hypotheses of this form reduce the risk of drawing incorrect conclusions about the likely sources of the significant effect, but do so at the cost of not being able to characterize precisely (i.e., focus on a single parameter) how population distributions differ (Serlin and Harwell, 2001). Because the methods herein are based on test statistics that employ mean differences, they are most sensitive to differences in location (Bradley, 1968). However, a significant result may also be attributable to difference in location, spread, and shape (Wilcox, 1993). Therefore, even if assumptions concerning identical distributions and homogeneous variances are not tenable, the researcher may still conclude that one group is stochastically dominant over the other (Vargha and Delaney, 1998).

Rank methods generalize to multiple group experiments; however, with multiple groups the multiple testing issues compound. For example, if there were $J = 4$ groups, there would be six possible pairwise comparisons, and thus, the number of tests conducted grows. We suggest that planned comparisons should be considered if multiple group designs are used. If more complicated designs that involve multiple factors and potential interaction terms (see Kerr and Churchill, 2000) are employed, then rank (Conover, 1980) and aligned rank testa (Salter and Fawcett, 1993) are available. Recently, mixed linear models have been suggested as a way to analyze data from microarray experiments (Guo et al., 2003). "Fully nonparametric" mixed linear model approaches (Brunner et al., 2002) and aligned rank approaches (Beasley, 2002) are available.

13.4.5 Cliff's δ Statistic

Similar to the WRS, Cliff's (1993) $\hat{\delta}$ statistic is a two-sample hypothesis test for ordinal data. In many applied situations, any variation in the difference between distributions implies that $P(y_j > x_i)$ is not equal to $\frac{1}{2}$ (Conover, 1980). Cliff's $\hat{\delta}$ method is a modification of the Fligner and Policello (1981) test for the Behrens–Fisher problem. Zumbo and Coulombe (1997) found that when testing the hypothesis of equal location parameters (i.e., under the IID[0, $\sigma^2_{\varepsilon(k)}$] assumption) for symmetric distributions, the Fligner–Policello test was conservative, whereas it performed inconsistently for skewed populations; however, studies have found Cliff's method to perform in a more stable manner (Long and Cliff, 1997). Thus, we restrict our discussion to Cliff's method.

Cliff's method was developed to directly test the null hypothesis of the probability that the response of a randomly selected member y_i in one population is higher than that of a randomly selected member x_j in the other, and is equal to the reverse probability:

$$H_{0(k)}\colon \delta = P(y_j > x_i) - P(y_j < x_i) = 0; \quad i = 1, \ldots, m; \quad j = 1, \ldots, n \qquad (13.14)$$

which is much more general than the null hypothesis for WRS and KW (Equation 13.13). Cliff (1993) shows that in the two-sample case, $\hat{\delta}$ is a linear transformation of the Mann–Whitney U statistic. If all $N = n + m$ scores are ranked and R_y are the ranks of the ny scores for the first group, then

$$\begin{aligned} U &= \Sigma R_y - n(n+1)/2 \\ \hat{\delta} &= 2U/nm - 1 \end{aligned} \qquad (13.15)$$

It can also be shown that it is identical to the form of Kendall's tau called Somers (1968) d. Thus, Cliff's $\hat{\delta}$ is related to the familiar WRS or U statistics. However, the inferential method for the WRS null hypothesis 13.13 is that the two distributions are identical. This is a limiting constraint in that the distributions could differ in shape or spread with and without $\delta = 0$ in Equation 13.14 true (see Cliff, 1996). In particular, δ is the population overlap, which does not depend on any of the assumptions regarding the population distributions. Therefore, the test of Equation 13.14 does not depend on randomization so the null hypothesis does not imply identical distribution functions. Since the Mann–Whitney U is linearly related to Cliff's d via $\hat{\delta} = 2U/nm - 1$ (Cliff, 1993) and U, WRS, and KW are statistically equivalent, for larger sample sizes WRS and Cliff's $\hat{\delta}$ are approximately statistically equivalent asymptotically.

Cliff's method computes the variance of $\hat{\delta}$ in different manner than the Fligner–Policello test. Statistical inference for Cliff's $\hat{\delta}$ involves constructing an asymmetric confidence interval for δ based on a sample estimate $(\hat{\delta})$ and a consistent estimate of the variance, $\hat{\sigma}^2$ (Long and Cliff, 1997). Let $\{d_{ij}\}$ be the dominance matrix defined by:

$$d_{ij} = \begin{cases} 1, & \text{if } y_i > x_j \\ 0, & \text{if } y_i = x_j \\ -1, & \text{if } y_i < x_j \end{cases} \tag{13.16}$$

Let $\hat{\delta}$ be the delta statistic given by $\hat{\delta} = \sum_{i=1}^{n} \sum_{j=1}^{m} d_{ij}/mn$. If $|\hat{\delta}| < 1$, an asymmetric $(1 - \alpha)100\%$ confidence interval for $\delta = P(y_i > x_j) - P(y_i < x_j)$ is given by

$$\mathrm{CI}(1-\alpha) = \frac{\hat{\delta} - \hat{\delta}^3 \pm z_{\alpha/2}\hat{\sigma}\sqrt{(1-\hat{\delta}^2)^2 + z_{\alpha/2}^2\hat{\sigma}^2}}{1 - \hat{\delta}^2 + z_{\alpha/2}^2\hat{\sigma}^2} \tag{13.17}$$

where $z_{\alpha/2}$ is the $(1 - \alpha/2)$th quantile of the standard normal distribution, $\hat{\sigma}^2 = ((n-1)s_{i.}^2 + (m-1)s_{.j}^2 + s_{..}^2)/mn$ an unbiased estimate of the variance of $\hat{\delta}$ with $s_{i.}^2 = \sum_{i=1}^{n}(\bar{d}_{i.} - \hat{\delta})^2/(n-1)$, $s_{.j}^2 = \sum_{j=1}^{m}(\bar{d}_{.j} - \hat{\delta})^2/(m-1)$, $\bar{d}_{i.} = \sum_{j=1}^{m} d_{ij}/m$ the column marginals of $\{d_{ij}\}$, and $\bar{d}_{.j} = \sum_{i=1}^{n} d_{ij}/n$ the row marginals of $\{d_{ij}\}$.

For $|\hat{\delta}| < 1$, the p-value from the Cliff's δ method is defined via the confidence interval $\mathrm{CI}(1-\alpha)$. If zero is not included in $\mathrm{CI}(1-\alpha)$ then the p-value is smaller than or equal to α. If $|\hat{\delta}| = 1$ then $\mathrm{CI}(1-\alpha)$ is not defined because the denominator $1 - \hat{\delta}^2 + z_{\alpha/2}^2\hat{\sigma}^2$ of $\mathrm{CI}(1-\alpha)$ is zero. This is because $1 - \hat{\delta}^2 = 0$ and $|\hat{\delta}| = 1$ if and only if all elements of the dominance matrix $\{d_{ij}\}$ are 1 or all elements of this matrix are -1, which implies that $|\hat{\delta}| = 1$ if and only if $\hat{\sigma}^2 = 0$. For $|\hat{\delta}| = 1$, which happens when both samples are nonoverlapping (i.e., all values in one sample are smaller than all values in the other) the asymmetric confidence interval is not defined. However, p-value can be assigned based on the permutation distribution that underlies the data. There are extensions to factorial designs and repeated measures (see Cliff, 1996).

13.4.6 Kolmogorov–Smirnov Statistic

The Kolmogorov test was modified by Smirnov to serve as a test of the null hypothesis that two samples come from identical populations (Massey, 1951). The test is given as:

$$\mathrm{KS}_{(k)} = \max |\dot{P}[x < a] - \dot{P}[y < a]| \tag{13.18}$$

where $\dot{P}[x < a]$ is the cumulative probability of sample x and $\dot{P}[y < a]$ is the cumulative probability of sample y. This statistic can also be framed in terms of cumulative distribution functions. Thus, this statistic tests a null hypothesis that is similar to the null hypotheses for the rank-based test (13.13) and Cliff's δ method (13.14). Furthermore, Hajek (1969) showed that the $\mathrm{KS}_{(k)}$ test can be expressed in terms of the ranks. An important consequence of the fact the value of $\mathrm{KS}_{(k)}$ depends only on the ranks of the observations is that its null distribution is determined by permutation distribution. However, the permutation distribution of $\mathrm{KS}_{(k)}$ is cruder (i.e., has fewer possible p-values) than the permutation distribution for the WRS or KW. Therefore, is likely to be less powerful than the KW or Cliff's statistics. Furthermore, since the KW is especially sensitive to location difference, the $\mathrm{KS}_{(k)}$ will tend be less powerful under the particular alternative hypothesis that the treatment somehow increases (or decreases) the gene expression level.

13.5 CHEBBY CHECKER METHODS

Permutation and bootstrap methods have a noted limitation when sample sizes are small. Specifically, in many situations, it is impossible to obtain "$p < .05$," let alone the far smaller p-values required if the significance level is corrected for multiple testing if a permutation, bootstrap, or any other resampling-based statistic is employed. Nevertheless, even with small sample sizes, at times, a group difference is so large that common sense suggests that the observed sample difference is significant regardless of the fact that nonparametric or bootstrap methods cannot yield significant p-values or the fact that normality assumptions are virtually impossible to assess. Real-life examples of huge differences in gene expression are readily found in the literature (see Beasley et al., 2003). One example is the effect of knocking out the interleukin-6 (IL-6) gene, which causes roughly a 35-fold change in expression of IGFBP-1 gene (Li et al., 2001). Such results sometimes lead biologists state that they do not need statisticians to tell them a difference is "real" and if statisticians say the difference is not significant, something must be wrong with statisticians or their methods.

Beasley et al. (2003) developed the *Chebby Checker* methods based on variants of Chebyshev's inequality. These methods are nonparametric and yet theoretically capable of yielding p-values that are continuous on the interval $0 < p \leq 1$ with *any* $N > 4$ cases (i.e., a balanced two-group experiment with $[n = m] > 2$). Three different versions of the Chebby checker method are presented. Each of these methods provides an upper bound for a p-value based on Chebyshev's original inequality or a variant of this inequality.

13.5.1 Chebby Checker Method 1: $CC_{(1)}$

The upper bound for a p-value of this method is based on Chebyshev's original inequality $P(|(\tau - \mu_\tau)/\sigma_\tau| \geq T) \leq 1/T^2$, where τ is any stochastic variable and μ_τ and σ_τ are the mean and standard deviation of τ. Let $\tau = |t/\sigma_t|$, with t being the ordinary t-test statistic given in Equation 13.11 and $\sigma_t = \sqrt{(m+n)/(m+n-2)}$ is the standard deviation of t under normality. Then the p-value for $CC_{(1)}$ is $p_{CC(1)} = 1/\tau^2$.

13.5.2 Chebby Checker Method 2: $CC_{(2)}$

The upper bound for a p-value from the $CC_{(2)}$ method is based on the modified Chebyshev's inequality $P(|(\tau - \mu_\tau)/\sigma_\tau| \geq T) \leq 1/3T^2$ proved by DasGupta (2000) in the case of a normal distribution of τ and obtained by multiplying the right-hand side of the original Chebyshev's inequality by 1/3. The resulting upper bound of the p-value for $CC_{(2)}$ is $p_{CC(2)} = 1/3\tau^2$.

13.5.3 Chebby Checker Method 3: $CC_{(3)}$

The upper bound for a p-value from the $CC_{(3)}$ method is based on a combination of the modified Chebyshev's equality $P(|(\tau - \mu_\tau)/\sigma_\tau| \geq T) \leq 1/((m+n)T^2)$ resulting from combining the inequality proved by Saw et al. (1984) with the original Chebyshev's inequality, and the ordinary t-test p-value. Let $p_{(C)} = 1/((m+n+1)\tau^2)$ and $p_{(t)} = 2P(t_{m+n-2} > \tau)$ be the p-value of the ordinary t-test (13.11). The $CC_{(3)}$ p-value is given by $p_{CC(3)} = \max\{p_{(C)}\ p_{(t)}\}$.

13.6 DISCUSSION

The small sample sizes and large number of hypothesis tests common to microarray research present a conundrum to microarray researchers as well as statisticians attempting to develop methods for microarray analysis. This is because it creates a situation wherein the researcher is trying to reject a null hypothesis at an extremely small alpha (due to correction for multiple testing) with a small sample, thus leading to a situation with extremely low statistical power. This problem is further compounded with the fact that microarray data frequently demonstrates departures from extreme departures normality (Hunter et al., 2001) with no known distributional form. This leads to a violation of the normality assumption underlying parametric tests. With small sample sizes, the asymptotic and robustness properties of parametric procedures may not apply. Furthermore, parametric procedures show a reduction in power as data become more skewed. In such cases, statisticians often suggest rank, nonparametric, permutation, or bootstrap procedures as robust alternatives to testing the difference in means (location parameters) of the expression level across conditions.

Ranks may be considered preferable to the actual expression level data because rank-transformed data will provide a robust adjustment for normalizing expression level data. That is, ranks are likely to be less sensitive to outliers and model misspecifications than other normalization procedures (Tsodikov et al., 2002). However, the effectiveness of rank-based statistics applied to gene expression data from microarray studies is debatable. The probability theory of statistics based on ranks is relatively straightforward and does not depend on the distribution in many cases (Bradley, 1968). When sample sizes are larger, these methods often rely on asymptotic theory and use a known statistical distribution (e.g., chi-square) as a referent distribution. However, for more complicated test statistics, the approximation of the estimated referent distribution ($\hat{T}_0$) to a known statistical distribution is often questionable, even with larger sample sizes.

Rank-based as well as other permutation-based tests have severe limitations in the context of sample sizes and small alpha. Consider the study by Lee et al. (2000), which used three mice per group. With $n = 3$ per group, conventional nonparametric tests for comparing two groups cannot possibly yield two-tailed p-values $<.10$. This is because conventional rank-based nonparametric tests (see Section 13.4.3) are based on the number of group combinations of ranks and there are a limited number of unique combinations that exist for finite data sets (Di Bucchiano, 1999). This is also the case for the Kolmogorov–Smirnov test (Section 13.4.6), Cliff's δ statistic (Section 13.4.5), and the permutation-based test (Section 13.4.1). Thus, in many situations, it is impossible to obtain "$p < .05$," let alone the far smaller p-values required if the significance level is corrected for multiple testing if a rank-based or any other permutation-based statistic is employed (see Equation 13.11).

Bootstrap techniques are often suggested as an alternative (Kerr and Churchill, 2001) because they need not assume normality or homogeneity of variance (Good, 1999) and are therefore less restrictive. If one chooses the bootstrap as an alternative, a similar complication arises when resampling from very few cases (see Equation 13.12). If sample sizes are very small (e.g., $n < 5$), p-values will be affected by the discreteness of the bootstrapped distribution and there will be a limited number of "distinct" p-values. Furthermore, the validity of bootstrap methods as robust alternatives to testing differences in mean expression levels when parametric assumptions are violated has come into question.

Thus, with small sample sizes, these procedures have discrete distribution and low statistical power, thus nullifying any expected power advantages. Table 13.1 shows the minimum two-tailed p-values of the KS test, WRS (KW) test, and bootstrap technique for several two-group balanced designs with small sample sizes. Beasley et al. (2003) developed the *Chebby Checker* methods (Section 13.5) as nonparametric procedures that are theoretically capable of yielding p-values that are continuous on the interval $0 < p \leq 1$ with *any* $N > 4$ cases. However, they admittedly demonstrate extremely low power (Brand et al., 2003).

Procedures such as EB and SAM attempt to correct the problem of discrete referent distributions by creating an empirical permutation test based on information from all genes in the microarray and adding a correction factor for this pooling. This procedure would seem to add more statistical power because the permutation distribution would be more continuous and thus could yield smaller p-values as compared to permutation or bootstrap tests based on sample size alone. However, the EB and SAM procedures use the unequal variance t-test with a correction factor for the denominator. The unequal variance t-test was designed to test the null hypothesis that both samples have equal population means (13.5); however, simulation studies have shown that this procedure is not robust to unequal variances between the two groups if the error components are not normally distributed (Algina et al., 1994). Similarly, EB and SAM were designed to test differences in means (13.5) and thus may be not be robust to differences in variances or shape across conditions, especially when there are small and unequal sample sizes.

Yet, there is no strong or empirically supported reason to suspect or assume that gene expression level data will have a constant variance and identical shapes across treatment conditions. Therefore, tests for detecting stochastic heterogeneity in the form of differences in distributional shape (13.14) or population overlap (13.15) may be much more appropriate (see Cliff, 1996). Unfortunately, tests for identical distributions (e.g., KW) or no population overlap (Cliff's δ) may have little to no power when sample sizes are very small due to the finite nature of their permutation distributions. Therefore, modifications of the EB or SAM procedures that use methods similar to KW or Cliff's δ should be a fruitful area for future statistical developments in microarray analysis. This would give these nonparametric tests a more continuous referent distribution and thus enhance statistical power. For skewed data, such a nonparametric modification of SAM (or EB) may provide more statistical power than the original methods based on parametric statistics.

REFERENCES

Agresti, A. and Pendergast, J. (1986) Comparing mean ranks for repeated measures data. *Comm. Stat. — Theor. Method*, 15, 1417–1433.

Akritas, M.G. and Arnold, S.F. (1994) Fully nonparametric hypotheses for factorial designs I: multivariate repeated-measures designs. *J. Am. Stat. Assoc.*, 89, 336–343.

Akritas, M.G., Arnold, S.F., and Brunner, E. (1997) Nonparametric hypotheses and rank statistics for unbalanced factorial designs. *J. Am. Stat. Assoc.*, 92, 258–265.

Algina, J., Oshima, T.C., and Lin, W.Y. (1994) Type I error rates for Welch's test and James's second-order test under nonnormality and inequality of variance when there are two groups. *J. Educ. Behav. Stat.*, 19, 275–291.

Allison, D.B. and Beasley, T.M. (1998) Method and computer program for controlling the family-wise alpha rate in gene association studies involving multiple phenotypes. *Genet. Epidemiol.*, 15, 87–101.

Allison, D.B. and Coffey, C.S. (2002) Two-stage testing in microarray analysis: what is gained? *J. Gerontol. A Biol. Sci. Med. Sci.*, 57, B189–B192.

Allison, D.B., Gadbury, G.L., Heo, M., Fernández, J.R., Lee, C.-K., Prolla, T.A., and Weindruch, R. (2002) A mixture model approach for the analysis of microarray gene expression data. *Comp. Stat. Data Anal.*, 39, 1–20.

Beasley, T.M. (2000) Nonparametric tests for analyzing interactions among intra-block ranks in multiple group repeated measures designs. *J. Educ. Behav. Stat.*, 25, 20–59.

Beasley, T.M. (2002) Multivariate aligned rank test for interactions in multiple group repeated measures designs. *Multivariate Behav. Res.*, 37, 197–226.

Beasley, T., Page, G., Brand, J.P.L., Gadbury, G.L., and Allison, D.B. (2003) Chebyshev's inequality for non-parametric testing with small N and α in microarray research. *J. R. Stat. Soc. Ser. C Appl. Stat.*, 53, 95–108.

Benjamini, Y. and Hochberg, Y. (2000) On the adaptive control of the false discovery rate in multiple testing with independent statistics. *J. Educ. Behav. Stat.*, 25, 60–83.

Boneau, C.A. (1960) The effects of violations of assumptions underlying the t test. *Psychol. Bull.*, 57, 49–64.

Box, G.E.P. and Watson, G.S. (1962) Robustness to non-normality of regression tests. *Biometrika*, 49, 93–106.

Bradley, J.V. (1968) *Distribution-Free Statistical Tests*. Prentice-Hall, Englewood Cliffs, NJ.

Brody, J.P., Williams, B.A., Wold, B.J., and Quake, S.R. (2002) Significance and statistical errors in the analysis of DNA microarray data. *Proc. Natl Acad. Sci., USA*, 99, 12975–12978.

Brunner, E. and Langer, F. (2000) Nonparametric analysis of ordered categorical data in designs with longitudinal observations and small sample sizes. *Biomet. J.*, 42, 663–675.

Brunner, E. and Munzel, U. (2000) The nonparametric Behrens–Fisher problem: Asymptotic theory and a small-sample approximation. *Biomet. J.*, 42, 17–25.

Brunner, E., Domhof, S., and Langer, F. (2002) *Nonparametric Analysis of Longitudinal Data in Factorial Experiments*. John Wiley & Sons, New York.

Cliff, N. (1993) Dominance statistics: ordinal analyses to answer ordinal questions. *Psychol. Bull.*, 114, 494–509.

Cliff, N. (1996) *Ordinal Methods for Behavioral Data Analysis*. Erlbaum, Mahwah, NJ.

Conover, W.J. (1980) *Practical Nonparametric Statistics*. John Wiley & Sons, New York.

Corominola, H., Conner, L.J., Beavers, L.S., Gadski, R.A., Johnson, D., Caro, J.F., and Rafaeloff-Phail, R. (2001) Identification of novel genes differentially expressed in omental fat of obese subjects and obese type 2 diabetic patients. *Diabetes*, 50, 2822–2830.

Davison, A.C. and Hinkley, D.V. (1997) *Bootstrap Methods and Their Application*. Cambridge University Press, Cambridge, UK.

DasGupta, A. (2000) Best constants in Chebychev inequalities with various applications. *Metrika*, 51, 185–200.

Di Bucchianico, A. (1999) Combinatorics, computer algebra and the Wilcoxon–Mann–Whitney test. *J. Stat. Plann. Inference*, 79, 349–364.

Dudoit, S., Yang, Y.-H., Callow, M.J., and Speed, T.P. (2002) Statistical methods for identifying differentially expressed genes in replicated cDNA microarray experiments. *Statistica Sinica*, 12, 111–139.

Dunn, O.J. (1961) Multiple comparison among means. *J. Am. Stat. Assoc.*, 56, 52–64.

Efron, B. and Tibshirani, R.J. (1993) *An Introduction to the Bootstrap*. Chapman & Hall, New York.

Efron, B. and Tibshirani, R.J. (2002) Empirical Bayes methods and false discovery rates for microarrays. *Genet. Epidemiol.*, 23, 70–86.

Esteban, M.D., Castellanos, M.E., Morales, D., and Vajda, I. (2001) Monte Carlo comparison of four normality tests using different entropy estimates. *Commun. Stat. Simul., Comp.*, 30, 761–785.

Fisher, N.I. and Hall, P. (1991) Bootstrap algorithms for small samples. *J. Stat. Plann. Inference*, 27, 157–169.

Fligner, M.A. and Pollicello, G.E. (1981) Robust rank procedures for the Behrens–Fisher problem. *J. Am. Stat. Assoc.*, 76, 162–168.

Gadbury, G.L., Page, G.P., Moonseong, H., Mountz, J.D., and Allison, D.B. (2003) Randomization tests for small samples: an application for genetic expression data. *J. Roy. Statist. Soc. Ser. C Appl. Statist.*, 52, 365–376.

Gibson, G. (2002) Microarrays in ecology and evolution: a preview. *Mol. Ecol.*, 11, 17–24.

Good, P.I. (2000) *Permutation Tests: A Practical Guide to Resampling Methods for Testing Hypotheses*. 2nd ed. Springer-Verlag, London.

Guo, X., Qi, H., Verfaillie, C.M., and Pan, W. (2003) Statistical significance analysis of longitudinal gene expression data. *Bioinformatics*, 19, 1628–1635.

Hajek, J. (1969) *A Course in Nonparametric Statistics*. Holden-Day, San Francisco.

Hodges, J.L. and Lehmann, E. (1956) The efficiency of some nonparametric competitors of the *t*-test. *Ann. Math. Stat.*, 27, 324–335.

Hollander, M. and Wolfe, D.A. (1973) *Nonparametric Statistical Methods*. John Wiley & Sons, New York.

Hotelling, H. (1961) The behavior of some standard statistical tests under non-standard conditions. In Neyman, J. (ed.), *Proceedings of the Fourth Berkeley Symposium on Mathematical Statistics and Probability*, Vol. 1. University of California Press, Berkeley, CA, pp. 319–359.

Hunter, L., Taylor, R.C., Leach, S.M., and Sirmon, R. (2001) GEST: a gene expression search tool based on a novel Bayesian similarity metric. *Bioinformatics*, 17 (Suppl 1), S115–S122.

Kayo, T., Allison, D.B., Weindruch, R., and Prolla, T.A. (2001) Influences of aging and caloric restriction on the transcriptional profile of skeletal muscle from rhesus monkeys. *Proc. Natl Acad. Sci., USA*, 98, 5093–5098.

Kerr, M.K. and Churchill, G.A. (2000) Analysis of variance for gene expression in microarray data. *J. Comp. Biol.*, 7, 819–837.

Kerr, M.K. and Churchill, G.A. (2001) Bootstrapping cluster analysis: assessing the reliability of conclusions from microarray experiments. *Proc. Natl Acad. Sci., USA*, 98, 8961–8965.

Koch, G.G. (1969) Some aspects of the statistical analysis of "split-plot" experiments in completely randomized layouts. *J. Am. Stat. Assoc.*, 64, 485–506.

Kruskal, W.H. and Wallis, W.A. (1952) Use of ranks on one-criterion variance analysis. *J. Am. Stat. Assoc.*, 48, 907–911.

Lee, C.K., Weindruch, R., and Prolla, T.A. (2000) Gene-expression profile of the ageing brain in mice. *Nat. Genet.*, 25, 294–297.

Lehmann, E.L. (1998) *Nonparametrics: Statistical Methods Based on Ranks*, Revised 1st ed. Prentice-Hall, Upper Saddle River, NJ.

Long, J.D. and Cliff, N. (1997) Confidence intervals for Kendall's tau. *Br. J. Math. Stat. Psychol.*, 50, 31–41.

Lönnstedt, I. and Speed, T. P. (2002). Replicated microarray data. *Statistica Sinica*, 12, 31–46.

Mann, H.B. and Whitney, D.R. (1947) On a test of whether one of two random variables is stochastically larger than the other. *Ann. Math. Stat.*, 18, 50–60.

Massey, F.J. (1951) The Kolmogorov–Smirnov test for goodness-of-fit. *J. Am. Stat. Assoc.*, 46, 68–78.

Mountz, J.D. and Zhang, H.G. (2001) Regulation of apoptosis of synovial fibroblasts. *Curr. Dir. Autoimmun.*, 3, 216–239.

Pan, W. (2003) On the use of permutation in and the performance of a class of nonparametric methods to detect differential gene expression. *Bioinformatics*, 19, 1333–1340.

Pan, W., Lin, J., and Le, C.T. (2002) How many replicates of arrays are required to detect gene expression changes in microarray experiments? A mixture model approach. *Genome Biol.*, 3, 1–10.

Salter, K.C. and Fawcett, R.F. (1993) The ART test of interaction: a robust and powerful test of interaction in factorial models. *Comm Stat: Sim & Comp*, 22, 137–153.

Saw, J.G., Yang, M.C.K., and Mo, T.C. (1984) Chebyshev inequality with estimated mean and variance. *Am. Stat.*, 38, 130–132.

Serlin, R.C. and Harwell, M.R. (2001) *A review of nonparametric test for complex experimental designs in educational research*. Paper presented at the American Educational Research Association, Seattle, WA.

Šidák, Z. (1967) Rectangular confidence regions for the means of multivariate normal distributions. *J. Am. Stat. Assoc.*, 62, 626–633.

Somers, R.H. (1962) A new asymmetric measure of association for ordinal variables. *Am. Socio. Rev.*, 27, 799–811.

Troyanskaya, O.G., Garber, M.E., Brown, P.O., Botstein, D., and Altman, R.B. (2002) Nonparametric methods for identifying differentially expressed genes in microarray data. *Bioinformatics*, 18, 1454–1461.

Tsodikov, A., Szabo, A., and Jones, D. (2002) Adjustments and measures of differential expression for microarray data. *Bioinformatics*, 18, 251–260.

Tusher, V.G., Tibshirani, R., and Chu, G. (2001) Significance analysis of microarrays applied to the ionizing radiation response. *Proc. Natl Acad. Sci., USA*, 98, 5116–5121.

Vargha, A. and Delaney, H.D. (1998) The Kruskal–Wallis test and stochastic homogeneity. *J. Educ. Behav. Stat.*, 23, 170–192.

Welch, B.L. (1947) The generalization of "Student's" problem when several different populations are involved. *Biometrika*, 34, 28–35.

Wilcox, R.R. (1993) Robustness in ANOVA. In Edwards, E. (eds.), *Applied Analysis of Variance in the Behavioral Sciences*, Marcel Dekker, New York, pp. 345–374.

Wilcoxon, F. (1945) Individual comparisons by ranking methods. *Biometrics*, 1, 80–83.

Wu, T.D. (2001) Analysing gene expression data from DNA microarrays to identify candidate genes. *J. Pathol.*, 195, 53–65.

Xu, X.L., Olson, J.M., and Zhao, L.P. (2002) A regression-based method to identify differentially expressed genes in microarray time course studies and its application in an inducible Huntington's disease transgenic model. *Hum. Mol. Genet.*, 11, 1977–1985.

Zhao, Y. and Pan, W. (2003) Modified nonparametric approaches to detecting differentially expressed genes in replicated microarray experiments. *Bioinformatics*, 19, 1046–1054.

Zimmerman, D.W. (1996) A note on homogeneity of variance of scores and ranks. *J. Exp. Educ.*, 64, 351–362.

Zumbo, B.D. and Coulombe, D. (1997) Investigation of the robust rank-order test for non-normal populations with unequal variances: the case of reaction time. *Can. J. Exp. Psychol.*, 51, 139–149.

14 Bayesian Analysis of Microarray Data

Jode W. Edwards and Pulak Ghosh

CONTENTS

14.1 Introduction 267
14.2 Probability of True Differential Expression 268
14.3 Estimating the Null Distribution 269
14.4 Estimating the Evidence 271
14.5 Estimating the Prior Probability of Nondifferential Expression 272
14.6 Hierarchical Models 273
  14.6.1 Gamma–Gamma and Normal–Normal Models 274
  14.6.2 B-Statistics 279
  14.6.3 Hierarchical Linear Models 282
References 286

14.1 INTRODUCTION

The main objective of many microarray experiments is to identify which of the genes are differentially expressed. In a classical statistical framework, this problem translates into testing thousands of null hypotheses of the form:

$$H_0: \quad \mu_1 = \mu_2$$
$$H_A: \quad \mu_1 \neq \mu_2$$

Numerous classical approaches have been described in the literature to test this hypothesis from a classical viewpoint, that is, to obtain a classical "p-value" [1–16]. Biological investigators are primarily interested in determining if a gene of interest is differentially expressed. Classical p-values alone do not directly answer this question. These provide the probability of observing a difference as large (or larger than) as those actually observed if, in fact, the gene was not differentially expressed. If the probability (the p-value) is very low, it is concluded by deduction that the gene was probably differentially expressed. The Bayesian approach to statistical inference can provide a more direct answer to the question "is a gene differentially expressed?" Bayesian inference is predicated on defining a parameter as a random variable and obtaining a probability distribution for the parameter given the data and the prior distribution of the parameter; this is in stark contrast to the frequentist approach to statistical inference, which is based on the distribution of observed data given a fixed constant value for the parameter. In Bayesian approaches to microarray analysis, the parameter of greatest interest is a binary one, say δ_i (for gene i), which takes a value of one if a gene is differentially expressed, and zero otherwise. According to Bayes law:

$$p(\delta|y) = \frac{p(\delta)p(y|\delta)}{p(y)}$$

The probability function for the parameter given the data, $p(\delta|y)$, provides a direct answer to an investigator, namely, what is the probability that the gene is differentially expressed given my observed data? The Bayesian probability of differential expression is a function of the likelihood of the data given the parameter value, $p(y|\delta)$, and the prior probability of differential expression, $p(\delta)$. The denominator, $p(y)$, is a normalizing constant so that $p(\delta|y)$ integrates (or sums) to unity, so it is a probability by definition. The objective of this chapter will be to outline a number of approaches described in the literature for estimating the Bayesian probability of differential expression. The methods covered range from the simple to the very complex, but all share the common objective, namely, estimation of the probability that a gene is truly differentially expressed given the observed data.

14.2 PROBABILITY OF TRUE DIFFERENTIAL EXPRESSION

Efron and colleagues [17] utilized Bayes law to obtain the probability of true differential expression in a very straightforward manner; many Bayesian methods are hierarchical in nature with complex hierarchical probability models representing the data. Efron [17] applied Bayes law only to reduced data. The data for each gene was reduced to a single summary statistic z, which was a function of the amount of differential expression. Any existing structure or ancillary information present in the data was removed prior to the application of Bayes law.

Differential or nondifferential expression can be represented by a binary parameter δ, which takes values of zero for nondifferential expression and one for differential expression. The probability of true differential expression, p_1, is the probability that the parameter δ is equal to one, that is, $P(\delta_i = 1) = p_1$. Conversely, the probability of nondifferential expression, p_0, is the probability that the parameter δ is equal to zero, that is, $P(\delta_i = 0) = p_0$. The conditional densities of the reduced data are

$$f_0(Z) = f(Z|\delta = 0) = \text{density of } Z \text{ under nondifferential expression}$$
$$f_1(Z) = f(Z|\delta = 1) = \text{density of } Z \text{ under differential expression}$$

The reduced data, z, has been replaced here with the random variable Z to emphasize that reduced data z are being treated as *exchangeable* random variables, drawn from probability density functions f_0 and f_1. The exchangeability assumption means that permutation of gene-specific subscripts, (which have been omitted from z for precisely this reason) would have no effect on the joint distribution of a vector of z values for multiple genes. The distributions of z conditional on differential or nondifferential expression can be combined as in a mixture distribution to obtain an expression for the marginal distribution of the data as $f(z) = p_0 f_0(z) + p_1 f_1(z)$. Then, combining the prior distribution of δ with the conditional distribution of z under differential expression and the marginal distribution of the data provides the posterior probability of differential expression:

$$P(\delta = 1|z) = \frac{p_1 f_1(z)}{f(z)} \tag{14.1}$$

Estimating the empirical sampling distribution of data for differentially expressed genes, $f_1(z)$ from only a single reduced summary statistic for each gene is unachievable without parametric assumptions that may be unreasonable [17]. However, $f_0(z)$ can be estimated under much less stringent assumptions than required for $f_1(z)$. Eliminating $f_1(z)$ from the expression for the probability of true differential expression we obtain [17]:

$$P(\delta = 1|z) = 1 - P(\delta = 0|z) = 1 - \frac{p_0 f_0(z)}{f(z)} \tag{14.2}$$

The denominator of Equation 14.2, $f(z)$, is much less difficult to estimate than $f_1(z)$ because it is directly observed from the data without knowledge of differential or nondifferential expression. The numerator, $f_0(z)$, is often referred to as the "null-distribution" [12,17] in the microarray literature because it is the expected distribution of z under the null hypothesis of no differential expression. Each of the three components of Equation 14.2 will now be discussed in detail.

14.3 ESTIMATING THE NULL DISTRIBUTION

The most common method for estimating $f_0(z)$ in the microarray literature is by permutation [12,13,17,18]. These approaches are based on computing z-scores, that is, standardized gene expression differences, on permuted data. The resulting z-scores computed from permuted data are called "null-scores" because they have an expected value of zero independently of the true level of differential expression. More importantly, if the null-scores are constructed to have the same expected dispersion as the true z-scores under the null hypothesis of no differential expression then the distribution of the null-scores provides an empirical approximation of the null-distribution, $f_0(z)$.

Efron et al. [17] based their original work on a study of the effects of ionizing radiation with two treatment conditions, irradiated and nonirradiated, applied to each of two cell lines. Two replicate RNA samples were observed within each treatment and cell line for a total of eight chips. Following data preprocessing steps, four contrasts between irradiated chips and nonirradiated chips within cell lines could be constructed. Because there were two replicate chips of each treatment within each cell line, the contrasts could be constructed in two arbitrary directions, that is, sample "A" could be contrasted with sample "A" or sample "B" for the nonirradiated treatment. The average of the four contrasts, $\bar{D}_i$, and the standard error of the contrasts, S_i, were used to compute summary statistics $z_i = \bar{D}_i/(a_0 + S_i)$. The constant a_0 was an arbitrarily chosen one used to correct the highly skewed distribution of sample variances, which can be substantially underestimated with very small samples. Adding a small constant to the denominator eliminates a small proportion of z_i values from being artificially large [14,17]. Null-scores were computed from differences between samples within cell lines and treatments in a randomly chosen direction (i.e., sample "A" minus sample "B" with probability $\frac{1}{2}$ and sample "B" minus sample "A" with probability $\frac{1}{2}$). From the four permuted differences for each gene, a null-score was computed by the same formula used for the z_i.

Permutations used by Efron et al. [17] were balanced with respect to treatments and cell lines. Pan [12] examined two general permutation methods by simulation and demonstrated that a balanced permutation method better approximates the true null-distribution. A permutation is balanced with respect to a factor if data are only reordered within levels of that factor but not between levels. In the example from Efron et al. [17], data permutations were balanced with respect to treatment and cell line because data were only reordered within a level of both cell line and treatment, which left only two possible orders of the data. Based on the conventions given in Reference 12, standardized null-scores can be generated by the following procedure:

1. Randomly permute observations within a treatment group
2. Divide observations within a treatment into two equal (or nearly equal in the case of an odd number) sets of observations and compute the sums of each of the randomly chosen sets of observations

Let x_{ij} be the mean of observations for a random group i in treatment j, and v_{ij} be the variance of x_{ij} $(i = 1, 2; j = 1, 2)$. The null-score, z_0, and the corresponding test statistic, z_1, for the difference in gene expression are [12]:

$$z_0 = \frac{((x_{11} + x_{12})/2) - ((x_{21} + x_{22})/2)}{\sqrt{((v_{11} + v_{12})/4) + ((v_{21} + v_{22})/4)}}, \quad z_1 = \frac{((x_{11} - x_{12})/2) + ((x_{21} - x_{22})/2)}{\sqrt{((v_{11} + v_{12})/4) + ((v_{21} + v_{22})/4)}}$$

The test statistic, z_1, is a difference between the average value of the random subsamples from treatment one and the average of the subsamples from treatment two. The null-score is a sum of within-treatment differences, which have expected value of zero under symmetric distributions. Both z_1 and z_0 have exactly the same standard error.

Pan's [12] procedure provides a generalized framework for obtaining test statistics, that is, reduced data, and corresponding null-scores used to estimate $f_0(z)$. However, other procedures could be used provided an appropriate data reduction is used to obtain a summary statistic for each gene and the null-distribution corresponding to the summary statistics can be computed. For example, Efron and Tibshirani [18] use a Wilcoxon rank-sum statistic. The null-distribution is assumed to be the known null-distribution for the Wilcoxon rank-sum statistic. These procedures for reducing data and finding an appropriate null-distribution, $f_0(z)$, do not have strong parametric assumptions because they do not rely on parametric probability distributions. There are, however, very important assumptions such as the exchangeability across genes and symmetry of error distribution. Furthermore, with limited sample sizes, the number of distinct permutations may be quite limited thereby providing unreliable estimates of true sampling distributions.

Estimating the null-distribution can be accomplished in numerous ways if one is willing to make parametric assumptions such as normally distributed errors. For example, Lee et al. [9] assumed that differences in gene expression were normally distributed with mean zero for nondifferentially expressed genes, and a Weibull distribution for genes that were either up- or down-regulated. A mixture distribution of a normal distribution with mean zero and two Weibull distributions, one with mean less than zero and another greater than zero, was fit to the data. From within this parametric framework, the estimated null-distribution, $f_0(z)$, and the conditional data distribution under differential expression, $f_1(z)$, were easily obtainable. Bayes law could then be applied to obtain gene-specific probabilities of differential expression.

14.4 ESTIMATING THE EVIDENCE

Because the evidence in the denominator of Bayes' law, $f(z)$, can be directly obtained from the data, there is little need for further manipulation in obtaining $f(z)$. The empirical data densities $f_0(z)$ and $f(z)$ are combined in Bayes' law with the prior probability of nondifferential expression to obtain the probability that a gene is not differentially expressed. Once observed distributions are obtained, smooth functions representing these empirical distributions are desired. Efron et al. [17] recommended a further simplification by estimating the ratio $f_0(z)/f(z)$ by logistic regression. Observed summary statistics were coded as successes and null-scores as failures. The collection of data and null-scores were ranked and logistic regression was used to estimate the ratio of probabilities of 'success' and 'failure' given the summary statistic z. Efron and Tibshirani [18] used a slight variation on this theme and applied a Poisson regression to the empirical frequency distribution of Wilcoxon rank-sum statistics. The empirical frequency distribution was simply a tabulation of the number of genes with each possible value of the Wilcoxon rank-sum statistic. The Poisson regression was applied to the empirical frequency distribution and the expected frequency distribution for Wilcoxon rank-sum statistics.

Several authors [9,11,13,19] have advocated the use of mixture models for microarray analysis in various contexts because of their flexibility in describing distributions that do not closely fit standard parametric distributions. Such a mixture would provide explicit parametric densities for both $f_0(z)$ and $f(z)$ that could be used in the application of Bayes law as an alternative to applying logistic or Poisson regression as done by Efron et al. [17].

14.5 ESTIMATING THE PRIOR PROBABILITY OF NONDIFFERENTIAL EXPRESSION

A potential limitation to the general empirical Bayes approach outlined here is that the prior probability, p_0, cannot be directly estimated without a parametric model, that is, in the words of Efron et al. [17], it is "unidentifiable without strong parametric assumptions." Efron et al. [17] outlined a somewhat conservative approach by setting an upper bound for p_0. Both Efron et al. [17] and Efron and Tibshirani [18] pointed out that because p_0 and p_1 are probabilities, both should be contained in the interval (0,1). However, if a large value of p_0 is chosen, it is possible for the posterior probability of the differential expression to be negative for values of Z near zero. Hence, Efron et al. [17] proposed p_0 equal to an upper bound that will maintain values of $p_1(z)$ greater than 0:

$$\hat{p}_{0,\max} = \min_z\{\hat{f}(Z)/\hat{f}_0(Z)\}$$

An alternative application of Bayes' law in the context outlined in Efron et al. [17] is available through mixture modeling of p-values [20,21]. In this approach, a standard statistical test is applied to each gene in an experiment producing a frequentist p-value, which is used as the summary statistic, z. Under the global null hypothesis that no genes are differentially expressed, the p-values are expected to follow an uniform distribution. Under the alternative hypothesis of differential expression, p-values tend to be closer to zero and thus can be well approximated with one or more beta distributions [20]. The marginal distribution of p-values takes on a mixture distribution of the form:

$$f(P) = \lambda_0 U(0,1) + \sum_{i=1}^{g} \lambda_i \beta(r_i, s_i)$$

where λ_0 is the proportion of p-values sampled from uniform distribution, $U(0,1)$ is the uniform distribution on the interval (0,1), λ_i is the proportion of p-values sampled from ith mixture component, $\beta(r_i, s_i)$ is the ith component beta density, and g is the number of beta components.

The p-values arising from a beta component of the mixture were assumed to represent truly differentially expressed genes so that $1 - \lambda_0$ is a parametric estimate of the proportion of genes that are truly differentially expressed (λ_0 is not interpreted as the proportion of genes not differentially expressed because for very small differences in gene expression, p-values may be closer to a uniform distribution). The null distribution is the uniform distribution, which is the expected distribution of p-values under no differential expression. The conditional distribution under differential expression, $f_1(z)$, is the mixture of beta distributions. Plugging the three components into Bayes' law results in a straightforward posterior probability of true differential expression [17]:

$$p(\delta = 1|p) = \frac{\lambda_1 \sum_{i=1}^{g} \beta(p; r, s)}{\lambda_0 U(p) + \lambda_1 \sum_{i=1}^{g} \beta(p; r, s)}$$

A similar application of Bayes' law can be used to obtain a probability that a randomly selected gene with a p-value less than some threshold, α, was truly differentially expressed [20]. The probability of obtaining a p-value less than α if a gene is not differentially expressed is equal to α, while the probability of obtaining a p-value less than α for genes that were truly differentially expressed is obtained by integrating the beta-mixture from zero to α, that is,

$$\text{Prob}(p < \alpha) = \int_0^{\alpha} \sum_{i=1}^{g} \lambda_i \beta(p; r_i, s_i) \partial p$$

The conditional probability that a gene was truly differentially expressed given that the p-value is less than the threshold α was obtained from the mixture model as the conditional probability that the p-value was sampled from the beta-mixture given that the p-value was less than the threshold:

$$\text{Prob}(\delta = 1 | p < \alpha) = \frac{\int_0^{\alpha} \sum_{i=1}^{g} \lambda_i \beta(p; r_i, s_i) \partial p}{\alpha + \int_0^{\alpha} \sum_{i=1}^{g} \lambda_i \beta(p; r_i, s_i) \partial p}$$

This probability was derived by an application of Bayes law, but was developed in a frequentist hypothesis testing paradigm. In particular, the "data" on which the parameter is conditioned is the probability of obtaining a p-value less than some threshold in the frequentist sense. Two formulas have been presented here to obtain a probability of differential expression conditional on p-values as data, but it is left to the reader to decide on which is preferable, with a word of caution that the two will have slightly different interpretations.

14.6 HIERARCHICAL MODELS

The approaches presented so far depend only on a single summary statistic and a single binary parameter that conditions the expected distribution of the summary statistic. Much information may be lost by reducing the entire prior distribution to a binary parameter taking on values zero or one. Hierarchical Bayesian methods provide a much richer class of models that can be used to utilize more of the information present in the original data. Robert [22] formally defines a hierarchical model as one in which "the prior is decomposed in conditional distributions." Hierarchical models describe variation in random processes at multiple levels of observational units [23]. In the approaches described so far, priors had only a single-level hierarchy with a term for differential vs. nondifferential gene expression. However, to this one level, we can add information at the levels of genes, chips, and treatments to describe the distribution of the data. Out of necessity, hierarchical models will be more complex and will rely on stronger parametric assumptions.

Three types of hierarchical models for microarray data have been described in the literature thus far. Newton et al. [24,25] and Kendziorski et al. [26] developed a generalized two-level hierarchical approach that included random variation in gene expression means and in gene expression observations conditional on gene expression means. Lonnstedt and Speed [27] and Gottardo et al. [28] developed "B-statistics" for testing whether the logarithm of cDNA gene expression ratios differs from zero. These statistics are based on a normal distribution at the observation level with a normal prior of log-ratios for differentially expressed genes and an inverse gamma prior for variances. The third approach is a more fully Bayesian approach due to Tadesse et al. [29] and Ibrahim et al. [30] that resembles a hierarchical linear model with normal priors on gene expression means and inverse gamma priors on variances. In addition, these authors have accounted for censoring of the data for considering upper and lower limits on expression values.

14.6.1 GAMMA–GAMMA and NORMAL–NORMAL MODELS

Newton et al. [24,25] and Kendziorski et al. [26] developed a series of hierarchical approaches to model gene expression data in hierarchical normal–normal models and gamma–gamma models. The first approach was developed to detect changes in gene expression on a single two-channel cDNA slide using a hierarchical gamma–gamma model [26]. Despite the current trend toward increasing replication in microarray research, the approach for unreplicated arrays warrants detailed discussion because it is illustrative and later work is built upon it. The state of differential or nondifferential expression for gene k was coded by a binary variable, z_k, which took a value of $z_k = 1$ for true differential expression with prior probability of differential expression of p. True, unobservable gene expression levels for individual genes, denoted by θ, were assumed to be exchangeable samples from a gamma distribution with shape parameter a_0 and scale parameter ν, that is, $\theta \sim \text{Gamma}(a_0, \nu)$ (see Figure 14.1 for example, Gamma distributions). Observed gene expression levels for the red and green channels, denoted by "r_k" and "g_k" for gene k respectively, were conditioned on the value of z_k and on true expression level θ. Specifically, if the gene was not differentially expressed (the true unknown value of z_k was 0), r_k and g_k were treated as exchangeable samples from a gamma distribution with shape parameter a and scale parameter θ, that is, $r \sim \text{Gamma}(a, \theta)$ and $g \sim \text{Gamma}(a, \theta)$. If the gene was differentially expressed (true value of z_k was 1), r_k and g_k were treated as samples from two independent gamma distributions with common shape parameter a but with independent scale parameters θ_r and θ_g that were exchangeable samples from a gamma distribution with parameters a_0 and ν, that is, $\theta_r \sim \text{Gamma}(a_0, \nu)$ and $\theta_g \sim \text{Gamma}(a_0, \nu)$. The interesting exchangeability assumption in this model is that true gene expression levels, θ, for individual genes and channels are all equally exchangeable samples from a gamma distribution across genes and across channels for those genes that are differentially expressed.

Estimation of hyperprior parameters a, a_0, ν, and p was accomplished by first integrating true gene expression levels, θ_r and θ_g, from the joint distributions of observations r_k and g_k given parameters $\theta_r, \theta_g, a, a_0, \nu$. The resulting marginal distributions of gene expression observations were denoted $p_0(r_k, g_k)$ for nondifferentially expressed genes and $p_A(r_k, g_k)$ for differentially expressed genes. According to Bayes law, the posterior probability of true differential expression was:

$$P(z_k = 1|r_k, g_k, p) = \frac{p \cdot p_A(r_k, g_k)}{p \cdot p_A(r_k, g_k) + (1 - p) \cdot p_0(r_k, g_k)}$$

Newton et al. [24] estimated parameters by using a Beta(2, 2) prior distribution on the hyperprior parameter p and used the Expected Maximization (EM) algorithm to estimate hyperprior parameters a, a_0, and ν.

Two very important features of this model merit further discussion. First, in contrast to the first approach presented due to Efron et al. [17], there are two random components, an observational or error component and variation among "true" gene expression values. Separate estimation of two independent, but confounded (because of lack of replication) random processes is made possible by "information borrowing" [31]. The term information borrowing is used to reflect information coming from other genes through prior distribution. The prior distribution provides a parametric probability model for both true expression levels and observed gene expression observations, given true expression levels. Given the parametric form of the prior and estimates of prior hyperparameters, a prior expectation is available for both the true expression level and the observed data. The prior information is thus described as "borrowed" from other genes.

The approach of Newton et al. [24] has also been extended to replicated chips and to a wider range of distributional forms [26]. The approach of Newton et al. [26] is quite similar to Newton et al. [24], but the Bernoulli component at the highest level of the hierarchy was treated as a mixture distribution, though consequences of this treatment are largely semantic (the primary difference seems to be in the definition owing to whether or not there is replication; without replication, each gene may be thought of as a Bernoulli trial). Data distributions were defined in general terms as $f(z|\mu_g)$, where z is an observed gene expression value for a single gene on a single chip. The true expression level of the hierarchy, μ_g, was further assumed to be an exchangeable sample from a (prior) probability distribution of true gene expression levels, $\pi(\mu_g)$. The marginal null data distribution for replicated measurements with common true expression level integrated out was expressed as [26]:

$$f_0(z_1, \ldots, z_n) = \int \left(\prod_{i=1}^{n} f(z_i|\mu_g) \right) \pi(\mu_g) \partial \mu_g$$

where the designation $f_0(z_1, \ldots, z_n)$ represents a set of exchangeable samples (nonvectors in this case) with common mean. In the case of nondifferential expression between groups, the data vectors x_g and y_g were samples from a population with a common mean value μ_g, which was expressed as $f_0(x_g, y_g)$. The notation is meant to show that all observations in vectors x_g and y_g have a common expected value. In the case of differential expression it was assumed that true gene expression levels for groups 1 and 2 were exchangeable samples from the common distribution of gene expression means, $\pi(\mu_g)$. Hence, the marginal data distribution for x_g and y_g, assuming differential expression, was expressed as $f_1(x_g, y_g) = f_0(x_g) f_0(y_g)$ to show that the marginal distribution under differential expression is the result of two independent data distributions (with independent means) for the two groups.

The marginal distribution of data, including the mixture distribution for differential and nondifferential expression was given as [26]:

$$f(x_g, y_g) = pf_1(x_g, y_g) + (1 - p)f_0(x_g, y_g)$$

where p is the probability of differential expression, $f_0(x_g, y_g)$ is the joint distribution of x_g and y_g given no differential expression, and $f_1(x_g, y_g)$ is the joint distribution of x_g and y_g given differential expression.

Two parameterizations of the model were described, a gamma–gamma model and a lognormal–normal model. In the gamma–gamma model, the observational and gene level distributions were identical to those defined in Newton et al. [24] with a small change in notation (the parameter θ in Newton et al. [24] was changed to λ in Kendziorski et al. [26]). In the lognormal–normal model, individual log-transformed observations were modeled as exchangeable samples with mean μ_g and variance σ^2. Mean expression levels for individual genes or individual conditions within genes, μ_g, were assumed to be normally distributed with prior mean μ_0 and variance τ_0^2. Both the gamma–gamma model and lognormal–normal model were put into the mixture framework to obtain a posterior probability of differential expression or posterior odds ratio, for which an analytical solution was given in Kendziorski et al. [26]. Parameter estimation was carried out using the EM algorithm [32] using the marginal likelihood of the full data over all genes, g:

$$l(\theta) = \sum_g \log[pf_1(x_g, y_g) + (1 - p)f_0(x_g, y_g)]$$

The authors recommended using any "off-the-shelf" optimization routine to obtain maximum likelihood estimators of all parameters [26].

A logical outcome of the models in Newton et al. [24,25] and Kendziorski et al. [26] is that differences in gene expression for truly differentially expressed genes follow a parametric distribution as either a difference between two independent gamma variates or a difference between two normal variates. To relax this parametric assumption, Newton et al. [25] introduced a semiparametric model based on the gamma–gamma model of Kendziorski et al. [26]. Observational and true expression levels of the hierarchy were modeled as distributed according to the gamma distribution as in Newton et al. [24] and Kendziorski et al. [26]. To relax the parametric assumption for differences in gene expression, three differential expression hypotheses were declared for the mean expression levels in group 1 and group 2 for gene g, equivalent expression, overexpression in group 1, and overexpression in group 2, that is,

$$H_{g,0}\colon\ \mu_{g,1} = \mu_{g,2}$$
$$H_{g,1}\colon\ \mu_{g,1} > \mu_{g,2}$$
$$H_{g,2}\colon\ \mu_{g,1} < \mu_{g,2}$$

True gene expression levels for each gene were conditional on these hypotheses. If gene expression levels were equal, as in $H_{g,0}$, all observations were derived from

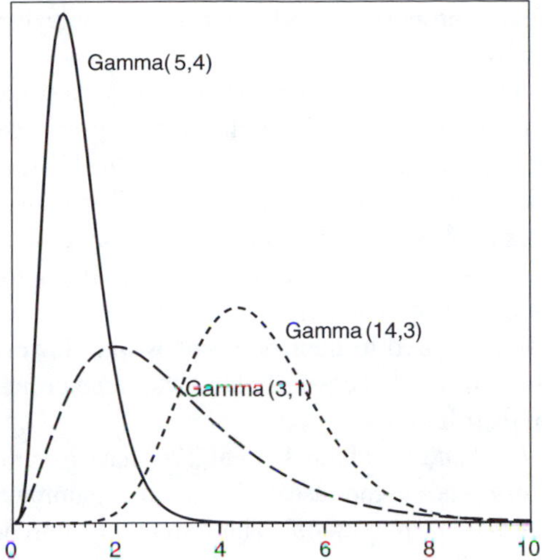

FIGURE 14.1 Example gamma distributions with shape and scale parameters indicated.

a common gamma distribution with a common scale parameter. Under hypotheses $H_{g,1}$ and $H_{g,2}$, observations were derived from independent gamma distributions with unique shape parameters for conditions 1 and 2 as follows:

$$f(x_{g,i}|a_1, \mu_{g,1}) \sim \text{Gamma}(a_1/\mu_{g,1}, a_1)$$
$$f(y_{g,i}|a_2, \mu_{g,2}) \sim \text{Gamma}(a_2/\mu_{g,2}, a_2)$$

The distribution of true expression levels was an inverse gamma distribution with scale parameter a_0x_0 and shape parameter a_0 (providing a grand mean of the distribution of x_0):

$$k(\mu) \sim \text{Inverse Gamma}(a_0x_0, a_0)$$

The parameterization with parameters a_0x_0 and a_0 was made so that under the null hypothesis of no differential expression, x_0 provided a measure of center of the marginal data distribution. The marginal distribution of gene expression means across the three hypotheses is a discrete mixture, specified as follows:

$$f(\mu_g) = p_0f_0(\mu_{g,1}, \mu_{g,2}) + p_1f_1(\mu_{g,1}, \mu_{g,2}) + p_2f_2(\mu_{g,1}, \mu_{g,2})$$

where

$$f_0(\mu) = k(\mu_1)1[\mu_1 = \mu_2]$$
$$f_1(\mu) = 2k(\mu_1)k(\mu_2)1[\mu_1 < \mu_2]$$
$$f_2(\mu) = 2k(\mu_1)k(\mu_2)1[\mu_1 > \mu_2]$$

p_0, p_1, and p_2 are mixing proportions and 1[expr] is an indicator equal to one if expr is true or zero otherwise.

With the indicator function part of each density, the density takes on a value of zero if the inequality in square brackets is false, that is to say, the likelihood for a pair of values of means is zero if the pair of values is not contained in the hypothesis specification. The authors found that the resulting posterior distribution of gene expression differences from an example dataset did not resemble any low-dimensional parametric family of distributions. The authors observed that the semiparametric mixture specification did not constrain gene expression differences to follow a smooth parametric distribution as compared to their previous work. However, in general, other semiparametric models may not necessarily have the same property because they can lead to smoothing of distributions.

Newton et al. [24,25] and Kendziorski et al. [26] have assumed a constant coefficient of variation across genes in the case of the gamma–gamma model and a constant variance in the lognormal–normal model. The remaining approaches to be discussed relax this assumption and allow variances to be heterogeneous among genes while at the same time combining information across genes to improve estimation of the variances.

Combining information across genes has been advocated in the literature in approaches involving t-tests because a small proportion of t-test statistics in microarray experiments have been observed to be quite large because of very small denominators [14,17]. Student's t-distribution is characterized by long tails, which are caused by frequent underestimation of the variance at small sample sizes. The sampling distribution of sample variances is highly skewed with small samples, so that the sample variance is underestimated more than 50% of the time. In fact, when a large number of sample variances are estimated with small sample sizes, the sample variance can be substantially underestimated a high proportion of the time (Figure 14.2).

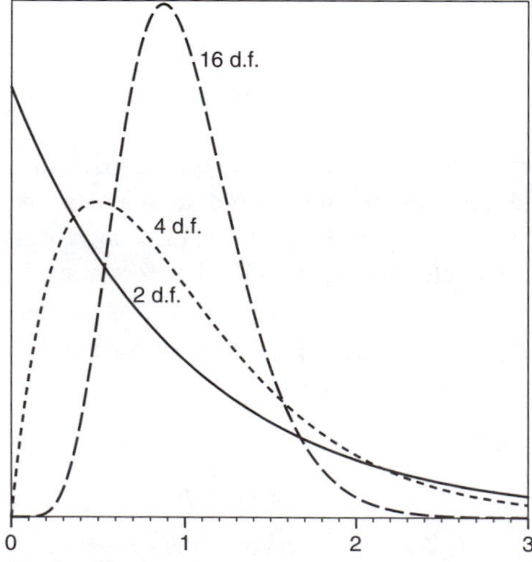

FIGURE 14.2 Sampling distribution of the sample variance estimator s^2 assuming 16, 4, and 2 degrees of freedom and a true variance of 1.0.

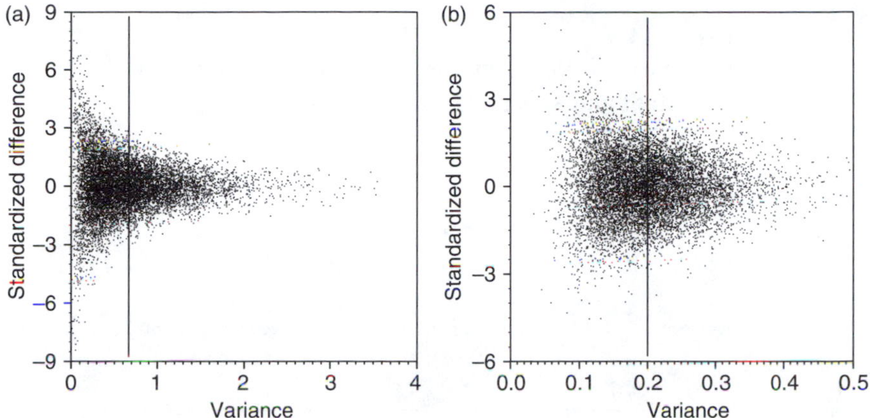

FIGURE 14.3 Simulated gene expression differences divided by the estimated standard error of the difference (Standardized difference) for 10,000 genes with sample sizes per group of 3 and 10. All genes were assumed to have identical expression between groups and error variances equaled one. Vertical reference lines represent the expected variance of the difference (0.67 and 0.2).

For example, at sample size five (four degrees of freedom), the estimated variance is less than one-fourth of its true value 13.9% of the time. To further demonstrate effects of small sample size in this context, we simulated differences in gene expression for 10,000 genes with sample sizes of three and ten. Observations were assumed to be normally distributed with mean of zero and variance one. Standardized mean differences vs. the variance of the difference are shown graphically in Figure 14.3. With a sample size of three, it is evident that many genes had very small estimates of the variance of the difference. Among those genes with the smallest estimates, some had very large standardized differences. At a larger sample size of 10, there is less underestimation of the variance, and fewer large standardized differences. An obvious "solution" to this problem might be to pool variances across genes and use a single estimator. However, it is also apparent that several authors [1,10,27–30] find the assumption of common variance untenable. Thus, a solution is needed that is between complete pooling of variances and computing many individual variances from very small sample sizes. Several authors have accomplished sharing of information across genes to accomplish shrinkage estimation of the variance by using hierarchical models within the Bayesian framework.

14.6.2 B-Statistics

Lonnstedt and Speed [27] and Gottardo et al. [28] have proposed a set of "B-statistics" to test for differential expression, though with slightly different parameterizations. The B-statistics incorporate a hierarchical probability model for log base-2 transformed ratios of gene expression from two-channel arrays. The objective is to obtain the probability that the gene expression ratio between two treatment groups differs from unity. Lonnstedt and Speed [27] and Gottardo et al. [28] developed hierarchical models that generate nearly identical probability distributions for the data. However,

TABLE 14.1
Summary of Notation for Hierarchical Models in Lonnstedt and Speed [27] and Gottardo et al. [28]

	D_g	Lonnstedt	Gottardo
Observations	0	$\mathrm{N}(\mu_i, \sigma_i^2)$	$\mathrm{N}(\mu_i, \tau_i)$
	1	$\mathrm{N}(\mu_i, \sigma_i^2)$	$\mathrm{N}(\mu_i, \tau_i)$
Means	1	$\mu_i \sim \mathrm{N}(\mu_a, \kappa\tau_i)$	$\mu_i \sim \mathrm{N}\left(0, c\frac{na}{2\tau_i}\right)^*$
	0	$\mu_i \equiv 0$	$\mu_i \equiv 0$
Variances	0	$\tau_i \sim \mathrm{Gamma}(\nu, 1)$	$\tau_i \sim \mathrm{InvGamma}(\nu_0, \tau_0)$
	1	$\tau_i \sim \mathrm{Gamma}(\nu, 1)$	$\tau_i \sim \mathrm{InvGamma}(\nu_a, \tau_a)$

* $\sigma_i^2 = na/2\tau_i$

the near equivalence of their models may not be apparent because of differences in parameterization. A parameter D_g (I_i in Lonnstedt and Speed [27]) is assigned to each gene to indicate true differential expression ($D_g = 1$) or nondifferential expression, with prior probability of differential expression equal to the hyperparameter p, that is, $D_g \sim \mathrm{B}(1, p)$. Conditional on D_g, observations are assumed to be normally distributed conditional on gene-specific means and variances. Gene-specific means were modeled as exchangeable samples from a normal distribution and gene-specific variances were modeled as exchangeable samples form an inverse gamma distribution. This hierarchical probability structure provides for information sharing across genes for log-ratio means and for variances of log-ratio observations. The hierarchical probability model is summarized in Table 14.1 for both Lonnstedt and Speed [27] and Gottardo et al. [28].

At the observation level, the two models are equivalent (Table 14.1). At the level of the gene-specific log-ratio means (μ_i), both models equate the mean to zero for nondifferentially expressed genes. However, for differentially expressed genes Gottardo et al. [28] assign a nonzero hyperprior to gene-specific log-ratio means whereas Lonnstedt and Speed [27] fix the mean at zero. The nonzero log-ratio mean allows an average change in gene expression in the direction of either up- or down-regulation. Allowing average gene expression to differ between treatments may have advantages, but should only be interpreted to the extent allowed by experimental design. Average differences between treatments could be the result of experimental variables that are confounded with the experimental design if these are not properly controlled.

The parameterization of variances appears quite different between models, but in fact it can be shown by transformation that the distribution of variances of gene expression observations conditional on true expression levels follow an inverse gamma distribution. Lonnstedt and Speed [27] provide a parametric gamma model for the parameter τ_i and define the observational variance, σ_i^2, as a function of τ_i. By transforming from τ_i to σ_i^2, it can be shown that σ_i^2 follows an inverse gamma distribution with shape parameter υ and scale parameter $na/2$. Gottardo et al. [28] parameterized the variance directly, which they denote as τ, as an inverse gamma distribution with shape parameters υ_0 and υ_a and scale parameters τ_0 and τ_a for nondifferentially and differentially expression genes, respectively (Table 14.1). Thus, the scale parameters, υ_0 and υ_a, in Gottardo et al. [28] have the same interpretation as the quantity $na/2$ in Lonnstedt and Speed [27]. The variance of gene-specific means is a multiple of the observation variance determined by hyperparameters c [27] and κ [28] (Table 14.1).

The goal of the hierarchical models of Lonnstedt and Speed [27] and Gottardo et al. [28] is to arrive directly at a probability or odds ratio of true differential expression. Lonnstedt and Speed [27] provided a log odds-ratio of differential expression in the form of (see Table 14.1 for parameter definitions):

$$B_g = \log \frac{p}{1-p} \frac{1}{\sqrt{1+nc}} \left[\frac{a + s_g^2 + M_{g\bullet}^2}{a + s_g^2 + (M_{g\bullet}^2/(1+nc))} \right]^{\nu+(n/2)}$$

where p is the probability of true differential expression, n is the number of observed ratios for gene g, s_g^2 is the sample variance, $M_{g\bullet}^2$ is the sample mean squared, and a, c, υ are hyperparameters (Table 14.1).

Lonnstedt and Speed's [27] B_g statistic is equivalent to the log odds-ratio of true differential expression, that is,

$$\log \frac{\Pr(D_g = 1|(Y_{ij}))}{\Pr(D_g = 0|(Y_{ij}))}.$$

Gottardo et al. [28] expressed their statistic as a probability of differential expression as follows:

$$\mathrm{Prob}(D_g = 1) = \left(1 + \frac{p}{1-p}\right) \sqrt{2} \frac{\Gamma(\nu_a)\Gamma(\nu_0 + n_2/2)\tau_0^{\nu_0}(\tau_a + ((n_2-1)/2)s_g^2)^{\nu_a + n_2/2}}{\Gamma(\nu_0)\Gamma(\nu_a + n_2/2)\tau_a^{\nu_a}(\tau_0 + (n_2/2)s_{g0}^2)^{\nu_0 + n_2/2}}$$

where p is the probability of true differential expression, n_2 is the number of observed ratios for gene g, s_g^2 is the sample variance, and s_{g0}^2 is the sum of squared observations (log-ratios), and υ_0, υ_a, τ_0, τ_a are hyperparameters (Table 14.1).

An advantage of the specific approaches of Gottardo et al. [28] and Lonnstedt and Speed [27] as compared to previously described approaches is the information sharing among genes for both the mean of the log-ratio and the variance of the log-ratio. Both means and variances are allowed to vary freely among genes, but a prior distribution is placed on these values that provides extra information about what values may be expected for them.

Each of the approaches requires estimates of hyperpriors. For the proportion of differentially expressed genes, p, Lonnstedt and Speed [27] used a fixed value under the assumption that the true proportion of differentially expressed genes is small (1 to 2%). Gottardo et al. [28] recommended a simple iterative algorithm to estimate p. A starting value of p was chosen and probabilities of true differential expression were computed for all genes. The proportion of genes with probabilities of true differential expression above .95 was taken as a new starting value, and the algorithm repeated until convergence. Both sets of authors recommended using method of moment estimators for remaining hyperparameters. Parameters specific to differentially expressed genes were estimated by using a proportion p of genes with the largest probability of differential expression or largest log odds-ratio of differential expression. The remaining proportion $1 - p$ genes were used to estimate hyperparameters specific to nondifferentially expressed genes. An exception was the scale parameters for the variance of means of differentially expressed genes (c [27] and κ [28]). These parameters serve nearly identical functions in the models, but the authors recommended quite different estimates for them. Gottardo et al. [28] fixed κ at a value of one over the sample size effectively making the variance of means a fixed scalar function of the variance at the observation level. Lonnstedt and Speed [27] used an empirical approach to obtain an estimate, namely they compared the histogram of means of differentially expressed genes to the histogram of means of nondifferentially expressed genes to find the difference in variances. These approaches were therefore highly empirical in nature in that hyperpriors were not defined for hyperparameters, but rather plain point estimates were substituted for hyperparameters [27,28]. It would seem that some room is left for readers to develop their own estimates of hyperparameters if better estimates are desired.

Gottardo et al. [28] extended their work beyond the one sample case to test three additional hypotheses incorporating a control group. The three testable hypotheses outlined were (1) treatment and control log-ratios have different means (B_2), (2) treatment and control log-ratios have different variances (B_3), and (3) treatment and control log-ratios have different means or different variances (B_4). The same hierarchical probability structure as that described in Table 14.1 was used, but with additional hyperparameters to accommodate both a treatment and a control log-ratio (Table 14.2). Additional extensions of this approach to include multiple treatments have been presented by Smyth [33].

14.6.3 Hierarchical Linear Models

All of the approaches described so far have utilized point estimates of at least one hyperparameter, and in some cases several. The use of point estimates for hyperparameters introduces additional error into the models that is not accounted for in posterior distributions and probabilities. The approaches by Ibrahim et al. [30] and Tadesse et al. [29] are intended to be more fully Bayesian. In their approaches, all priors have hyperprior distributions on them that can be specified to be quite noninformative (flat). Both models use a hierarchical probability model quite similar to Lonnstedt and Speed [27] and Gottardo et al. [28], that is, normally distributed gene expression values, normally distributed gene expression means, and inverse-gamma distributed variances. However, some additional elements are added to these models to reflect certain observed properties of microarray data.

Microarray data, particularly from Oligonucleotide arrays (such as Affymetrix® arrays, Affymetrix, Inc., Santa Clara, CA), tends to be censored on the lower end of the scale. Ibrahim et al. [30] formally model the censoring by modeling observations as samples from a mixture distribution of a constant value, c_0, for nonexpressed genes, and values sampled from a lognormal distribution, y, for expressed genes. The

TABLE 14.2
Summary of Probability Model for Gottardo's [28] B-Statistics

	D_g	B_2	B_3	B_4
Data				
Control	0	$N(\mu_i, \tau_i)$	$N(\mu_{1i}, \tau_i)$	$N(\mu_i, \tau_i)$
	1	$N(\mu_{1i}, \tau_i)$	$N(\mu_{1i}, \tau_{1i})$	$N(\mu_{1i}, \tau_{1i})$
Treatment	0	$N(\mu_i, \tau_i)$	$N(\mu_{2i}, \tau_i)$	$N(\mu_i, \tau_i)$
	1	$N(\mu_{2i}, \tau_i)$	$N(\mu_{2i}, \tau_{2i})$	$N(\mu_{2i}, \tau_{2i})$
Means				
Control	0	$\mu_i \sim N(\mu_0, \kappa_0\tau_i)$	$\mu_{1i} \sim N(\mu_{a1}, \kappa_1\tau_i)$	$\mu_i \sim N(\mu_0, \kappa_0\tau_i)$
	1	$\mu_{1i} \sim N(\mu_{a1}, \kappa_1\tau_i)$	$\mu_{1i} \sim N(\mu_{a1}, \kappa_1\tau_{1i})$	$\mu_{1i} \sim N(\mu_{a1}, \kappa_1\tau_{1i})$
Treatment	0	$\mu_i \sim N(\mu_0, \kappa_0\tau_i)$	$\mu_{2i} \sim N(\mu_{a2}, \kappa_2\tau_i)$	$\mu_i \sim N(\mu_0, \kappa_0\tau_i)$
	1	$\mu_{2i} \sim N(\mu_{a2}, \kappa_2\tau_i)$	$\mu_{2i} \sim N(\mu_{a2}, \kappa_2\tau_{2i})$	$\mu_{2i} \sim N(\mu_{a2}, \kappa_2\tau_{2i})$
Variances				
Control	0	$\tau_i \sim \text{IG}(\nu_0, \tau_0)$	$\tau_i \sim \text{IG}(\nu_0, \tau_0)$	$\tau_i \sim \text{IG}(\nu_0, \tau_0)$
	1	$\tau_i \sim \text{IG}(\nu_a, \tau_a)$	$\tau_{1i} \sim \text{IG}(\nu_{a1}, \tau_{a1})$	$\tau_{1i} \sim \text{IG}(\nu_{a1}, \tau_{a1})$
Treatment	0	$\tau_i \sim \text{IG}(\nu_0, \tau_0)$	$\tau_i \sim \text{IG}(\nu_0, \tau_0)$	$\tau_i \sim \text{IG}(\nu_0, \tau_0)$
	1	$\tau_i \sim \text{IG}(\nu_a, \tau_a)$	$\tau_{2i} \sim \text{IG}(\nu_{a2}, \tau_{a2})$	$\tau_{2i} \sim \text{IG}(\nu_{a2}, \tau_{a2})$

observed expression level for gene g in treatment j, and subject i expressed as:

$$x_{jig} = \begin{cases} c_0 & \text{with probability } p_{jg} \\ c_0 + y_{jig} & \text{with probability } 1 - p_{jg} \end{cases}$$

where

$$y_{jig} \sim \text{LN}(\mu_{jg}, \sigma_{jg}^2)$$

The primary objective of Ibrahim et al. [30] was to estimate the posterior distribution of the ratio of expression values and determine which of the genes are differentially expressed. The ratio was constructed from the ratio of posterior expectations of true gene expression for the two treatment groups. The posterior expectation of true gene expression (ψ_{jg}) for treatment j and gene g was:

$$\psi_{jg} = c_0 p_{jg} + (1 - p_{jg})(c_0 + \mathrm{e}^{\mu_{jg}+\sigma_{jg}^2/2})$$

where $\mathrm{e}^{\mu_{jg}+\sigma_{jg}^2/2}$ is the posterior expectation of y_{jg}.

The posterior gene expression ratio (fold change) was computed from the posterior of the ratios $\xi_g = \psi_{2g}/\psi_{1g}$.

The true expression level for each treatment of each gene depends on three gene and treatment specific parameters, μ_{jg}, σ_{jg}, and p_{jg}, each of which are random variables that come from a prior distribution:

$$\mu_{jg}|\mu_{j0},\sigma_{jg}^2 \sim N\left(\mu_{j0},\frac{\tau_0\sigma_{jg}^2}{\bar{n}_j}\right)$$

$$\sigma_{jg}^2 \sim \mathrm{IG}(a_{j0},b_{j0})$$

$$\mathrm{logit}(p_{jg}) = \log\left(\frac{p_{ig}}{1-p_{jg}}\right) = e_{jg} \sim N(u_{j0},k_{j0}w_{j0}^2$$

The term $\bar{n}_j$ in the prior on μ_{jg} is the average sample size, which is defined as the average number of subjects for which a gene is expressed ($x_{jig} > c_0$) for the jth treatment (averaged over all genes), that is, $\bar{n}_j = \frac{1}{G}\sum_{g=1}^{G}\left(n_j - \sum_{i=1}^{n_j}\delta_{jig}\right)$, where δ_{jig} is an indicator taking a value of one if $x_{jig} > c_0$ and zero otherwise. From the priors on μ_{jg}, σ_{jg}, and p_{jg}, hyperprior distributions were declared for the hyperparameters as follows:

$$\mu_{j0} \sim N(m_{j0},\nu_{j0}^2)$$

$$b_{j0} \sim G(q_{j0},t_{j0})$$

$$u_{j0} \sim N(\hat{u}_{j0},h_{j0}w_{j0}^2)$$

The prior and hyperprior parameters τ_{j0}, a_{j0}, k_{j0}, w_{j0}^2, m_{j0}, ν_{j0}^2, q_{j0}, t_{j0}, $\hat{u}_{j0}$, and h_{j0} are all predetermined values based on some type of prior knowledge (or data). Recommendations were given by Ibrahim et al. [30] for using the data as a "guide" to choose hyperparameter values. The choice of values for prior parameters was important as sensitivity analysis revealed that priors that were "too informative" or "too vague" resulted in a poor model fit, as determined by correspondence between observed data and predicted observations from the model.

Expression values for all genes within a treatment were modeled by two random values in the hierarchical prior structure, a gene specific value, μ_{jg}, and a random mean expression level for the treatment, m_{j0}. Because of the common, random mean expression level for all genes within a treatment there is a covariance between expression levels within tissues of ν_{j0}^2. Likewise, gene-specific variances and probabilities of expression have common random means for each tissue, which induces similar covariances within tissues for these parameters as well. Random scale parameters for treatments were included in the hierarchical prior because they induced covariances between genes within treatments, which the authors claimed to be a desirable property of the model.

Tadesse et al. [29] extended the approach of Ibrahim et al. [30] with a more sophisticated treatment of censoring and an expanded linear model that included a factorial structure with covariates. The linear model for observed gene expression observations, x_{jki}, was

$$x_{jki} = \mu_{jk} + \lambda_k' z_{ji} + \varepsilon_{jki}$$

where μ_{jk} was the mean expression level (intercept) of gene k in treatment j, z_{ji} is the covariable for subject i in treatment j, λ_k is the regression parameter for gene k, and ε_{jki} is the error term. The mean expression level for each gene and treatment combination was expressed using a factorial model such that $\mu_{jk} = \alpha_j + \beta_k + \gamma_{jk}$, where α_j is the effect of group j, β_k is the effect of gene k, γ_{jk} is the gene by treatment interaction. Because some of the observed expression values were considered outside the range of detection of the technology, the data likelihood was developed to include the probability of a censored outcome. Specifically, the likelihood was a function of censored values, y_{jki}, where

$$y_{jki} = \begin{cases} c_l & \text{if } x_{jki} < c_l \\ c_r & \text{if } x_{jki} > c_r \\ x_{jki} & \text{if } c_l < x_{jki} < c_r \end{cases}$$

and c_l and c_r are lower and upper detection limits and x_{jki} is the observed gene expression value. For each value, y_{jki}, indicator variables, l_{jki} and r_{jki}, were defined as:

$$l_{jki} = I(x_{jki} < c_l)$$
$$r_{jki} = I(x_{jki} > c_r)$$

where $I(\cdot)$ equals one if the expression is true and zero otherwise.

The likelihood function for censored observations is a product of the Gaussian likelihood for observations within the range of the technology and tail area probabilities for observations outside the detection limits of the technology:

$$L_{jki} = \left[\Phi\left\{\frac{y_{jki} - (\mu_{jk} + \lambda_k' z_{ji})}{\sigma}\right\}\right]^{l_{jki}} \left[1 - \Phi\left\{\frac{y_{jki} - (\mu_{jk} + \lambda_k' z_{ji})}{\sigma}\right\}\right]^{r_{jki}}$$
$$\times \left(\frac{1}{\sqrt{2\pi\sigma^2}} \exp\left[-\frac{1}{2\sigma^2}\{y_{jki} - (\mu_{jk} + \lambda_k' z_{ji})\}^2\right]\right)^{(1-l_{jki})(1-r_{jki})}$$

Each scale parameter α_j, β_k, γ_{jk}, and λ_{kq}, was modeled as an exchangeable sample from a hierarchical normal–normal prior distribution, with a normal prior for the parameter, and a second level normal distribution for the mean of the normal prior on the parameter. The two-level hierarchy resulted in numerous covariances being built into the model so that all the random effects in the linear model had compound symmetry covariance structures. For example, gene by treatment interaction effects, γ_{jk}, had a normal prior distribution with mean γ_{0k} and variance σ_γ^2, that is, $\gamma_{jk}|\gamma_{0k}, \sigma_\gamma^2 \sim N(\gamma_{0k}, \sigma_\gamma^2)$. The gene-specific mean of the prior distribution, γ_{0k}, was itself an exchangeable sample from a normal hyperprior with gene-specific mean c_{0k}, and variance ν_γ^2, that is, $\gamma_{0k}|c_{0k}, \nu_\gamma^2 \sim N(c_{0k}, \nu_\gamma^2)$. Hence, for each gene x the treatment interaction is a sum of two random variables, a gene-specific scale parameter for interactions, c_{0k}, and a gene and treatment-specific interaction effect, γ_{0k}. Gene by treatment interactions had variance $\sigma_\gamma^2 + \nu_\gamma^2$ and covariance within genes of ν_γ^2.

The primary goal of this approach was to estimate (and draw inference from) differential gene expression between groups j and j'. A logical quantification of differential expression would be the contrast between interaction terms, $\eta_k = \gamma_{j'k} - \gamma_{jk}$. However, Tadesse et al. [29] found that this function was not estimable. The lack of estimability of η_k was likely due to the redundancy of effects in the model because of the numerous random variables in the hierarchical prior. For example, average gene by treatment interaction, γ_{0k}, appeared to be confounded with the average gene effect, β_k. Because the hierarchical prior contained two effects for the mean value of each gene, these parameters are unestimable. In order to solve the estimability problem, the authors propose an alternative estimator of gene expression that compares the gene expression difference between two genes, namely $\xi_k = \gamma_{j'k} - \gamma_{jk} - \gamma_{jr} + \gamma_{j'r}$, which compares gene expression differences between genes k and r. Gene r could be chosen to be a control gene, that is, some gene that is expected or known to be not differentially expressed. The lack of estimability was evident in Gibbs sampling chains because the ξ_k did appear to have fully identifiable location parameters. However, other parameters had identifiable locations and appeared to have very stable posterior distributions (i.e., good mixing). As in Ibrahim et al. [30], sensitivity analysis demonstrated that priors that were too informative over informed data and nearly determined the values of gene expression differences.

The approaches of Ibrahim et al. [30] and Tadesse et al. [29] were comprehensive modeling methods intended to represent many aspects of microarray data including censoring of the data at low and high expression values, factorial structure, covariates, and heterogeneous error variances. Results of both approaches clearly depended on choice of hyperprior parameters, which is discussed by the authors in greater detail. Thus, while these fully Bayesian approaches have the potential to provide the best representation of the most features of microrarray data, caution is required in choosing appropriate prior parameters.

REFERENCES

1. P. Baldi and A.D. Long. A. Bayesian framework for the analysis of microarray expression data: regularized t-test and statistical inferences of gene changes. *Bioinformatics* 17: 509–519, 2001.
2. T.M. Beasley, G.P. Page, and J.P.L. Brand. Chebyshev's inequality for nonparametric testing with small N and alpha in microarray research. *Journal of the Royal Statistical Society Series C—Applied Statistics* 53: 95–108, 2004.
3. Y. Chen, V. Kamat, E.R. Dougherty, M.L. Bittner, P.S. Meltzer, and J.M. Trent. Ratio statistics of gene expression levels and applications to microarray data analysis. *Bioinformatics* 18: 1207–1215, 2002.
4. Y. Chen, E.R. Dougherty, and M.L. Bittner. Ratio-based decisions and the quantitative analysis of cDNA microarray images. *Journal of Biomedical Optics* 2: 364–374, 1997.
5. T.M. Chu, B. Weir, and R. Wolfinger. A. systematic statistical linear modeling approach to oligonucleotide array experiments. *Mathematical Biosciences* 176: 35–51, 2002.
6. X. Cui and G.A. Churchill. Statistical tests for differential expression in cDNA microarray experiments. *Genome Biology* 4: 210, 2003.
7. M.K. Kerr, M. Martin, and G.A. Churchill. Analysis of variance for gene expression microarray data. *Journal of Computational Biology* 7: 819–837, 2000.

8. M.K. Kerr and G.A. Churchill. Statistical design and the analysis of gene expression microarray data. *Genetic Research* 77: 123–128, 2001.
9. M.L. Lee, W. Lu, G.A. Whitmore, and D. Beier. Models for microarray gene expression data. *Journal of Biopharmaceutical Statistics* 12: 1–19, 2002.
10. A.D. Long, H.J. Mangalam, B.Y. Chan, L. Tolleri, G.W. Hatfield, and P. Baldi. Improved statistical inference from DNA microarray data using analysis of variance and a Bayesian statistical framework. Analysis of global gene expression in *Escherichia coli* K12. *Journal of Biological Chemistry* 276: 19937–19944, 2001.
11. W. Pan. A comparative review of statistical methods for discovering differentially expressed genes in replicated microarray experiments. *Bioinformatics* 18: 546–554, 2002.
12. W. Pan. On the use of permutation in and the performance of a class of nonparametric methods to detect differential gene expression. *Bioinformatics* 19: 1333–1340, 2003.
13. W. Pan, J. Lin, and C.T. Le. A mixture model approach to detecting differentially expressed genes with microarray data. *Functional and Integrative Genomics* 3: 117–124, 2003.
14. V.G. Tusher, R. Tibshirani, and G. Chu. Significance analysis of microarrays applied to the ionizing radiation response. *Proceedings of the National Academy of Sciences, USA* 98: 5116–5121, 2001.
15. M.A. van de Wiel. Significance analysis of microarrays using rank scores. *Kwantitative methoden* 71: 25–37, 2004.
16. Y.H. Yang, M.J. Buckley, S. Dudoit, and T.P. Speed. Comparison of methods for image analysis on cDNA microarray data. *Journal of Computational and Graphical Statistics* 11: 108–136, 2002.
17. B. Efron, R. Tibshirani, J.D. Storey, and V. Tusher. Empirical Bayes analysis of a microarray experiment. *Journal of American Statistical Association* 96: 1151–1160, 2001.
18. B. Efron and R. Tibshirani. Empirical Bayes methods and false discovery rates for microarrays. *Genetic Epidemiology* 23: 70–86, 2002.
19. P. Broet, S. Richardson, and F. Radvanyi. Bayesian hierarchical model for identifying changes in gene expression from microarray experiments. *Journal of Computational Biology* 9: 671–683, 2002.
20. D.B. Allison, G.L. Gadbury, M.S. Heo, J.R. Fernandez, C.K. Lee, T.A. Prolla, and R. Weindruch. A mixture model approach for the analysis of microarray gene expression data. *Computational Statistics and Data Analysis* 39: 1–20, 2002.
21. S. Pounds and S.W. Morris. Estimating the occurrence of false positives and false negatives in microarray studies by approximating and partitioning the empirical distribution of p-values. *Bioinformatics* 19: 1236–1242, 2003.
22. C.P. Robert. *The Bayesian Choice*. New York: Springer-Verlag, 2001.
23. A. Gelman, J.B. Carlin, H.S. Stern, and D.B. Rubin. *Bayesian Data Analysis*. 1st ed. New York: Chapman & Hall, 1995.
24. M.A. Newton, C.M. Kendziorski, C.S. Richmond, F.R. Blattner, and K.W. Tsui. On differential variability of expression ratios: improving statistical inference about gene expression changes from microarray data. *Journal of Computational Biology* 8: 37–52, 2001.
25. M.A. Newton, A. Noueiry, D. Sarkar, and P. Ahlquist. Detecting differential gene expression with a semiparametric hierarchical mixture method. *Biostatistics* 5: 155–176, 2004.
26. C.M. Kendziorski, M.A. Newton, H. Lan, and M.N. Gould. On parametric empirical Bayes methods for comparing multiple groups using replicated gene expression profiles. *Statistical Medicine* 22: 3899–3914, 2003.

27. I. Lonnstedt and T. Speed. Replicated microarray data. *Statistica Sinica* 12: 31–46, 2002.
28. R. Gottardo, J.A. Pannucci, C.R. Kuske, and T. Brettin. Statistical analysis of microarray data: a Bayesian approach. *Biostatistics* 4: 597–620, 2003.
29. M.G. Tadesse, J.G. Ibrahim, and G.L. Mutter. Identification of differentially expressed genes in high-density oligonucleotide arrays accounting for the quantification limits of the technology. *Biometrics* 59: 542–554, 2003.
30. J.G. Ibrahim, M.H. Chen, and R.J. Gray. Bayesian models for gene expression with DNA microarray data. *Journal of American Statistical Association* 97: 88–99, 2002.
31. B.P. Carlin and T.A. Louis. *Bayes and Empirical Bayes Methods for Data Analysis*, 2nd ed. New York: Chapman & Hall/CRC, 2000.
32. A.P. Dempster, N.M. Laird, and D.B. Rubin. Maximum likelihood from incomplete data via the EM algorithm. *Journal of Royal Statistical Society Series B* 39: 1–38, 1977.
33. G.K. Smyth. Linear models and empirical Bayes methods for assessing differential expression in microarray experiments. *Statistical Applications in Genetics and Molecular Biology* 3: Article 3, 2004.

15 False Discovery Rate and Multiple Comparison Procedures

Chiara Sabatti

CONTENTS

15.1 Multiple Comparison in Microarrays 289
15.1.1 The Problem of Multiple Comparison 289
15.1.2 Microarrays 291
15.2 Multiple Testing 292
15.2.1 Global Error Rates and the Definition of FDR 293
15.2.2 Controlling for Multiple Testing without Resampling 294
15.2.3 Controlling for Multiple Testing with Resampling 296
15.2.4 Corrections for Multiple Testing in Microarrays 298
15.2.5 Bayesian Alternatives 300
15.3 Simultaneous Inference — Beyond Testing 301
References 302

15.1 MULTIPLE COMPARISON IN MICROARRAYS

15.1.1 The Problem of Multiple Comparison

The problem of multiple comparison is the title of a monograph by Tukey [1] that, while unpublished till recently in the collection of his complete works, has been extremely influential in the development of this area of research. Interestingly, the reader that searched for a definition of the problem would hardly find one there. In a typical Tukey style, however, one finds a series of very thought-provoking examples and observations. Let me quote from the opening page:

> Having measured [...] the frequencies of certain blood groups among the residents of six areas [...] and knowing something of his accuracy of measurement, the naive, *unspoiled* investigator wishes, at the very least, to compare each of the six determinations with every other, and to consider what he has learned about the differences between many, if not all, pairs. Until now, this natural, and the writer believes *wise*, desire has been in conflict with the desire of an increasing number of investigators to attach some definite statistical meaning to their statements. An investigator with both desires has been likely

> to be told that what he wishes to do involves non-independent tests of significance, with the implication that *such things are not spoken of in polite society* (italics is mine).

I find the above quotation instructive in that it is both dated and very actual. Generally, what we define as a problem of multiple comparison has to do with a natural question that an *unspoiled* researcher would want to ask and to which an intellectually lazy mainstream statistician would react as *unspoken of in polite society*. It takes guts as Tukey's to recognize that the question is indeed *wise* and similar intellect to devise an appropriate strategy to achieve an answer. This assessment of the problem of multiple comparison appears to me as extremely actual. What makes the example dated is that the investigator is dealing with measurements on only six quantities and the problem of multiplicity arises because all the pairwise comparisons are considered. Contemporary data sets are typically measurements on thousands of variables and for the researcher it is very natural to behave in an *unspeakable of* manner, even if pairwise comparisons are avoided. In Dohono's words [2], we are spectators of a "Data deluge," and there is no point in pretending that such data should not be thoroughly investigated, even if current statistical methodology cannot always provide the appropriate tools. If the data is there and there is a meaningful question to ask, somebody will do it — in an appropriate or not so appropriate manner. The history of the term data mining is quite illustrative. Statistician coined the expression to designate a kind of excessive data-snooping that would not lead to valuable results in terms of prediction. Today, data mining designates one of the "hottest" areas in computational sciences and statistics. Fortunately, there is a growing body of research that addresses the multiplicity issues that arise from mining large datasets. Model selection when the space of possible explanatory variables is large; de-noising of sparse signals; control of global errors in multiple testing are some of the challenges that statisticians have started to take up.

Measurements from gene expression arrays are a typical example of the kind of massive datasets that characterize contemporary research. Their analysis encounters all the problems described above and, in some ways, has motivated methodological development or promoted the popularization of some approaches that raise to the challenge of data-deluge. The definition of False Discovery Rate and the outline of strategies to control it in presence of multiple testing is due to Benjamini and Hochberg in 1995 [3]. I think it is a fair statement that this clever approach owes a considerable part of its present popularity to the fact that it can be fruitfully applied to the analysis of microarrays. It is, then, particularly interesting to review its applications in this context. Before "diving" into the heart of the matter, however, we need to clarify what we are going to consider as the scope of "multiple comparison." Because of the nature of the present work, which aims at reviewing available approaches and methods, rather than presenting novel contributions, we will let the present literature define the extent of the problem of multiple comparison in microarrays. This will translate in most of the discussion being devoted to multiple testing. However, we do want to point out to the reader that this is only one aspect of the corrections for multiplicity that are possible and appropriate in the context of microarrays analysis. We then devote the next section to a survey, from a multiplicity standpoint, of the issues encountered in the study of gene expression.

15.1.2 Microarrays

The nature of microarray experiments is fundamentally exploratory. Indeed, they have played an important role in the recent shift in biomedical research from hypothesis-driven investigation to hypothesis formulation. With a microarray, one gathers noisy measurements on the expression levels of tens of thousands of genes under one specific cell condition. The goal is to obtain some insights on how the expression of genes is affected by the specific treatment/genetic modification/cell cycle conditions under study. *A priori*, there is either no information on which genes experience regulation in expression, or information on the behavior of a very small subset of genes (order of magnitude 1–10). It is not known which genes are expected to have similar behavior. Microarray experiments are used in the scientific community as a "screening tool." Once a series of scientific hypothesis are formulated, the investigators will use other, less throughput, more costly, and more specific experimental techniques for verification.

Given this scientific background, the statistical analysis of microarray data has to be substantially exploratory rather than confirmatory. Within this framework, variety of statistical tools have been applied. In almost all cases, issues of multiplicity arise. Some examples of the most common statistical methods applied to the analysis of array data will serve as illustration.

Often, a series of microarray experiments is conducted under different conditions on the same genes, in order to identify which genes may be co-regulated or part of the same biological pathway. In such cases, cluster analysis is a natural tool: thousands of genes are clustered on the basis of their expression values in a small set of experiments (around 10). Typically, "good-looking clusters" contains tens of genes. One of the questions that statisticians have to address is how seriously should such clusters be taken: with so few experiments, and so many genes, could not one expect to find small groups of similar ones just by chance?

Another very common setting is the comparison of expression values for genes under two conditions of interest: a reference and a study sample. In some cases, investigators look for genes whose expression provides best prediction of the conditions. A variety of classification tools have been used in such contexts. All of these have to deal with the fact that the total number of possible predictors is much higher than the number of observations: appropriate strategies are needed to avoid overfitting. In other cases, investigators would like to individuate all the genes that experience a difference in regulation under the two conditions being studied. If this is the case, statistical tests of the null hypothesis that genes have constant expression are a natural tool to use. However, it is clear that we face issues of multiple testing.

Given that there is a problem of multiple comparison, what is the appropriate framework within which to address it? Again, I think a quotation from Tukey [1] is helpful in clarifying the goals of statistical analysis of microarray.

> A statistical analysis is made for *action*, if its purpose is to affect a single particular given decision, whether to be made now or at some time in the future.
>
> A statistical analysis is made for *indication* (or benchmarking) if its purpose is to make a value (or limits for a value) available for future use in unspecified decisions.
>
> A statistical analysis is made for *sanctification* if its purpose is to throw an aura of statistical reliability about some statement or appearance. (Italics as in the original.)

Given that microarray is an exploratory tool, it is seldom clear *a priori* what will be the nature of the subsequent step. There is no prespecified action item for which the statisticians could provide guidance. In general, sanctification is not very interesting, but in the case of microarray data is clearly inappropriate. Scientists are well aware that any statement based on results of array experiments need validation with different techniques. It would be presumptuous and useless for statisticians to try to substitute these. Indication is hence the domain of statistics in microarray analysis. When correcting for multiplicity, we have to take into account that the primary goal is to help the scientist to focus on patterns, appearances that have a certain degree of credibility. A quantification of such credibility, the identification of limits for the values of interest are the primary goals. Given these premises, let us start our journey among the procedures presented in the literature to correct for multiple comparisons in the context of microarray analysis.

15.2 MULTIPLE TESTING

By far, the context within which the issues of multiplicity have been discussed in more detail is the one of multiple testing. Let m be the total number of genes for which we have measurements. Let then $T_1, \ldots, T_m$ be a set of test statistics for testing the hypotheses $\{H_1, \ldots, H_m\}$, where H_i is true if gene i does not experience a change in expression in the two compared conditions. Let H_0 be the hypothesis that corresponds to each of the H_i being true $H_0 = \cap_{i=1}^{m} H_i$. When we conduct tests of these hypotheses, we try to answer two types of questions (1) can H_0 be rejected? (2) If H_0 is rejected, which of the H_i should be rejected? The statistical technique used to answer the first question is called a global test, while the one addressing the second is called a multiple test procedure. In standard textbooks, practically only one multiple comparison procedure is documented: the Bonferroni correction. In order to clarify the issues involved, it is then useful to refer to the Bonferroni procedure, well known to the practitioner. It offers an easy answer to both the global test and the multiple comparisons problem. Let $p_1, \ldots, p_m$ be the p-values associated with each of the tests statistics and let $p_{(1)}, \ldots, p_{(m)}$ be their ordered counterpart. According to Bonferroni, one can reject H_0 if $p_{(1)} < \alpha/m$, where α is the desired level for the test of H_0. As for the second question, all hypotheses H_i, for which $p_i < \alpha/m$, will be rejected. This is a very general procedure, which has the advantage of being very simple. However, it is not the only possible approach to multiple comparisons and it is based on specific choices that may not always be appropriate. Three characteristics, in particular, of the Bonferroni approach are worth discussing.

1. If the test statistics are positively dependent, the Bonferroni correction is much too conservative. This is best exemplified by looking at the extreme case where all the tests are the same

$$\begin{aligned}\Pr(\text{Reject } H_0 | \text{when } H_0 \text{ is true}) &= \Pr(p_{(1)} \leq \alpha/m | H_0)\\ &= \Pr(p_1 \leq \alpha/m | H_0) = \alpha/m\end{aligned}$$

The actual level of the test is now α/m; in other words, we did not need any correction.

2. The Bonferroni procedure is very conservative as it is a single-step method, that is all the p-values from the various statistics are compared to the same benchmark value. In contrast, stepwise methods generally work on ordered sets of p-values and have a different cutoff values for each $p_{(i)}$. The idea behind these methods is that once the H_i corresponding to $p_{(1)}$ has been rejected, we should believe that it is false. We are then "fishing" among only $m - 1$ hypothesis, so that the appropriate cutoff for the second significant result should be $\alpha/(m - 1)$.
3. The Bonferroni procedure controls a measure of global error that is known in statistics as familywise error rate, FWER. This is defined as the probability to wrongly reject at least one H_i. Controlling the FWER has been the dominant statistical paradigm for analyzing the problem of multiple comparisons. The FWER is a natural extension of the significance level of a single test to the context of multiple tests, that is, it represents the "p-value of the p-value," it gives the probability of observing a p-value as low as the minimum one when all the null hypotheses are true. This similarity with the p-value concept has contributed to the popularity of this conservative criterion, although its shortcomings with respect to loss of power have been well documented (see, e.g., [4]). Having so clarified the nature of the problem, we will consider what are the available alternatives as they have been applied in the analysis of microarray data. However, before proceeding further in this direction, it is worth mentioning that other approaches to multiple comparison are possible. For example, building on some notions introduced by Fisher, [5,6] one can develop methods to combine evidence from some subset of hypothesis, rather than considering each hypothesis separately — thus reducing the problem to a more manageable size. This approach, however, has not been extensively considered for microarrays and hence we will not review it further.

15.2.1 Global Error Rates and the Definition of FDR

As mentioned, the FWER is just one of the possible measures of global error that we may want to control with a multiple comparison procedure. A thorough review of some others error rates can be found in the review paper [7], which we invite the reader to consult. Here, we limit our scope to the comparison of FWER and FDR, which are the two alternatives considered in the analysis of microarrays.

Let us consider the set of m tested hypothesis. With respect to the true nature of the hypotheses and the tests results, they can be organized in the following table:

	# nonrejections	**# rejections**	
# true null	U	V	m_0
# false null	T	S	m_1
	$m - R$	R	m

The FWER is the probability that $V > 1$. The FDR is the expected fraction of mistakes among the rejected hypothesis $E(Q)$:

$$Q = \begin{cases} \frac{V}{V+S}, & \text{if } V + S > 0 \\ 0, & \text{otherwise} \end{cases}$$

FWER is an appropriate measure of error when there is an overriding reason to not make any incorrect rejections of the null hypothesis. However, in many cases, researchers are interested in measuring the overall error rate of multiple tests, rather than focusing on the presence of at least one error. For example, suppose that after testing 1000 hypotheses, one rejects 100, one of which was truly null and compare this situation with the one where it has two rejections, one of which is wrong. From the FWER point of view these two situations are equally bad, as we have one wrong rejection. However, in the first case, we have 1 wrong discovery against 99 right ones and in the second case, half of our discoveries are wrong. The FDR captures the idea that the first situation is much better than the second. If all the H_i are true, the FDR coincides with the probability of wrongly rejecting at least one hypothesis, so that FDR and FWER lead to similar global test conclusions. However, if there are a number of false null hypotheses, we become more lenient toward committing some false rejections when we are detecting them. It is the opinion of the writer that the definition of error rate provided by FDR is much more appropriate for an exploratory investigation like the one carried out with microarrays.

While the details of the theoretical discussion on the definition of FDR are beyond the scope of the present review, it is perhaps useful to the reader of microarray literature to notice that a slightly different definition of FDR has been proposed by Storey [8] and is implemented in some of the software dedicated to microarrays. Storey defines positive FDR the expected value $E[V/R|R > 0]$, and provides a Bayesian interpretation of this quantity appropriate in some cases. We will return to this point in later chapters. While the literature on FDR has grown considerably and its review is beyond the scope of this presentation, we cite for completeness few publications that have contributed to develop a theoretical understanding of the BH procedure and FDR controlling, using a point of view that is different from the one originally taken by Benjamini and Hochberg [3]: Storers [9] introduces a Bayesian interpretation of FDR, that is base for developments in Reference 10; a generalization of BH procedure can be found in Reference 11.

15.2.2 Controlling for Multiple Testing without Resampling

Defined an error rate, the meat of a multiple testing is in individuating a procedure that would control it. There are two definitions of control that correspond to two choices of the probability distributions with which the error rates can be calculated. The term *weak* control is used to define procedures that control the error rates under a null distribution satisfying the complete null hypothesis $H_0 = \cap_{i=1}^{m} H_i$. The term *strong* control refers to the control of error rates under any combination of true and false null hypotheses. In the case of expression arrays, it is generally preferable to opt for procedures that offer strong control, as it is expected that some of the null hypothesis will be false (albeit, without knowing which of these, in particular). Moreover, because it is expected that more than one hypothesis is false, it is typically preferable to opt for stepwise procedures that increase the power of detection. We will describe two such procedures that can be applied very easily to a set of ordered p-values $p_{(1)} < p_{(2)} < \cdots < p_{(m)}$. In the following, α is the desired level of control for the global error under consideration. The first procedure offers strong control of FWER [12]:

(Holm) Start with $i = 1$. If $p_{(i)} > \alpha/(m - i + 1)$ accept $H_{(i)}, \ldots, H_{(m)}$ and stop. Otherwise, reject $H_{(i)}$ and continue

As it can be easily seen, the first cutoff value of this rule is the same as the one suggested by Bonferroni. It is only for suggestive rejections that the criterion becomes less stringent. It is also clear that it will take a sizable amount of rejections for the cutoff value to substantially decrease. (Notice that there are other stepwise procedures to control FWER. Sidak's one is a good example. However, reviewing all of them is beyond the scope of this chapter, see [13].)

The following procedure, which strongly controls FDR, leads to a much faster lowering of the significance cutoff.

(BH) Proceed from $i = n$ to $i = n - 1$ and so on, until, for the first time, $p_{(i)} \leq i\alpha/(p_0 n)$. Where $p_0 = m_0/m$. If this quantity is unknown, set it to 1. Denote that i by k and reject all $H_{(i)}$ with $i = 1, \ldots, k$.

This step-down rule was proposed in Reference 3. At the time, the authors proved that it controlled FDR for independent tests. Subsequent work [4,8,10] showed that this is also true for a quite general class of dependent distributions. In Reference 4, another procedure is also proposed that compares $p_{(i)}$ with $i\alpha/\left(n \sum_{j=1}^{m} 1/j\right)$: this is guaranteed to strongly control FDR for any type of dependence, even if it may lead to a significant loss of power.

The two described procedures also lead to the definition of corresponding adjusted p-values $\tilde{p}_{(i)}$, which are a quite useful statistics to report as they enable other, future researchers to make informed decision on which effects should be considered significant.

(Holm) $\tilde{p}_{(i)}^{H} = \max_{k=1,\ldots,i}\{\min((m - i + 1)p_{(k)}, 1)\}$
(BH) $\tilde{p}_{(i)}^{BH} = \min_{k=i,\ldots,m}\{\min(m/k p_{(k)}, 1)\}$

A clear advantage of the above strategies is that a reasonable answer can be obtained easily. However, a possibly serious disadvantage here is the fact that in case of dependent tests these procedures are not quite as powerful as one could wish. While there are reasons to believe that the tests involved in microarray analysis may be dependent, there is no available model for such dependency. In this case, the only possibility of taking into account dependency between the test statistics to increase power lies in procedures based on resampling, which are reviewed in the following section.

Before considering resampling, let us spend a few words on control of the positive FDR. If all the hypotheses are true, pFDR $= 1$ by definition, so it is not possible to control it in the usual sense. The strategy suggested by Storey is to fix a rejection region and estimate its pFDR. This approach lends itself to the evaluation of the q-value: $q(t)$ is the minimum pFDR that can occur when rejecting a statistic with p-value t or bigger. The author suggests an intuitive estimation of this quantity. Moreover, Storey and colleagues stress the importance of estimating empirically the proportion p_0 of true null hypothesis. The implication of this procedure in terms of strong control is investigated in Reference 14. We will revisit it when discussing specific software and applications to microarrays.

15.2.3 Controlling for Multiple Testing with Resampling

The power of resampling-based methods to correct for multiple comparisons has been extensively put forward in Reference 13, in context of FWER control. The major underlying goal is to take advantage of the dependence between the tests in order to avoid excessive corrections. A very thorough review of their application is given in Reference 15. If a p-value for each hypothesis can be obtained analytically, it is rather simple to set up a resampling procedure to obtain FWER corrected p-values. Enumerating all permutations, or sampling a number of them (depending on the size of the dataset), or using bootstrap resampling to better tailor the specific hypothesis tested, one creates a collection B of fictional datasets generated under the null hypothesis (a discussion description of how to best do this can be found in Reference 16). For each of these permutation data sets, one can evaluate the p-values for each of the tested hypothesis. When looking at this collection of p-values for all the hypotheses in a series of permutation datasets, one can gather information on the distribution of the p-value under the complete null and hence derive global p-value. An efficient way of doing such evaluations that ensures monotonicity is presented in Reference 13. Let $o_1, \ldots, o_m$ define the order of the original hypothesis in terms of their uncorrected p-values p in the original dataset, so that $p(1) = p_{o_1}, p(2) = p_{o_2}$, and so forth. This order remains fixed during the simulation. Evaluations are performed one permutation at the time. That is, one generates a permuted dataset and obtains p-values $p^*_{o_1}, \ldots, p^*_{o_m}$ for all the hypothesis. Subsequently one proceeds to adjusting these p-values in order to obtain a set of values $u_{o_1}, \ldots, u^*_{o_m}$ that have the same monotonicity of the original p-values:

$$
\begin{aligned}
u_{o_m} &= p^*_{o_m} \\
u_{o_{m-1}} &= \min(q_{o_m}, p^*_{o_{m-1}}) \\
u_{o_{m-2}} &= \min(q_{o_{m-1}}, p^*_{o_{m-2}}) \\
&\vdots \\
u_{o_1} &= \min(q_{o_2}, p^*_{o_1})
\end{aligned}
$$

At this point, we are ready to compare these ordered permutation p-values $u_{o_1}, \ldots, u_{o_m}$ with the observed ones $p_{o_1}, \ldots, p_{o_m}$. The proportion of times, across the entire collection of B permutations, in which $p_{o_i} < u_{o_i}$ is going to be the resampling-based global p-value for hypothesis H_{o_i}, modulo a final monotonicity correction.

Problems arise when one cannot rely on analytic approximation of the p-value and permutations are needed to estimate them. In Reference 13 two rounds of permutations are described: one to get the p-values and one to obtain the global p-values. However, if the number of hypothesis evaluated is considerable, this strategy quickly becomes excessively time-consuming. This is an important point to consider in the analysis of microarray data. Typically, researchers are reluctant to engage in one specific distributional assumption for the expression measurements. This, together with the fact that few observations per genes are typically available, makes it impossible to analytically obtain p-values. The two round of simulation procedure described above, however, is computationally too intensive to be realistically performed.

One way of reducing the number of permutations to be evaluated involves in storing the matrix of permutations, and to directly estimate global p-values proceeding gene by gene. However, this translates in a memory burden that can be decreased with smart storing rules. One such procedure is suggested in Reference 15.

Resampling/permutation estimates of the False Discovery Rate, are trickier to obtain, even from a conceptual standpoint and forgetting, for a minute, computational complexity. The FDR definition depends not only on the number of false discovery, but also on the number of true discoveries. It is clear that a permutation resampling procedure can gain us some insights on the first one, but not on the second one. To date, two procedures with this goal have been proposed: they differ in how the number of true rejections is evaluated and also for the estimation of the ratio at the base of FDR (one suggesting an average of ratios and the other a ratio of averages). The first algorithm is due to Yekutieli and Benjamini [17], in an article that is unfortunately hard to read. The second suggestion is due to Storey and Tibshirani [18] and derives quite naturally from their definition of pFDR and idea of fixing the rejection regions. The method of Yekutieli and Benjamini suggests two possible estimators for the number of true rejections, one substantially based only on the original dataset, and one that takes into account permutation results. In both cases, once estimated, this number is fixed for all the permutations. A series of permutation datasets is created as described above and, for each dataset, counts of false rejections are recorded. The empirical FDR rate associated to that dataset is then evaluated as the ratio of such count and the sum of false rejection count and the fix number of true rejections. For detailed description, we invite the reader to consult the original paper. It is worth to point out that this estimator has desirable conservative properties only if the distribution of the number of false positive and true positive are independent — which is hard to evaluate. The algorithm in Reference 18 estimates the denominator of FDR using the total number of rejections in the original dataset. Resampling is used only to estimate the numerator, which can be done is a rather straightforward manner, doing resampling from the complete null hypothesis, averaging the number of rejections across resampled datasets and weighting it by an estimate of the proportion of true null that is obtained comparing the rejections is a selected region in the original sample and their average for that region across the resampled datasets.

As for the case described above, computational difficulties arise when one wants to evaluate the simple p-values using permutation techniques. We are not aware of any proposed shortcut at this time.

15.2.4 Corrections for Multiple Testing in Microarrays

As mentioned in the introduction, a substantial part of the recent interest in multiple testing procedure is linked to the necessity of analyzing microarray experiments. The reviews we have referred to References 7 and 15 are specifically dealing with the case of gene expression array studies. Many of the development proposed by Storey and co-authors are also associated with this type of data. Hence the literature on multiple testing in microarray is quite extensive, and much of it easily accessible to the nonspecialist, as addressed really to the wide community of statisticians that are or start to be involved in the study of these datasets. We hence urge the reader to consult it, with the only caveat that many of the papers are work-in-progress in the sense that is the result of the first years of work of the statistical community on these issues. It is easy to foresee that in a few years, a better consensus will be reached on the more appropriate procedure and, at the same time, algorithms that are more effective may be developed.

The purpose of this section is to make some of the above observations more specific to the case of microarrays and to point the reader to some software available at the time of the writing (end of 2003), that may facilitate initial analysis.

I believe that control of FDR is more appropriate than control of FWER in the context of gene expression studies conducted with microarrays. As I have already pointed out, these experiments are exploratory in nature. Their fundamental goal is to direct the researcher toward some genes whose expression may be affected by the condition under study. Unless the results are validated with other more-specific methods, they will be considered preliminary. While it is clear that there is no point in following thousands of false leads (hence the necessity of multiple comparison procedures), there is really no point in being excessively conservative. I have never interacted with a scientist who thought that bringing attention to one gene that subsequently turns out not to be interesting is a serious mistake. What is important is that a large fraction of the genes that are presented as candidates for further studies actually turn out to be of biological importance. This is precisely the quantity controlled (in expected value) by FDR.

Given that we decide to control FDR, one needs to select the type of procedure to use. One issue that comes up in making such a decision is the dependence/independence of the tests. There are two separate issues at stake (1) control of FDR and (2) power. (1) A method like BH, which controls FDR for independent tests may not control it for dependent ones. It has been proved that BH controls FDR for tests that are positive regression-dependent on a subset. This is quite a technical condition that, however satisfied in some note-worthy cases and can be loosely interpreted as "positive" dependence. In general, BH seems to be pretty robust to departure from dependency in terms of control of FDR. (2) Even if a method controls FDR for dependent tests, the fact that it does not capitalize on dependency may lead to loss of power. This is the main issue when developing resampling-based procedures.

Now, are tests for microarray dependent? Often it is noted that genes are co-regulated, that is, groups of genes are regulated by same elements, or genes are part of the same pathway, so that their levels of expression are linked. This is certainly the case. Levels of expression of genes are dependent. This may not necessarily translate in dependence of the test statistics under consideration that are under the null hypothesis. However, it appears that measurement errors are correlated, and also biological variability, that is often incorporated in the error in the analysis introduces dependency. Therefore, it is reasonable to assume that tests regarding the expression of different genes may be dependent. Stepwise procedures as Holm's control FWER with dependent tests and the modified BH procedure that compares $p_{(i)}$ with $i\alpha/\left(n\sum_{j=1}^{m}1/j\right)$ is guaranteed to control FDR. To increase power, one may want to resort to resampling-based controlling procedures. The point here is that we are often in the difficult situation described above where we want to use permutations to both assess the p-values and correct them. What are the approaches taken in practice? A way out consists in using an approximate analytical p-value, then avoiding one level of permutations. One other (related) assumption often made is that the statistics leading to the p-values are identically distributed, even if this is not known to be necessarily the case. In the case of FWER control, this assumption allows achieving corrections for multiple comparison working directly with the statistics and leads to procedures that are more robust to smaller number of permutations (for details, see Reference 15). In the case of FDR rate, this assumption is at the base of the algorithm described in Reference 18. Again, assuming that all the test statistics have the same distribution greatly facilitates the inferential process: instead of having few observations from each distribution (as many as the number of replicate experiments) we have a large amount from the same (as many as the number of genes times the number of experiments). This allows borrowing strength across genes, which is the theme of the next section. In summary, the researcher carrying out statistical analysis needs to decide which of the two is going to carry heavier consequences: omitting to consider the dependence structure between tests or the difference between distributions of the genes. The choice of multiple-comparison procedure, then, is tightly dependent on the choice of test statistics and available permutation schemes. It is then, worth reviewing some of the most common routes.

Assuming that all test statistics have the same distributions is rather common. This is easy to justify when a nonparametric test like Wilcoxon rank test is used. Otherwise, to date, there is really not enough data to satisfactorily test this hypothesis, as the number of measurements per gene with the same platform and under the same condition is very limited. It is nevertheless rather obvious that different genes, in different experiments, exhibit different variances. Indeed, most researchers construct standardized statistics like the t-statistics to carry out comparisons. Variations are in the evaluation of the standard deviation at the denominator of the statistics: since the number of observations per gene is very limited, one needs to regularize the variance estimate pooling toward an overall value that can be treated as tuning parameter [19,20], or can be evaluated pooling genes with similar expression values [21], or using calibration experiments [22,23]. Calibration experiments are sometimes available for two channel systems as cDNA spotted arrays: the same mRNA is labeled with the two dyes and it is hybridized onto the same slide. Any measurement of change in expression value is then to be attributed to technical variability. They provide an ideal setting to study noise. One can then assume that the distribution of these standardized difference statistics is reasonably approximated by a t-distribution (in which case, one may actually use different degrees of freedom for different genes, according to the amount of missing data in each sample). Another possibility is that one can assume that the distribution of the test statistics is the same for all genes and estimate it empirically, with the aid of calibration experiments or permutations/resampling techniques. The first option allows to conduct resampling-based multiple comparison corrections, while the second allows more precise evaluation of the row p-values. To implement the second option, one has to pay some attention to what is an appropriate empirical estimator of the null hypothesis. If one is working with two channel systems, calibration slides can be successfully used. Permutations of cell line labels are an option in one-channel systems such as Affimetrix, Codelink, and Agilent arrays to evaluate the distribution of statistics under the null hypothesis of equal distribution under the two conditions compared. If one want to test a more specific null hypothesis (e.g., equivalence of means rather than entire distributions), or one is working with cDNA spotted arrays, one needs to resort to resampling techniques as "bootstrap with surgery" [16].

In Reference 24, the same distribution for all the test statistics is assumed, evaluated using calibration experiments, and used to control FDR with a BH procedure. In the program statistical analysis of microarray (SAM) [19,20], the authors use a permutation estimate of the null distribution to identify potential cutoff rules whose FDR is then estimated using permutation procedures as in Reference 8. Examples of what can be achieved if the t-distribution is considered an acceptable choice for evaluating the row p-values, are the FDR resampling under dependence correction documented in References 18 and 25. Ge et al. [15] conduct a comparison study using permutations both to assess the row p-values and their correction.

In conclusion, it is appropriate to briefly mention some of the software sources that are available as of November 2003. From the website of Rob Tibshirani, it is possible to download SAM. John Storey provides an R program that calculates q-values starting form p-values. Benjamini distributes an R program that adjusts p-values using FDR controlling procedures. Within the R based project bioconductor, there is a package called multtest that performs multiplicity corrections (for FWER and FDR) for a set of test statistics (t, F, paired t, block F, Wilcoxon): some permutation procedures for evaluating adjusted p-values are available here.

15.2.5 Bayesian Alternatives

So far all the discussion has been conducted in a frequentist's framework. However, Bayesian approaches may be particularly useful in a microarray context and often lead to inferential results that are surprisingly close to what we have described so far. One of the themes emerged from the previous section has been "borrowing strength." Generally speaking this is achieved with empirical Bayesian procedures, which have indeed been applied to microarray data (see appropriate chapter in the book). One of the advantages of empirical Bayes procedures is that, in some sense, force the researcher to do simultaneous inference: all the genes are studied at the same time.

It is then easier to introduce some elements of inference that deal with multiplicity. Perhaps the best illustration of this is given in Efron's memorial article in honor of Robbins [26]. The setting is as follows: a mixture model is assumed for the test statistics y of each gene (the same for every gene):

$$f(y) = p_0 f_0(y) + p_1 f_1(y)$$

where p_0 is the probability that the gene does not experience a change in expression under the studied conditions and $f_0(y)$ the density of the distribution of the test statistics under the null. Analogously, $p_1 = 1 - p_0$, and f_1 is the density under the alternative. Bayes theorem leads to $p_0(y) = p_0 f_0(y)/f(y)$, probability that the null hypothesis is true. The density $f(y)$ can be estimated from data using a standard density estimation method (this is the crucial empirical Bayes, step). The density f_0 can either be known or estimated with permutations. The value of p_0 remains undetermined, and is unidentifiable without parametric assumptions on f_0, f_1. However, one can obtain meaningful results screening the implications of a range of values for p_0: from the most conservative one, to the minimum value of p_0 that makes positive all the posterior probabilities $p_1(y)$ of a change in expression. Therefore, this approach is practical. It is also remarkably close to FDR. Indeed, defining "Bayesian FDR" for the rejection rule $\{Y_i \leq y\}$ to be $\mathrm{Fdr}(y) \equiv p_0 F_0(y)/F_{(y)}$ (where F stays for cumulative distribution function) and using $\widehat{F}(y)$ to estimate $F(y)$ in $\widehat{\mathrm{Fdr}}$, it is possible to prove [26] that the BH rule is equivalent to rejecting all H_i having $Y_i \leq y_\alpha$, where $y_\alpha = \max\{\widehat{\mathrm{Fdr}}(y) \leq \alpha\}$. Note that since $0 \leq p_0 \leq 1$, the BH approach is, in general, more conservative than Efron's empirical Bayes approach.

Also based on a mixture model similar to the one above is the contribution of Reference 27. These authors work with the p-values rather that test statistics. The uniform distribution for p-values under the null is assumed and a parametric form for f_1 is chosen (mixture of beta distributions). After estimation of the parameters involved in the model, evaluations similar to the ones by Efron are possible. Interestingly, this model can be extended to p-values obtained with resampling procedures [28].

Another related approach is found in Reference 29. Here the authors adopt a Bayesian model selection framework in order to identify which genes may be changing expression value within an ANOVA framework. Interesting comparisons with BH results are carried out.

15.3 SIMULTANEOUS INFERENCE — BEYOND TESTING

While most of the discussion of multiplicity in the context of microarrays is carried out with regard to multiple testing, there are other domains where simultaneous inference plays a role. These are, in our opinion, quite important, as by no means the only use for microarray is comparative experiments. A very common goal is, for example, to individuate genes that are co-regulated: this is best achieved by analyzing multiple experiments together and applying some clustering techniques. Another major motivation for array experiments is the identification of novel regulatory proteins binding sites: in the most recently developed methods, one uses expression values, as measured with microarray, as a response in regression setting. In the first case, it is clearly important to be able to assign a measure of confidence to the obtained clusters. In the second case, it is important to obtain the best estimate of expression values for each gene, in order for the regression procedure to lead to relevant results. Both these problems involve consideration of multiplicity. It is beyond the scope of the present introduction to give a detailed account of the literature on these issues. However, we will mention some of the directions of research more closely related with our topic.

To date, there are a number of approaches proposed in the literature to obtain better estimates of expression values (or change in expression values) from microarray data. A considerable amount of these uses a Bayesian framework, with more or less "strength borrowing" across genes (e.g., [22,30,31]). The interest of these approaches is that they not only allow for testing null hypothesis on the expression of the genes, but also provide point estimates of it, and confidence intervals. Now, both point estimates and confidence intervals can be approached from the point of view of multiplicity. When dealing with point estimates, one needs to define a loss function that considers all the genes at the same time, rather than one at the time. This type of loss functions are at the base of shrinkage estimators, so that it is not surprising to obtain results in this direction in this context. One very interesting feature, is that if one makes sparsity assumptions on the vector of true expression change values, thresholding estimators become valuable options. The expression change is estimated to be zero, unless a threshold is attained — this threshold can be linked to FDR procedures (see [24,32]).

The issue of simultaneous confidence interval was particularly dear to Tukey. Unfortunately, not much progress has been made on this front in the case of microarrays. For example, the work of Tseng et al. is exemplary in stressing the importance of confidence intervals over testing, but the authors gloss over the issue of multiple comparisons. A promising route is the work in Reference 33 on FDR corrected confidence intervals that can be applied to the genes for which the null hypothesis has been rejected.

The problem of determining how many clusters are really present in a dataset or how well-defined are the clusters boundaries is of relevance for many other fields besides microarrays. In the specific context of gene expression studies, Pollard and van der Laan [34] had some interesting suggestions.

REFERENCES

1. Braun, H. (ed.) (1994) *The Collected Works of John W. Tukey. Vol. VIII Multiple Comparisons: 1948–1983*. Chapman & Hall, New York.
2. Donoho, D. (2001) Data! data! data! challenges and opportunities of the coming data deluge, Michelson Memorial Lecture Series (sponsored by the US Naval Academy).
3. Benjamini, Y. and Hochberg, Y. (1995) Controlling the false discovery rate: a practical and powerful approach to multiple testing. *Journal of the Royal Statistical Society B* 57: 289–300.
4. Benjamini, Y. and Yekutieli, D. (2001) The control of the false discovery rate in multiple testing under independence. *The Annals of Statistics* 29: 1165–1188.

5. Zaykin, D., Zhivotovsky, L., Westfall, P., and Weir, B. (2002) Truncated product method for combining *P*-values. *Genetic Epidemiology* 22: 170–185.
6. Dudbridge, F. and Koeleman, B. (2003) Rank truncated product of *P*-values with application to genomewide scans. *Genetic Epidemiology* 25: 360–366.
7. Dudoit, S., Popper Shaffer, J. and Boldrick, J. (2003) Multiple hypothesis testing in microarray experiments. *Statistical Science* 18: 71–103.
8. Storey, J.D. (2003) The positive false discovery rate: a Bayesian interpretation and the *q*-value. *Annals of Statistics* 31: 2013–2035.
9. Storey, J. (2002) A direct approach to false discovery rates. *Journal of the Royal Statistical Society B* 64: 479–498.
10. Genovese, C.R. and Wasserman, L., (2002) Operating characteristics and extensions of the false discovery rate procedure. *Journal of the Royal Statistical Society B* 64: 499–518.
11. Sarkar, S. (2002) Some results on false discovery rate in stepwise multiple testing procedures. *Annals of Statistics* 30: 239–257.
12. Holm, S. (1979) A simple sequentially rejective multiple test procedure. *Scandinavian Journal of Statistics* 6: 65–70.
13. Westfall, P. and Young, S. (1993) *Resampling-Based Multiple Testing*. John Wiley & Sons, New York.
14. Storey, J.D., Taylor, J.E., and Siegmund, D. (2004) Strong control, conservative point estimation, and simultaneous conservative consistency of false discovery rates: a unified approach. *Journal of the Royal Statistical Society, Series B*, 66: 187–205.
15. Ge, Y., Dudoit, S., and Speed, T. (2003) Resampling-based multiple testing for microarray data-analysis. *Test* 12: 1–77.
16. Pollard, K.S. and van der Laan M.J. (2002) Resampling-based multiple testing: asymptotic control of type I error and applications to gene expression data (June 2003). U.C. Berkeley Division of Biostatistics working paper series. Working Paper 121. http://www.bepress.com/ucbbiostat/paper 121.
17. Yekutieli, D. and Benjamini, Y. (1999) Resampling-based false discovery rate controlling multiple test procedures for correlated test statistics. *Journal of Statistical Planning and Inference* 82: 171–196.
18. Storey J. and Tibshirani, R. (2001) Estimating false discovery rates under dependence, with applications to DNA microarrays. Technical Report 2001-28, Department of Statistics, Stanford University, Standford, CA.
19. Tusher, V., Tibshirani, R., and Chu, G. (2001) Significance analysis of microarray applied to the ionizing radiation response. *Proceedings of the National Academy of Sciences, USA* 98: 5116–5121.
20. Storey, J.D. and Tibshirani, R. (2003) SAM thresholding and false discovery rates for detecting differential gene expression in DNA microarrays. In *The Analysis of Gene Expression Data: Methods and Software*, G. Parmigiani, E.S. Garrett, R.A. Irizarry, and S.L. Zeger (eds.). Springer-Verlag, New York.
21. Baldi, P. and Long, A.D. (2001) A Bayesian framework for the analysis of microarray expression data: regularized *t*-test and statistical inferences of gene changes. *Bioinformatics* 17: 509–519.
22. Tseng, G.C., Oh, M.-K., Rohlin, L., Liao, J.C., and Wong, W.H. (2001) Issues in cDNA microarray analysis: quality filtering, channel normalization, models of variation and assessment of gene effects. *Nucleic Acids Research* 29: 2549–2557.

23. Erickson, S. and Sabatti, C. (2005) Empirical Bayes estimation of a sparse vector of gene expression, UCLA stat preprint 413.
24. Sabatti, C., Karsten, S., and Geschwind, D. (2002) Thresholding rules for recovering a sparse signal from microarray experiments. *Mathematical Biosciences* 176: 17–34.
25. Reiner, A., Yekutieli, D., and Benjamini, Y. (2003) Identifying differentially expressed genes using false discovery rate controlling procedures. *Bioinformatics* 19: 368–375.
26. Efron, B. (2003) Robbins, empirical Bayes and microarrays. *Annals of Statistics* 31: 366–378.
27. Allison, D.B., Gadbury, G., Heo, M., Fernandez, J., Lee, C.-K. Prolla, T.A. and Weindruch, R. (2002) A mixture model approach for the analysis of microarray gene expression data. *Computational Statistics & Data Analysis* 39: 120.
28. Gadbury, G.L., Page, G.P., Heo, M., Mountz, J.D., and Allison, D.B. (2003) Randomization tests for small samples: an application for genetic expression data. *Journal of the Royal Statistical Society: C (Applied Statistics)* 52: 365–376.
29. Ishwaran, H. and Rao, J. (2003) Detecting differentially expressed genes in microarrays using Bayesian model selection. *Journal of American Statistical Analysis* 98: 438–455.
30. Newton, M.A., Kendziorski, C.M., Richmond, C.S., Blattner, F.R., and Tsui, K.W. (2001) On differential variability of expression ratios: improving statistical inference about gene expression changes from microarray data. *Journal of Computational Biology* 8: 37–52.
31. Ibrahim, J.G., Chen, M.H., and Gray, R.J. (2002) Bayesian models for gene expression with DNA microarray data. *Journal of the American Statistical Association* 97: 88–99.
32. Abramovich, F., Benjamini, Y., Donoho, D., and Johnstone, I. (2000) Adapting to unknown sparsity by controlling the false discovery rate. Stanford Statistics Department, Technical Report # 2000-19.
33. Benjamini, Y. and Yekutieli, Y. (2005) False discovery rate controlling confidence intervals for selected parameters. *Journal of American Statistical Association* 100: 71–80.
34. Pollard, K.S. and van der Laan, M.J. (2002) A method to identify significant clusters in gene expression data. In *Proceedings of SCI*, Vol. II, pp. 318–325.

16 Using Standards to Facilitate Interoperation of Heterogeneous Microarray Databases and Analytic Tools

Kei-Hoi Cheung

CONTENTS

16.1 Introduction 305
16.2 Using Standards to Tackle the Heterogeneity Problem 307
16.2.1 Minimum Information about a Microarray Experiment 307
16.2.2 Standard Vocabularies and Ontologies 308
16.2.2.1 MGED Ontology 308
16.2.2.2 Gene Ontology 309
16.2.2.3 Other Bio-Ontologies 310
16.2.3 XML-Based Standards and Technologies 310
16.2.3.1 Biological Data Representation 311
16.2.3.2 Web Services 312
16.2.3.3 Web Service Choreography Language 312
16.2.3.4 Other Web Service Issues 313
16.3 Future Directions 314
Acknowledgments 316
References 317

16.1 INTRODUCTION

The high volume and complexity of data generated by DNA microarrays has created both opportunities and challenges for bioinformaticians and computational biologists as the large-scale analysis of expression data requires advanced database, computational, and statistical approaches. One such approach seeks the seamless interoperation of different data sources and analytic tools that are used for analyzing microarray data. Despite the advent of integration technologies such as web technology and database connectivity (DBC) technologies (e.g., Open DBC and Java DBC), the heterogeneous nature of these data sources, analytic tools, and microarray data measurement methods poses a major challenge to their interoperation. In general, there are two types of heterogeneities, namely syntactic heterogeneities and semantic heterogeneities.

1. *Syntactic heterogeneities.* They refer to (i) the different ways in which data are represented, formatted and accessed, and (ii) the different ways in which software applications are written and executed. For example, data may be represented using different data models (e.g., relational model vs. object-oriented model). Different database platforms may provide different query languages for users to access and query the data. For example, different relational database vendors (e.g., Oracle and Sybase) implement the (SQL) query standard slightly differently. In terms of data dissemination, a variety of formats may be used (e.g., tab-delimited, ASN.1, and eXtensible Markup Language [XML] formats). Different analytic software tools may be written using different programming languages (e.g., C, Java, Perl, etc.) with different user interfaces (e.g., command-line interface vs. graphical user interface). These programs may also run on different computer platforms (e.g., Unix, Windows, and Mac).
2. *Semantic heterogeneities.* These have to do with what the data mean (e.g., what type of data it is, who produced them on what date using what types of experimental methods, etc.) and what a program does (what kind of analysis it performs, who wrote the program, what version it is, what input parameters it takes, what kind of output it produces, etc.). One common source of semantic heterogeneities in the biosciences domain is the use of synonymous terms for describing the same concept. In gene nomenclature, the same gene (e.g., dopamine receptor D2) may have synonymous symbols (e.g., DRD2, D2, D2DR, etc.). In microarray data measurement, the level of gene expressions can be described using synonymous terms (e.g., up-regulation/down-regulation vs. over-expression/under-expression). Besides synonyms, the same term (e.g., insulin) can be used to represent different concepts (e.g., gene, protein, drug, etc.). This problem can also occur at the level of data modeling. For example, the concept experiment in one microarray database (e.g., SMD [1]) may refer to a series of samples (corresponding to different experimental conditions) hybridized to different arrays. In another microarray database (e.g., RAD [2]), an experiment may refer to a single hybridization. In terms of data analysis, a particular category of analysis (e.g., clustering) may be implemented using different techniques or algorithms (e.g., hierarchical clustering, K-means, self-organizing map, etc. [3]). Another example is data normalization (or data scaling), which can be implemented based on a variety of methods such as global vs. local normalization methods [4]. Such a diversity of data acquisition, representation, and analysis methods makes comparison and integration of microarray gene expression data very difficult. Whether the interpretation

of microarray data is meaningful or not depends on the semantics of the expression values (i.e., how the data are acquired and preprocessed) and the semantics of the analysis programs used (i.e., the type of analysis involved and how the analysis is implemented). Without such semantic information, gene expression data (and hence their interpretation) becomes meaningless.

16.2 USING STANDARDS TO TACKLE THE HETEROGENEITY PROBLEM

Heterogeneity is unavoidable in a young field like DNA microarrays. Different methodologies need to be developed, explored, compared, and evaluated to determine which one would work better under what circumstances. However, uncontrolled multiplicity of methodologies will hamper research advance and the maturity of a scientific field necessitates the use of standards. Good science requires its data to be verifiable and reproducible, which is impossible to achieve without using standardized methods. In addition, reduced heterogeneity in data measurement, analysis, and representation methods helps lower the complexity of interoperation of the data and analytic tools involved. In the following sections, a number of standard-based approaches to microarray data and tool interoperability are described.

16.2.1 Minimum Information about a Microarray Experiment

As a first step toward tackling the problem of heterogeneity posed by microarray technology, a standard was proposed by the microarray gene expression database (MGED) consortium (http://www.mged.org/) to provide the minimum information about a microarray experiment (MIAME) [5] that is needed to meaningfully compare the results of microarray data stored in different databases. As the amount of publicly accessible microarray data grows, new methods are being developed to cross-validate the data. For example, a statistical approach has been developed to translate experimental results across different microarray platforms (cDNA- vs. oligonucleotide-based arrays) [6]. MIAME serves as a standard guideline or specification on how to annotate the entities involved in the process of generating and preprocessing microarray data. Broadly speaking, these entities include: samples, array design, experiments, hybridizations, measurements, and normalization controls. As a result of this standardization effort, a number of MIAME-compliant/supportive microarray databases have been developed. While some of these databases (e.g., RAD [2], GeneX [7], LAD [8], YMD [9], SMD [1], etc.) are intended for use by individual laboratories, centers, or institutions with their own specific needs, others (e.g., ArrayExpress [10] and GEO [11]) are designed to serve as public repositories of microarray data. Making microarray data available electronically in MIAME format has become a requirement made by a number of scientific journals for publishing papers that describe the results of microarray experiments.

To make the MIAME standard computer-usable, an object-oriented data model microarray gene expression object model (MAGE-OM [12]) was built using the unified modeling language (UML). In addition, the model was translated into an XML-based common language microarray gene expression markup language (MAGE-ML [12]) to facilitate automatic exchange of microarray data between analytic tools and databases. A software toolkit (MAGE-STK) was developed for a variety of programming languages (including Perl, Java, C++, and Python) to ease the integration of MAGE-OM and MAGE-ML into end users' systems. The applications of MAGE-ML have been shown in a number of microarray software packages such as Rosetta Inpharmatics' Revolver (http://www.rosettabio.com/products/resolver/default.htm). In addition, public data repositories such as ArrayExpress accept microarray data submission in MAGE-ML format. A MAGE-ML-based message broker framework has been developed using simple object access protocol (SOAP) for exchange and integration of microarray data [13].

16.2.2 Standard Vocabularies and Ontologies

While MIAME helps interoperability at the database structure level, interoperability of data contents can be facilitated by the use of standard terms (vocabularies) and ontologies. An ontology consists of a well-defined set of terms and a well-defined set of relationships between the terms. More progress has been made in the field of medical informatics than in the field of bioinformatics in terms of developing and using standard vocabularies and ontologies. For example, the unified medical language system (UMLS [14]) has been developed to integrate a variety of clinically related standard vocabularies including SNOMED, MESH, ICD9, etc. The UMLS has been frequently used to facilitate clinical data interoperability [15], medical decision support [16], and natural language processing [17,18]. As the amounts and diversity of genome data (e.g., gene expression data) increase, there will be a growing demand for developing and utilizing standard vocabularies/ontologies to realize data unification. As a step toward meeting this demand, an Open Biological Ontologies or OBO (http://obo.sourceforge.net) has been established as an umbrella project for creating well-structured controlled vocabularies/ontologies for shared use across different biological and biomedical domains. In the following sections, a number of bio-ontologies (some of which are included in OBO) will be discussed.

16.2.2.1 MGED Ontology

While MIAME specifies what constitutes the minimum information about a microarray experiment, it does not place any restriction on the information content. For example, MIAME recommends that experiment type be specified as part of the experimental description, but it does not control what go in as actual values (it only gives representative examples of experiment types such as dose response, time course, etc.). To avoid the proliferation of synonymous data values, the MGED ontology working group (http://mged.sourceforge.net/ontologies/) was established, which is charged with developing standard terms (controlled vocabulary) for annotating microarray experiments including descriptions of biological samples. Such standard vocabularies or ontologies will ease cross-database queries and development of software tools that mine the databases. Indeed, they are expected to evolve especially as new applications of microarray technology arise that require descriptive terms. To minimize the impact of ontological evolution on software development, a core MGED ontology is being developed, which will not change. A second layer of the ontology, the extended MGED ontology, will contain all additional terms that are logically consistent with the core ontology.

16.2.2.2 Gene Ontology

Given a mixture of known and unknown genes exhibiting a similar expression pattern, it can be hypothesized that the biological function of the unknown genes may be similar to those that are known. Moreover, the function of the unknown genes may be deduced from the known homologous genes (of different organisms) identified by sequence alignment programs such as the Basic Local Alignment Search Tool (BLAST) [19]. It is a very daunting task to compile the knowledge of the known genes from the literature since the current systems of nomenclature for describing the biological roles of genes and their products are divergent. The term *biological function* has broadly been used in the biomedical literature to describe the biochemical activities, biological roles, and cellular structure of genes or proteins. For example, the function of a protein such as *tubulin* has been described as *GTPase* or *constituent of the mitotic spindle*. The main goal of the Gene Ontology (GO) is to standardize the gene (protein) description in a more precise way across different species. As a result, there are three broad categories in GO.

1. *Biological process* refers to the biological objective to which the gene or gene product contributes.
2. *Molecular function* is defined as the biochemical activity of a gene product. One or more ordered assemblies of molecular functions can accomplish a biological process.
3. *Cellular component* refers to the place in the cell where a gene product is active.

The GO also facilitates interoperability by providing cross-references between many databases (e.g., Pfam, Flybase, Swiss-Prot, etc.). As pointed out by Karp [20], GO can be enhanced by adding more ontological features to it (e.g., causal transitivity, temporal ordering, assertion of disjoint subconcepts, etc.). This is particularly true as the size of GO has expanded significantly (spanning a wider variety of organisms) in recent years and its utility in microarray data analysis has been on the rise (e.g., Onto-Express [21], GoMiner [22], DAVID [23], FatiGO [24], etc.). Given its curatorial success, GO has also been used as a gold standard for testing or evaluating literature-mining approaches (e.g., [25]) that compute biological functions. Recently, GO has undergone the following extensions.

1. GO Slim. GO possesses a fine granularity of abstraction of concepts that are restricted to the molecular biology domain. Although this fine level of granularity is needed for understanding a particular molecular mechanism controlling a specific biological pathway (e.g., "S phase of mitotic cell cycle"), it may be too detailed for someone interested in a broader biological

context (e.g., "cell replication"). To address this issue, GO Slim (ftp://ftp.geneontology.org/pub/go/GO_slims/), which consists of a set of high-level terms under each of the three GO ontologies, was introduced to cover most aspects of each of these three ontologies without overlapping in paths in the GO hierarchy.

2. GONG. The Gene Ontology Next Generation (GONG) Project [26] is developing a staged methodology to evolve the current representation of the Gene Ontology into a knowledgebase using the knowledge representation language, DAML+OIL [27], in order to take advantage of the richer formal expressiveness and the reasoning capabilities of the underlying description logic. Each stage provides a step level increase in formal explicit semantic content with a view to supporting validation, extension, and multiple classification of the GO.

16.2.2.3 Other Bio-Ontologies

This section describes other bio-ontologies that are also useful in the integrated analysis of microarray data. While GO is an organism-independent ontology, there are ontologies that are organism-specific, focusing on a different category of biological data such as pathway data. For example, EcoCyc [28] is a database that involves using a frame-based ontology [29] to classify and relate concepts for construction of the biological pathways of the bacterium *Escherichia coli*. It demonstrates the benefits of using a formal and expressive ontology to compute biological functions. As an extension to EcoCyc, MetaCyc [30] is a collection of metabolic pathways and enzymes from a wide variety of organisms with extensive references to literature citations. It also describes the reactions, chemical compounds, and genes. In addition, it includes a set of software tools that supports querying, visualization, curation, and prediction of pathways. Metacyc (as well as EcoCyc) is a member of BioCyc (http://BioCyc.org), which represents a larger collection of pathway/genome databases.

The Kyoto Encyclopedia of Genes and Genomes (KEGG) [31] is a suite of databases and associated software, integrating our current knowledge on molecular interaction networks in biological processes. The KEGG Ontology (KO) captures aspects of biochemistry such as the relationships between reactants, catalysts, substrates, and products. Initially, its focus was on enzyme metabolic pathways. To overcome the problems inherent in the enzyme nomenclature, KO extends ortholog IDs based on computational analysis, as well as manual curation, of decomposing all genes in the complete genomes into sets of orthologs. Here, two genes are considered as orthologs, or belonging to the same KO group, when they are mapped to the same KEGG pathway node. KO can be used to explore unknown pathways as well as to characterize all known pathways.

Tools (e.g., AraCyc [32] and PathMAPA [33]) are available to give an overlay of expression data on the biochemical pathway overview diagrams. These pathway diagrams may be constructed using experimental data or literature data (e.g., GeneWays [34]).

16.2.3 XML-Based Standards and Technologies

Thanks to the World Wide Web, a large variety of types of genomic data are now publicly and easily accessible to many genome researchers over the Internet. Using a client browser (e.g., Netscape, Firefox, and Internet Explorer), a researcher can access a remote genome website to query and download the data of his/her interest. Very often, such query results (e.g., in HTML format) or downloaded datasets (e.g., in tab-delimited format) are massaged and analyzed by one or more programs. For example, a gene expression dataset for a particular microarray experiment may be downloaded from a microarray database and reformatted for input to a clustering program. For further analysis, the genes within a cluster of interest may be linked to an annotation resource such as GO for functional categorization (this may involve writing yet another program to extract from the cluster output the gene identifiers within the cluster). One problem noted in this data analysis pipeline is that its automation is difficult due to the following:

1. The data are formatted differently (e.g., HTML vs. flat file format). These individual formats need to be processed by different parsing programs.
2. The data semantics are not captured explicitly in a format that is machine-readable. Very often the meaning of the data is described in the file header or in a separate README file that can only be understood by human beings. Formats such as HTML are designed for data display and hypertext linking and they are not machine-friendly either.

To address these problems, the XML was used as a standard data representation format for exchanging information between different software applications (including databases) that can be accessed over the Internet. XML allows each data value to be described by a meaningful tag (element). For example, the value "3" can be tagged with an element labeled "ratio" indicating that it is a ratio expression value. An XML element can have its own attributes and it can relate to other elements hierarchically. This hierarchical feature provides a straightforward way to design data structures that meet a wide range of application needs. However, some critics have pointed out that the XML approach is more verbose than the traditional data representation approach. Despite the criticisms, the benefits brought by the XML technology outweigh its costs.

16.2.3.1 Biological Data Representation

The popularity of XML has grown rapidly in the bioinformatics domain due to the great need for a common format that can be used to represent and integrate diverse types of genomic and proteomic data (e.g., gene and protein microarray data). Also, XML has a number of advantages including its self-describing nature, ease of use, validity, and a large base of open software support. A wide variety of types of biological data have been represented using XML. Examples include MAGE-ML for describing gene expression data [12], BIOML for describing biopolymer data [35], BSML for describing biological sequence data (http://www.bsml.org/), PEML for describing proteomics experiments [36], SBML for describing biochemical network models [37], and so forth. Although XML has widely been adopted as a common language for data representation, this alone does not solve the heterogeneity problem. In fact, the same problem will occur if multiple XML formats are used to describe the same type of data. Efforts have been underway to either consolidate existing XML formats or come up with a consensus on how to standardize a new XML format for representing a certain type of data. For example, MAGE-ML is the result of merging two different XML formats (GEML and MAML) that were used to describe microarray gene expression data. Having learned from the experience with MAGE, emerging standards such as Proteomics Experiment Markup Language (PEML) [36] has been progressing in a much more concerted way. More recently, the semantic web technology (http://www.w3.org/2001/sw) has emerged in the bioinformatics and life sciences communities as standard way of using XML to represent bio-ontologies.

16.2.3.2 Web Services

Besides data representation, XML has been used to interoperate bioinformatics data and software applications. The SOAP (http://www.w3.org/TR/SOAP/) is an XML-based standard for exchanging XML-formatted messages between applications that are known as *web services*. SOAP is not bound to any transport protocol, but is in general built on top of the HyperText Transfer Protocol (HTTP), which makes it firewall friendly in contrast to protocols used by Common Object Request Broker Agent (CORBA), for example. A web service is a published interface to a type of data or computation. In addition to SOAP, two standards have emerged that comprise the web services model.

1. The Web Service Description Language (WSDL; http://www.w3.org/TR/2003/WD-wsdl12-20030611/) is an XML language that describes a web service in an abstract manner by defining the web service interface and the exchange of messages between the provider and requester.
2. Universal Description, Discovery and Integration (UDDI; http://www.uddi.org/specification.html) is a standard protocol designed to publish details about an organization and the web services. It provides a description and definition of web services in a central repository, which functions as yellow pages for web services.

The web service approach is considered as a type of distributed computing as they are run on different computer hosts and can be integrated (e.g., one web service can invoke another) to perform a larger computing task. Although other distributed computing approaches such as CORBA have been used to facilitate data and tool interoperation in the genome context [38], the web service approach is considered to be easier and simpler to implement without introducing a heavy system overhead. Despite their differences, these alternative distributed computing approaches can coexist (e.g., web services can exist within a CORBA framework). According to Stein [39], the "bioinformatics nation" can be unified using web services. Examples of bioinformatics web services include XEMBL [40], SOAP-HT-BLAST [41], and Distributed Annotation Server (DAS) [42].

16.2.3.3 Web Service Choreography Language

While interoperating between web services is relatively simple, in reality, applications such as different microarray analysis programs need to work together in order to offer more advanced functionality. For example, the analysis may involve a coordinated sequence of processing steps such as filtering, normalization, cluster analysis, access to a variety of genome annotation sources, and so forth. To compound the problem of interoperation, there are different ways of analyzing the same dataset. For example, some microarray analyses such as clustering are exploratory in nature while others are more specific to the nature of the biological problem at hand (e.g., genetic network modeling [43]). Moreover, there are different ways to perform the same type of analysis (e.g., there are different clustering approaches such as hierarchical methods vs. nonhierarchical methods).

To make full use of the advantages that the web service model offers, such as advanced automation and application integration to group a set of applications together, we need more than a simple point-to-point connection. That is, we need a language that allows us to create a chain of interdependent web services that easily operate together. This concept of interrelated web services that are linked together in a functionally coherent and repeatable process is called a "Web Services Choreography."

The World Wide Web Consortium (W3C) has published a draft of its specification for web services choreography language (http://www.w3.org/2002/ws/chor/). Currently, there are two leading emerging standards: the Business Process Execution Language for Web Services (BPEL4WS; http://www-106.ibm.com/developerworks/webservices/library/ws-bpel/) and the Business Process Markup Language (BPML; http://www.bpmi.org/bpml.esp) are the leading emerging standards. They are compatible languages and both offer a rich set of workflow options. BPEL4WS seems most promising at this moment to emerge as a widely supported standard for web services choreography since it has the broad industry support from companies like IBM, Microsoft, and BEA, while SUN recently also joined the Technical Committee. In addition, BPEL4WS is well documented and tools such as editors and choreography engines are already available. Both BPEL4WS and BPML allow complex scenarios of choreography such as concurrent processes, synchronous and asynchronous messaging, roll back mechanisms, data manipulation, and error handling. These features benefit certain classes of biomedical informatics applications that involve sensitive data and the use of parallel programming to speed the complex analyses of large datasets. While web services and web services choreography are still undergoing consolidation and evolution, ongoing projects such as BioMoby [44] and myGrid [45] use such standard web service choreography languages with some customization to create and manage the workflow of bioinformatics web services.

16.2.3.4 Other Web Service Issues

Other important issues of web services include the following:

1. *Security.* While the web service approach is firewall friendly, it does not mean that security can be comprised. For example, some microarray experiments may involve the use of confidential DNA samples of individuals (e.g., HIV patients). When such sensitive data are exchanged between web services, we need to ensure their confidentiality and security. Groups including

IBM and Microsoft have been working on the extension of the current SOAP specifications to incorporate into web services security features (e.g., data signature and encryption) that are provided by current technologies including the Public Key Infrastructure (PKI) and Secure Sockets Layer (SSL).

2. *Compatibility.* Web services are intended to be platform-independent. In reality, web services that are generated using different methods (e.g., Microsoft.NET and Java-based Axis) are not as easily interoperable as one may imagine. To address this, the Web Service Interoperability (WS-I) organization (http://www.ws-i.org/) has proposed Web Service Profiles (http://www.ws-i.org/docs/WS-I_Profiles.pdf) to ensure the future compatibility of web service products put out by different nonprofit groups or commercial vendors.
3. *Compression.* One trade-off of encoding data semantics in XML format is the significant increase in data size. Microarray-related web services may involve exchanging large amounts of gene expression data in XML formats such as MAGE-ML. Transmitting such large expression data files over network and processing them can become a major performance bottleneck. To address this problem, approaches to compressing large XML files have been developed. They include conversion of XML tags into binary tokens (http://xml.coverpages.org/xmlandcompression.html). Some methods such as XML-Xpress (http://www.ictcompress.com/products_xmlxpress.html) can achieve a better compression ratio by exploiting the schema of XML files.

16.3 FUTURE DIRECTIONS

As software tools are being developed to offer new or better ways of analyzing microarray data, they need to be easily accessible to the users (including both experimentalists and bioinformaticians). Currently, most of the tools (some of which include source code) can be downloaded from different microarray software resources including SMD (http://genome-www5.stanford.edu/), TIGR (http://www.tigr.org/software/tm4/), and Bioconductor (http://www.bioconductor.org/) and installed on local computers. This approach has the following limitations:

1. Some of these analyses may run slowly on large microarray datasets. This creates a difficulty to users who neither have enough computing power nor bioinformatics skills to optimize these programs.
2. It is difficult to link these tools together as they may be written in different programming languages and run on different platforms.
3. Even for users who have computing power and bioinformatics expertise, this approach may still not be flexible especially when the tools evolve over time.

One solution we have implemented is to provide the users with an integrated platform (the Yale Microarray Database [YMD]; http://info.med.yale.edu/microarray) for both data storage and data analysis. In this approach, YMD is seamlessly linked to a variety of analysis methods including Expressyourself (an integrated platform for analyzing and visualizing microarray data) [46], GProcessor

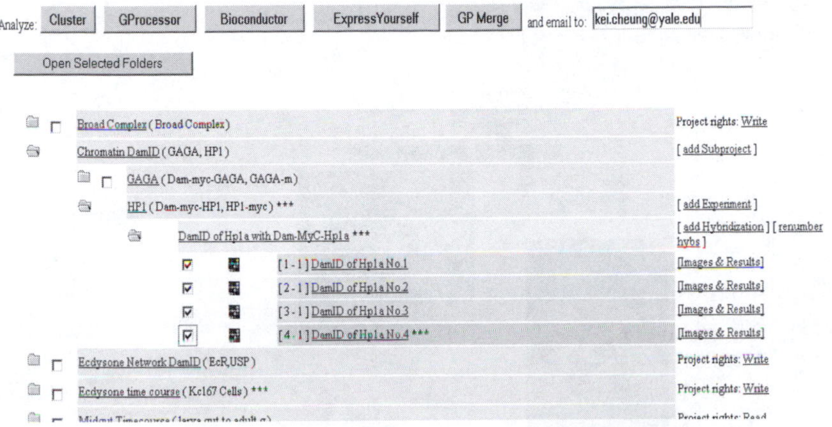

FIGURE 16.1 Tool linking through YMD's project management interface.

(a program that allows data to be normalized using nonlinear methods such as Lowess fit and Analysis of variance [ANOVA]; http://bioinformatics.med.yale.edu/software/readmegprocessorlinux.pdf), boxplot (that is available through the Bioconductor package), and a number of clustering methods. These programs are written in different programming languages (e.g., C and Java) and run on different computers. While we have installed some of these programs locally (e.g., Bioconductor and GProcessor), others (e.g., Expressyourself) are maintained and updated by the original developers and are running on their computers. In addition, some of the tools such as different clustering methods are implemented as web services (this is a collaboration between the Yale Center for Medical Informatics and a Computer Science Group in the University of Hong Kong). As shown in Figure 16.1, the user can select the datasets (hybridized arrays for a microarray experiment) he or she wants and click on a button (near the top of the screen) corresponding to an analysis method (Expressyourself). Once the datasets are submitted, the user can continue to do other tasks while the analysis is being performed on another computer. When the analysis is done, an e-mail message with a hypertext link to the analysis output page will be sent to the user. Figure 16.2 shows the analysis results generated by Expressyourself. The figure indicates that one of the datasets (arrays) was selected for array image recreation (one of the features of Expressyourself). The image spots within a selected block (subarray) are displayed alongside with a log–log intensity scatter plot (Cy3 vs. Cy5).

As indicated by Campbell [47], the central role of data analysis will become even more critical in the future as microarray data is successfully integrated with other massive datasets such as genome sequence data or data from high-throughput drug screening techniques. Currently, most microarray-related data can be accessed interactively via the web (e.g., through a web query form) or can be downloaded in bulk as text files. Although the web interface allows a data item provided by one resource to be linked to another data item provided by another, this approach, as pointed out by Karp [48], is not scalable to genome-wide data integration. The problem with the file download approach is that these files are heterogeneous in their formats requiring

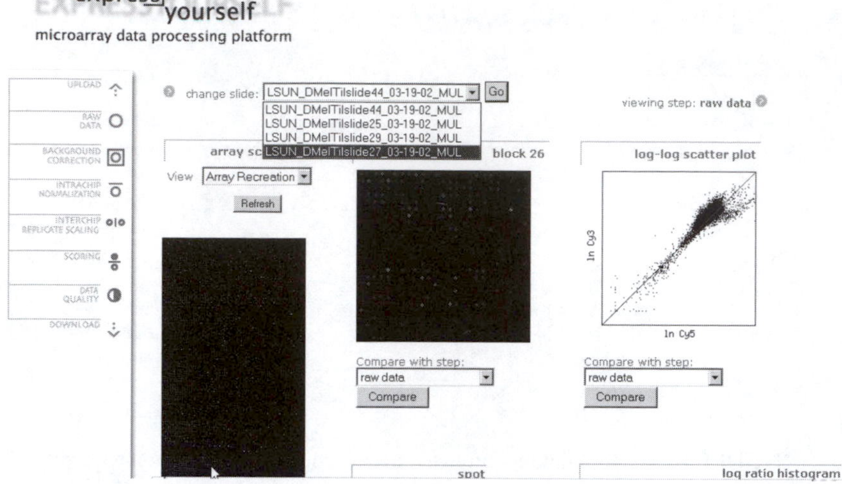

FIGURE 16.2 Analysis results of datasets submitted from YMD to Expressyourself.

specific scripts to parse them. To make it scalable and machine friendly, we need a standard (XML) format for describing the data structure (metadata). We also need a standardized interface easing programmatic access to the data.

If HTML has revolutionized the way genome scientists access information and tools over the Internet, XML has provided a standardized way for future bioinformatics software applications (in the form of web services) to interact automatically and intelligently. With the increasing use of XML in biological data representation and emerging bioinformatics web services, we are in the transition to the next generation of semantic web. In the semantic-web scenario, each resource (whether it is a data providing service or a data analysis service) will be registered centrally and described using a common framework language (e.g., an ontological language such as OWL and DAML+OIL [27]) making it easier to find the required service. There will be a mechanism that allows software developers to register their databases/tools and end users (including both software developers and microarray researchers) to query the registry to find the services they want. Once identified, these services can be put together easily by using a web service choreography engine to create a data analysis workflow that can be executed automatically (without human intervention) or semiautomatically (with some human intervention).

ACKNOWLEDGMENTS

This work is supported in part by NSF grants DBI-0135442, MCB-0090286, NIH grants K25 HG02378, and U24 DK58776. The author would like to thank Dr. Janet Hager and Dr. Michael V. Osier for their comments.

REFERENCES

1. Gollub J., Ball C., Binkley G., Demeter J., Finkelstein D., Hebert J., Hernandez-Boussard T., Jin H., Kaloper M., Matese J., Schroeder M., Brown P., Botstein D., and Sherlock G. The Stanford Microarray Database: data access and quality assessment tools. *Nucleic Acids Research*, 2003, 31: 94–96.
2. Stoeckert C., Pizarro A., Manduchi E., Gibson M., Brunk B., Crabtree J., Schug S., Shen-Orr S., and Overton G. A relational schema for both array-based and SAGE gene expression experiments. *Bioinformatics*, 2001, 17: 300–308.
3. Shannon W., Culverhouse R., and Duncan J. Analyzing microarray data using cluster analysis. *Pharmacogenomics*, 2003, 4: 41–51.
4. Quackenbush J. Microarray data normalization and transformation. *Nature Genetics*, 2002, 32 (Suppl): 496–501.
5. Brazma A., Hingamp P., Quackenbush J., Sherlock G., Spellman P., Stoeckert C., Aach J., Ansorge W., Ball C., Causton H., Gaasterland T., Glenisson P., Holstege F., Kim I., Markowitz V., Matese J., Parkinson H., Robinson A., Sarkans U., Schulze-Kremer S., Stewart J., Taylor R., Vilo J., and Vingron M. Minimum information about a microarray experiment (MIAME) — toward standards for microarray data. *Nature Genetics*, 2001, 29: 365–371.
6. Ferl G., Timmerman J., and Witte O. Extending the utility of gene profiling data by bridging microarray platforms. *Proceedings of the National Academy of Sciences, USA*, 2003, 100: 10585–10587.
7. Mangalam H., Stewart J., Zhou J., Schlauch K., Waugh M., Chen G., Farmer A., Colello G., and Weller J. GeneX: an Open Source gene expression database and integrated tool set. *IBM Systems Journal*, 2001, 40: 552–569.
8. Killion P., Sherlock G., and Iyer V. The Longhorn Array Database (LAD): an open-source, MIAME compliant implementation of the Stanford Microarray Database (SMD). *BMC Bioinformatics*, 2003, 4: 32.
9. Cheung K., White K., Hager J., Gerstein M., Reinke V., Nelon K., Masiar P., Srivastava R., Li Y., Li J., Li J.M., Allison D., Snyder M., Miller P., and Williams K. YMD: a microarray database for large-scale gene expression analysis. In *Proceedings of the American Medical Informatics Association 2002 Annual Symposium*, 2002, pp. 140–144.
10. Brazma A., Parkinson H., Sarkans U., Shojatalab M., Vilo J., Abeygunawardena N., Holloway E., Kapushesky M., Kemmeren P., Lara G.G., Oezcimen A., Rocca-Serra P., and Sansone S.A. ArrayExpress — a public repository for microarray gene expression data at the EBI. *Nucleic Acids Research*, 2003, 31: 68–71.
11. Edgar R., Domrachev M., and Lash A. Gene Expression Omnibus: NCBI gene expression and hybridization array data repository. *Nucleic Acids Research*, 2002, 30: 207–210.
12. Spellman P., Miller M., Stewart J., Troup C., Sarkans U., Chervitz S., Bernhart D., Sherlock G., Ball C., Lepage M., Swiatek M., Marks W., Goncalves J., Markel S., Iordan D., Shojatalab M., Pizarro A., White J., Hubley R., Deutsch E., Senger M., Aronow B., Robinson A., Bassett D., Stoeckert C., and Brazman A. Design and implementation of microarray gene expression markup language (MAGE-ML). *Genome Biology*, 2002, 3: 1–9.
13. Tjandra D., Wong S., Shen W., Pulliam B., Yu E., and Esserman L. An XML message broker framework for exchange and integration of microarray data. *Bioinformatics*, 2003, 19: 1844–1845.

14. Cimino J. Review paper: coding systems in health care. *Methods of Information in Medicine*, 1996, 35: 273–284.
15. Ingenerf J., Reiner J., and Seik B. Standardized terminological services enabling semantic interoperability between distributed and heterogeneous systems. *International Journal of Medical Informatics*, 2001, 64: 223–240.
16. Cimino J., and Sideli R. Using the UMLS to bring the library to the bedside. *Medical Decision Making*, 1991, 11 (Suppl): S116–S120.
17. Weeber M., Vos R., Klein H., Berg LDJ-VD., Aronson A., and Molema G. Generating hypotheses by discovering implicit associations in the literature: a case report of a search for new potential therapeutic uses for thalidomide. *Journal of American Medical Informatics Association*, 2003, 10: 252–259.
18. Hahn U., Romacker M., and Schulz S. How knowledge drives understanding — matching medical ontologies with the needs of medical language processing. *Artificial Intelligence in Medicine*, 1999, 15: 25–51.
19. Altschul S., Gish W., Miller W., Myers E., and Lipman D. Basic local alignment search tool. *Journal of Molecular Biology*, 1990, 215: 403–410.
20. Yeh I., Karp P., Noy N., and Altman R. Knowledge acquisition, consistency checking and concurrency control for Gene Ontology (GO). *Bioinformatics*, 2003, 19: 241–248.
21. Khatri P., Draghici S., Ostermeier G., and Krawetz S. Profiling gene expression using onto-express. *Genomics*, 2002, 79: 266–270.
22. Zeeberg B., Feng W., Wang G., Wang M., Fojo A., Sunshine M., Narasimhan S., Kane D., Reinhold W., Lababidi S., Bussey K., Riss J., Barrett J., and Weinstein J. GoMiner: a resource for biological interpretation of genomic and proteomic data. *Genome Biology*, 2003, 4: R28.
23. Dennis G., Jr., Sherman B., Hosack D., Yang J., Gao W., Lane H., Lempicki R. DAVID: database for annotation, visualization, and integrated discovery. *Genome Biology*, 2003, 4: 3.
24. Al-Shahrour F., Díaz-Uriarte R., and Dopazo J. FatiGO: a web tool for finding significant associations of Gene Ontology terms with groups of genes. *Bioinformatics*, 2004, 20: 578–580.
25. Raychaudhuri S., Chang J., Imam F., and Altman R. The computational analysis of scientific literature to define and recognize gene expression clusters. *Nucleic Acids Research*, 2003, 31: 4553–4560.
26. Wroe C., Stevens R., Goble C., and Ashburner M. A methodology to migrate the gene ontology to a description logic environment using DAML+OIL. In *Proceedings of the 8th Pacific Symposium on Biocomputing*, Lihue, Hawaii, USA., 2003.
27. Horrocks L. DAML+OIL: a reasonable web ontology language. In *Proceedings of the International Conference on Extending Database Technology*, Prague, Czech Republic, 2002.
28. Karp P., Riley M., Saier M., Paulsen I., Collado-Vides J., Paley S., Pellegrini-Toole A., Bonavides C., and Gama-Castro S. The EcoCyc database. *Nucleic Acids Research*, 2002, 30: 56–58.
29. Karp P. Pathway databases: a case study in computational symbolic theories. *Science*, 2001, 293: 2040–2044.
30. Krieger C.J., Zhang P., Mueller L.A., Wang A., Paley S., Arnaud M., Pick J., Rhee S.Y., and Karp P.D. MetaCyc: a multiorganism database of metabolic pathways and enzymes. *Nucleic Acids Research*, 2004, 32: D438–D442.
31. Kanehisa M., and Goto S. KEGG: kyoto encyclopedia of genes and genomes. *Nucleic Acids Research*, 2000, 28: 27–30.

32. Mueller L.A., Zhang P., and Rhee S.Y. AraCyc: a biochemical pathway database for arabidopsis. *Plant Physiology*, 2003, 132: 453–460.
33. Pan D., Sun N., Cheung K., Guan Z., Ma L., Holford M., Deng X., and Zhao H. PathMAPA: a tool for displaying gene expression and performing statistical tests on metabolic pathways at multiple levels for *Arabidopsis*. *BMC Bioinformatics*, 2003, 4: 56.
34. Krauthammer M., Kra P., Iossifov I., Gomez S.M., Hripcsak G., Hatzivassiloglou V., Friedman C., and Rzhetsky A. Of truth and pathways: chasing bits of information through myriads of articles. *Bioinformatics*, 2002, 18: 249S–257S.
35. Fenyo D. The biopolymer markup language. *Bioinformatics*, 1999, 15: 339–340.
36. Taylor C., Paton N., Garwood K., Kirby P., Stead D., Yin Z., Deutsch E., Selway L., Walker J., Riba-Garcia I., Mohammed S., Deery M., Howard J., Dunkley T., Aebersold R., Kell D., Lilley K., Roepstorff P., Yates J., Brass A., Brown A., Cash P., Gaskell S., Hubbard S., and Oliver S. A systematic approach to modeling, capturing, and disseminating proteomics experimental data. *Nature Biotechnology*, 2003, 21: 247–254.
37. Hucka M., Finney A., Sauro H., Bolouri H., Doyle J., Kitano H., Arkin A., Bornstein B., Bray D., Cornish-Bowden A., Cuellar A., Dronov E., Gilles E., Ginkel M., Gor V., Goryanin I., Hedley W., Hodgman T., Hofmeyr J., Hunter P., Juty N., Kasberger J., Kremling A., Kummer U., Novere N.L., Loew L., Lucio D., Mendes P., Minch E., Mjolsness E., Nakayama Y., Nelson M., Nielsen P., Sakurada T., Schaff J., Shapiro B., Shimizu T., Spence H., Stelling J., Takahashi K., Tomita M., Wagner J., and Wang J. The systems biology markup language (SBML): a medium for representation and exchange of biochemical network models. *Bioinformatics*, 2003, 19: 524–531.
38. Fumoto M., Miyazaki S., and Sugawara H. Genome Information Broker (GIB): data retrieval and comparative analysis system for completed microbial genomes and more. *Nucleic Acids Research*, 2002, 30: 66–68.
39. Stein L. Creating a bioinformatics nation. *Nature*, 2002, 417: 119–120.
40. Wang L., Riethoven J., and Robinson A. XEMBL: distributing EMBL data in XML format. *Bioinformatics*, 2002, 18: 1147–1148.
41. Wang J., and Mu Q. Soap-HT-BLAST: high throughput BLAST based on Web services. *Bioinformatics*, 2003, 19: 1863–1864.
42. Dowell R., Jokerst R., Day A., Eddy S., and Stein L. The distributed annotation system. *BMC Bioinformatics*, 2001, 2: 1–7.
43. Someren Ev, Wessels L., Backer E., and Reinders M. Genetic network modeling. *Pharmacogenomics*, 2002, 3: 507–525.
44. Wilkinson M., and Links M. BioMoby: an open source biological web services proposal. *Brief Bioinform*, 2002, 3: 331–341.
45. Stevens R., Robinson A., and Goble C. myGrid: personalised bioinformatics on the information grid. *Bioinformatics*, 2003, 19 (Suppl 1): I302–I304.
46. Luscombe N., Royce T., Bertone P., Echols N., Horak C., Chang J., Snyder M., and Gerstein M. Express Yourself: a modular platform for processing and visualizing microarray data. *Nucleic Acids Research*, 2003, 31: 3477–3482.
47. Campbell C. Editorial: new analytical techniques for the interpretation of microarray data. *Bioinformatics*, 2003, 19: 1045.
48. Karp P. A strategy for database interoperation. *Journal of Computational Biology*, 1995, 2: 573–586.

17 Postanalysis Interpretation: "What Do I Do with This Gene List?"

Michael V. Osier

CONTENTS

17.1 Introduction 321
17.2 Overview of Current Methods 322
17.3 Knowledgebase Approaches 323
  17.3.1 Gene Ontology Database 323
  17.3.2 GenMAPP 325
  17.3.3 FatiGO 325
  17.3.4 GoMiner 326
  17.3.5 GOMine 326
  17.3.6 Database for Annotation, Visualization, and Integrated Discovery 327
17.4 Supplementary Data Approaches 327
17.5 Tentative Function Assignment Approaches 328
17.6 Future Directions 328
  17.6.1 Database Integration 329
  17.6.2 "Plug-and-Play" Analysis 329
  17.6.3 Combining Analytical Methods 329
  17.6.4 Evaluating Statistical Significance 330
17.7 Conclusions 331
Acknowledgments 332
References 332

17.1 INTRODUCTION

You complete the molecular portion of your microarray experiment. You run the analysis most appropriate to your data set to see which genes are differentially expressed under the experimental conditions. Now you are left with the question: "What does the array data tell me?" Sometimes this can be an easy question to answer, with only a few genes standing out from the many you have probed. Sometimes the question is not so easy with dozens or hundreds of genes, or even a representation of clustering results, to wade through. In the former case, an expert biological researcher can decide how to interpret the results, come up with new questions, and design new experiments. In the latter case, however, a computational analysis of the identified genes would be considerably helpful and sometimes even necessary.

17.2 OVERVIEW OF CURRENT METHODS

A number of groups have realized there is a problem of information overload even after the amount of data to be examined has been statistically reduced by identifying which genes are differentially expressed. Multiple approaches are being developed to analyze the expression profiles and aid the researcher in the interpretation of the data. Some methods identify biological patterns, some methods help direct future research, some provide easy access to additional relevant information, and some do more than one of the above. Those identifying biological patterns often use outside data sources to provide context, helping the researcher to see the forest rather than the trees. Those helping to direct future research sometimes generate hypotheses and sometimes partially take on the role of Laboratory Information Management Systems (LIMS), helping the researcher decide and keep track of what experiments could be conducted next.

However, within this simplistic breakdown lies more depth in the categories of analysis. For example, several of the pattern identifying approaches harness the expertise represented by curated knowledge databases such as the gene ontology database (GODB; http://www.godatabase.org/dev/database/; [1]), GenMAPP (http://www.genmapp.org/; [2]), Kyoto encyclopedia of genes and genomes (KEGG; http://www.kegg.com; [3]), and the original literature as represented in Medline. Other pattern identification analyses use the actual sequence data to correlate gene expression profiles with sequence similarities and, presumably, biological function.

With increasing number of data sources comes a corresponding increase in the number of combinations by which one data source can be analyzed against another. The combination of rapid changes in the way one can statistically analyze a data set and how one wants the results presented has led to a large corresponding increase in the number of combinations of analytical methods and tools a researcher needs to choose. In the short term, it seems likely that this trend of increasing options will continue. Since this new field of research into postanalysis interpretation of microarray data is difficult to keep track of, we present a brief selection of examples that are a representation of a subset of analysis types. The author apologizes in advance for the methods and tools, both present and future, which are not described in the following pages; however, notes that researchers eagerly anticipate these new developments. In the long term, one can hope that a standard emerges, either de facto or by a group consensus, reducing the difficulty for the researcher in choosing an appropriate method and tool.

17.3 KNOWLEDGEBASE APPROACHES

Multiple tools are available that harness knowledgebases to aid the interpretation of microarray analysis results. Many of these, such as Onto-Express (http://vortex.cs.wayne.edu/Projects.html; [4]) utilize the GODB. Several tools also use KEGG, UniGene (http://www.ncbi.nlm.nih.gov/entrez/query.fcgi?db=unigene), and other databases for both interpretation and annotation. The common theme to these tools is that they use the knowledge placed in these databases by experts to help the researcher in the decision process when working with statistically analyzed microarray data. They generally do this by searching for correlations between the structures of the two data sources. Sometimes these are simple statistical summaries, and sometimes the actual statistical significance level of the result is examined. Most experts attempt to assign biological meaning to the data, while some try to facilitate the interpretation process through the presentation of supplementary data (e.g., database cross references) without making any inferences.

Among those that attempt to make an inference based on the data, two approaches appear to predominate. Some algorithms perform a statistical test for each term or category based on which genes are directly associated (DA) with that term, which is a method we will call "strict association" for the sake of simplicity. Other algorithms perform a statistical test for each term where the knowledgebase is hierarchical based on which genes are associated with that DA term or any term "above" or "parental" to that DA term, which we will call "inclusive association."

Both methods have strengths and weaknesses. Strict association reduces some statistical dependencies easing inference calculations. Inclusive association ensures that a gene is also associated with all terms that are supersets of the term(s) that the gene is directly associated with, thereby giving the gene weight with terms that can be indirectly associated with it. However, in both cases many tests are performed, with questions such as "what is an appropriate statistical correction?" This question is hard to answer especially for inclusive association, since there is more dependence between terms. Later sections of this chapter discusses on how to answer this question of multiple test correction.

17.3.1 Gene Ontology Database

The GODB appears to be the most commonly used knowledgebase for the interpretation of microarray data. GODB is a directed acyclic graph (DAG) of biological roles, named "terms," that is curated by the Gene Ontology (GO) Consortium. A considerable amount of annotation is connected directly and indirectly to these terms through the various tables of GODB, including gene names and symbols, database cross references, biological sequences for specific genes, and references to the evidence supporting the gene–term associations, and is simplified in Figure 17.1. Programs generally connect terms to microarray spots through gene identifiers (Figure 17.2). These identifiers can include gene names, gene symbols, and database cross references, among other references. In this way, GODB provides a powerful way to connect gene expression data to known biological information gathered through other experimental means.

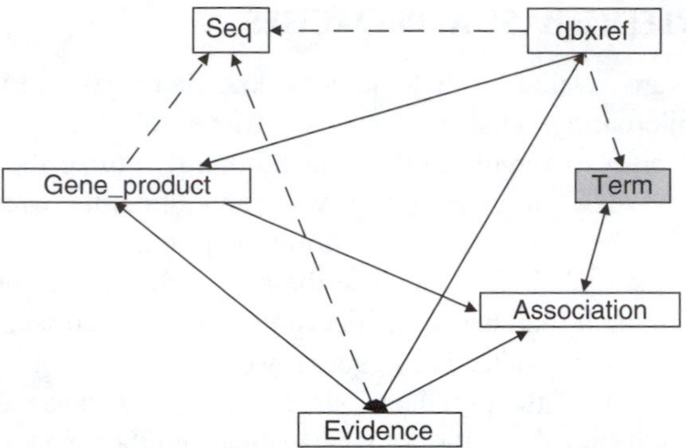

FIGURE 17.1 Simplified GODB schema. The core tables of GODB are linked in a complex structure. Tables connected directly are noted with solid arrows. Tables linked through other tables are noted with dashed arrows.

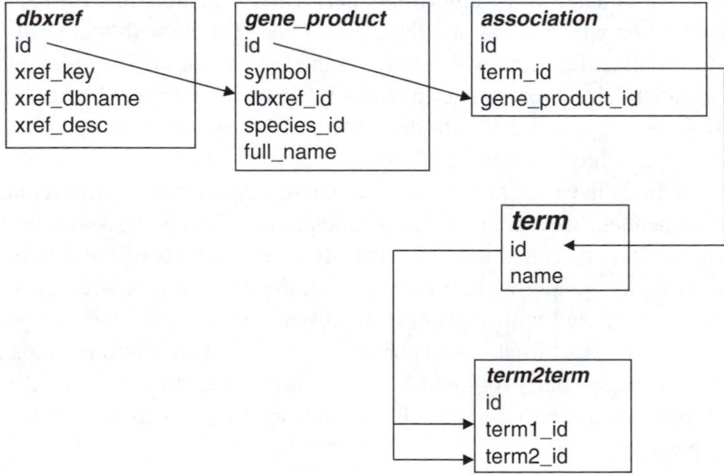

FIGURE 17.2 Gene–term associations in GODB. Genes can be connected to terms through a variety of tables and fields in GODB, including external database references and gene names/symbols.

The structure of the GODB term DAG adds to the power of the resource. The terms within GO are effectively rooted into three subontologies through three terms: biological process, cellular component, and molecular function. In the database, these three terms are rooted to the single term "Gene Ontology." This creates an interesting handle on the graph since the rooting makes the structure a directed, *rooted*, acyclic graph (rooted DAG). A single node in a DAG can have multiple parents, possibly in widely dispersed subtrees, which is the definition of a polyhierarchical tree. From the perspective of designing efficient algorithms around the GO data structure, traversing unrooted directed graphs can take $O(V + E)$ time where V is the number of vertices and E is the number of nodes, while traversing trees generally takes $O(\log n)$ time where n is the number of nodes [5]. Statistical analyses that traverse GO should have a performance somewhere between that of an unrooted directed graph and a tree. However, for a human, a polyhierarchical tree is easier to navigate and to visually represent than a DAG. The structure of GODB is therefore a well-balanced compromise of the two data structures, allowing both for flexibility of links between terms and for easy visualization and navigation by the user.

17.3.2 GenMAPP

While GO provides an ontology of how genetics terms are related to each other, data sources are still needed to describe how these components actually interact. For example, knowing that glucose-6-phosphatase and fructose-1,6-bisphosphatase are both involved in glycolysis does not tell us anything about their specific role within the pathway. GenMAPP fills this role with what it calls microarray pathway profilers (MAPPs). MAPPs, independent of GenMAPP, are representations of how the cell components such as genes and proteins, interact with each other. These interactions include concepts such as metabolic pathways, signal transduction pathways, and subcellular components. GenMAPP presents microarray data in the context of these MAPPs by showing which genes are up- or downregulated through color-coding of the genes within the MAPP file. This is a powerful way to visualize the effect of the experimental conditions within a set of predefined interactions.

The problem of using GenMAPP within the context of a microarray experiment, however, is determining which MAPPs may be of interest to the researcher without looking at all of them. The tool MAPPFinder 6 scores the microarray results in the context of GO terms. From the list of GO terms, the user can access specific MAPPs related to that GO term using GenMAPP. This allows quick navigation from microarray data, to GO terms, to the MAPP that shows the gene(s) in the context of their pathway.

17.3.3 FatiGO

One tool that takes advantage of the information within GODB is FatiGO [7]. FatiGO is a web-accessible tool that allows the user two modes of searching GO. In the simplest method, the user can search for GO terms associated with a single list of genes. This search is limited to a single subontology of GO and a single species. The user can choose from several formatting and analysis options including presenting the results in a table or tree, arranging terms by GO terms or gene/protein names, performing statistical analysis on only GO annotated genes or all genes, and determining what level of GO (how far the term is from the root of the subontology) should be examined. Alternately, the user can compare two lists of genes to examine the differential distribution of GO terms. This allows the user to quickly compare the two sets of genes. The graphical representation makes the results easier to interpret here, displaying a histogram of what percentage of the genes are associated with each of the GO terms with a different color for each of the gene lists.

The single gene list search alone provides a simple and rapid way to identify what GO terms the genes are associated with, at a given level of the subontology. This is also limited to a certain extent since the analysis requires that the user choose a single subontology and a single level of that subontology. As noted in the graphical summary in FatiGO under "total number of genes with GO but NOT at level 5 and ontology. . .," sometimes genes do not have associated GO terms at that level or within that subontology. In fact, they often do not. An analysis across multiple term levels and subontologies might give a better overall picture. This search within FatiGO provides a good, quick, overview of the input genes.

17.3.4 GoMiner

GoMiner [8] also provides a way for researchers to extract information content from GODB based on the results of microarray experiments. GoMiner is a stand-alone Java program, requiring the user to install Java on his or her system if it is not there already. The program inputs two sets of gene lists: all genes on the array and those genes the user flags as "interesting." The latter category is determined by a statistical analysis of gene expression levels through an independent program.

GoMiner displays the information from the input files to the user using different formats in the context of the GO terms. In the primary format, the user is presented with a list of genes and an expandable and collapsible tree of GO terms. Genes that are up- or downregulated are marked with color-coded symbols, making it easy for the researcher to note trends in gene expression. In the tree, the number of genes assigned to that term and the subtree below it are reported, as is a two-sided p-value calculated with an implementation of Fisher's exact test. This tree structure allows for easy traversal of the tree by the user, aiding in the exploration of gene–term relationships.

GoMiner also has the ability to create scalable vector graphic (SVG) images showing the graph of the subontologies and their terms. In this view, each term is represented by a box. Terms enriched for interesting genes are colored red; terms depleted for interesting genes are colored blue; all other terms are colored gray. By scrolling the mouse and clicking over the nodes, the user can get the name of the term and the list of genes associated with that term. This provides another easy way for the researcher to traverse the GO graph of terms looking for patterns of expression. Indeed, this is a general strength of GoMiner: connecting the GO graph with methods of determining the importance of a term for an experiment, such as expression levels or an disproportionate number of interesting genes in an easy-to-use interface.

17.3.5 GOMine

Another, more recently developed, tool that attempts to mine the information content of GODB for analyzing microarray data is GOMine [9]. The unfortunate similarity in names to GoMiner has been noted, and a change of name is being planned and should be anticipated. The current version of GOMine is a standalone Perl program, therefore Perl must be present on the user's local machine before the program can be used. A web accessible version of GOMine is also being planned for the near future, which will ease accessibility by the end user. In fact, GOMine presents the results of the analysis in a stand-alone HTML file, making this transition to a web interface trivial.

Like GoMiner, GOMine scores terms based on those genes associated with either the term itself or a term in the subtree below that term. Unlike GoMiner, GOMine only reports those terms that have an excess of GOI and not the full GO tree. This allows the researcher to focus specifically on the high scoring terms. The output is a single HTML file containing a simple overview of the input data set, a z-score rank-ordered list of terms that score at least as significant as a user-defined cutoff value, and a couple of tree representations of these high scoring terms. GOMine also archives all statistics used in that particular analysis of a microarray experiment. The specific statistical methods used by analysis programs are likely to change over time as new, more appropriate, statistical tools are implemented. Archiving the exact method of calculation gives the user an idea if it would be appropriate to compare two result sets.

17.3.6 Database for Annotation, Visualization, and Integrated Discovery

Several divisions of the National Institutes of Health (NIH) campus have together released the database for annotation, visualization, and integrated discovery (DAVID; [10]). DAVID is a web-accessible tool designed to annotate and aid in the interpretation of microarray data. Like the previously described programs, DAVID will analyze the array data in the context of GO terms using its GOCharts analysis. Unlike the other tools, however, DAVID does not attempt to assign significant values to the results. Instead it shows those terms, within a certain "level" of the GO hierarchy, representing the greatest percentage of genes. This approach is rapid; however, it does have some disadvantages. Terms with a greater number of genes associated with them may be disproportionately represented in the results. By limiting the search to a single level or depth of the GO hierarchy, the program could have the same problem noted for FatiGO, that sometimes genes do not have associated GO terms at that level or within that subontology.

The real strength of DAVID is the ability to move between tools without the need to reupload the input data set. Users can upload a gene list, perform a GOCharts analysis on these genes, then quickly pull up annotation for all the genes. With the addition of more analysis methods, DAVID can meet the ability of integrated discovery from its name that is needed in the microarray community. Currently, DAVID supports annotation across a number of external databases such as LocusLink (http://www.ncbi.nlm.nih.gov/LocusLink/; [11]), Genbank (http://www.ncbi.nih.gov/Genbank/; Reference [12]), UniGene, GO, and the affymetrix descriptions. DAVID also provides tools to visualize the distribution of genes in KEGG and National Center for Biotechnology Information's (NCBI) molecular modelling database (MMDB) for protein structure (http://www.ncbi.nlm.nih.gov/entrez/query.fcgi?db=Structure; [13]). This is an impressive initial set of tools, and the addition of other tools will make DAVID even more useful to the researcher.

17.4 SUPPLEMENTARY DATA APPROACHES

An alternate method of analyzing the results of a microarray experiment uses supplementary data from sources other than curated knowledgebases. Generally, these approaches can be thought of as another level of analysis that inputs the microarray data and a supplementary data set, to generate a new summary data set, which is a complex interpretation of the initial data. For example, the program regulatory element detection using correlation with expression; (REDUCE; [14]) takes the upstream sequence data to identify a set of common regulatory elements at the DNA level. These methodologies perform most of the monotonous work of interpreting the microarray results in light of the supplementary data. While these approaches probably should not be used to replace the researcher, since small but important effects (e.g., rare promoter elements) may be missed, they are extremely useful since they do reduce the number of possibilities that the researcher must examine. Therefore, these tools should rather be considered as powerful ways to ease the interpretation of microarray data.

17.5 TENTATIVE FUNCTION ASSIGNMENT APPROACHES

Tools are as well being developed to aid the researcher in deciding what experiments should be conducted next. These programs assign tentative functions to genes of unknown role, or assign new functions to previously annotated genes. Obviously, these tools also require supplementary data sources of some form, such as YPD (http://www.proteome.com/YPDhome.html; [15]) or the original literature, to make their correlations. Because of this, they can be thought of as a higher level of the data mining programs, facilitating another logical step in the interpretation of array data.

One example of this sort of tool is the shortest-path analysis method of Zhou et al. [16]. The authors took a graph of distances between genes in YPD and the information in GODB to assign tentative functions to genes. They did this by identifying "transitive" genes between two genes annotated by YPD as being in the same pathway. They then assigned functions to the intermediate genes based on prior knowledge. Genes that are transitively coexpressed are those that are not strongly correlated with each other, but are those wherein both are strongly correlated with one or more genes. The example they give is when genes *a* and *b* have strong expression correlation, as do genes *b* and *c*, but genes *a* and *c* do not have a strong expression correlation. In this case, genes *a* and *c* are transitively coexpressed. This is an excellent example of how knowledge about the function of genetic networks can be applied to streamline the process of forming new hypotheses. Similar analyses to automate gene annotation are being performed by other groups such as Wu et al. [17].

17.6 FUTURE DIRECTIONS

There are some obvious directions to improve the tools to aid the interpretation of microarray data. Tools need to harness the rapidly expanding databases, both in terms of the data set sizes and the numbers of databases. Analytical tools also need to harness the rapid change in analytical methods. The most efficient way to do this, from the perspective of a researcher, is by incorporating various analysis methods and data sources through a single interface. The common need here is "Integration." Additionally, there is a problem of correction for multiple testing that needs to be addressed when determining our confidence in the statistical significance assigned to results.

17.6.1 Database Integration

Since many biological databases are not currently integrated and new databases are being created for specific needs, it will be useful for tools to rapidly incorporate any relevant data sources for annotation and interpretation. If current trends continue, it is likely that many of these will be model organism-specific databases. The reason for this divergence of databases rather than convergence is that each model organism has its own set of information that is relevant to it and not to other organisms. Innovative approaches to seamlessly make these data sources available for analysis will be helpful to the researcher.

17.6.2 "Plug-and-Play" Analysis

Likewise, writing analysis tools so that they can be harnessed by integrated analysis systems may expand the reach and usefulness of new analytical methods. There are several ways to do this. Using commonly available and easily installed languages (such as Java, Perl, and C/C++) when writing the code will make it easier for the integrated systems to be able to run the code. The separation of algorithm and interface in these tools, by means such as a model-view-controller (MVC) approach, will allow the analysis methods to be run either as stand-alone applications or as components of a larger package. The use of more portable and flexible input and output formats, such as eXtensible Markup Language (XML), will allow the output of one algorithm to be piped into the input of another or to the end output for user visualization. The computer science tools for easy integration of algorithms are generally present, which simply need to be employed to increase the usefulness of the methods we design. By using these implementation tools, both data and data analysis can be made more portable for researchers.

17.6.3 Combining Analytical Methods

The approach with the greatest potential for the end user, the biologist, is the one that combines a number of aspects of the above analytical approaches. Tools built from this perspective can answer many of the questions asked by the researcher: Do we see an expression change in the genes that would be predicted a priori (both adding evidence of the known functions of individual genes and increasing confidence in the results of the particular microarray experiment)? Do we see expression changes in genes of known functions that would not be predicted (possibly revealing new biological functions for these genes or revealing errors in the current experiment)? Do we see expression changes in genes of unknown function (possibly allowing the tentative assignment of function)? Integrating different forms of analysis into a single integrated discovery package like DAVID will ease this postarray analysis process by putting as many "keys to the kingdom" as possible into the hands of the researcher.

Tools that can successfully link a number of approaches and data structures may have an advantage for a number of reasons, including reducing the overhead of learning how to use multiple independent tools and helping to manage the data and experimental design. To be maximally useful, however, these tools need to be efficient, accurate, reliable, and easy to use, which is a high bar to measure up to, but one that is attainable with current methodologies in all relevant fields.

Microarray databases, such as the Yale Microarray Database (YMD; http://info.med.yale.edu/microarray/; [18]) and the Stanford Microarray Database (SMD; http://genome-www.stanford.edu/microarray; [19]), can also play a crucial role in this consolidation since they are already at the center of the data. By allowing the researcher to pipeline their data into analysis tools external to the database interface, these projects can serve as facilitators of the entire analysis process, right from the gathering of the data to the interpretation of results. Indeed, the current version of YMD implements this feature for some analysis tools. In the near future, YMD will also link to other tools such as Bioconductor (http://www.bioconductor.org/; [20]) and GOMine to facilitate the last step in the analysis pathway interpretation.

17.6.4 Evaluating Statistical Significance

The authors of GoMiner quite appropriately note that "Overall, the *p*-values quoted should be considered as heuristic measures, useful as indicators of possible statistical significance, rather than as the results of formal inference." Given the large numbers of tests performed by many of the analysis methods presented here and the strong nonindependence of each gene/term, a Bonferroni correction for multiple tests is overly conservative to the point of being an impediment since few, if any, terms would remain significant. Resolving the question of "what is an appropriate yardstick for evaluating significance?" is one of the most difficult problems in the field.

Part of the problem is the high degree of interdependence in the data sets being examined. Within the microarray data, there are at least two levels of structure. The changes in gene expression levels are certainly not independent of each other. Additionally, the genes that are represented on the array are rarely chosen randomly. Generally, genes are chosen because they have a known function or they match some well studied idea of "what is a gene?" through predictive models. Genes, or concepts of genes, that are well studied are usually present and genes that are not well studied, or maybe even unknown, are usually not present. When mining the microarray data set against a knowledgebase, other types of structure add to the complexity. Knowledgebases such as GODB and GenMAPP have the same problem of genes/terms that are not independent of each other within their structure. When a structured microarray experiment is mined against a structured knowledgebase, the problem of finding an appropriate correction for multiple tests takes on the complexity of the "N-body problem" in physics. With a very small amount of structure, or two planetary bodies, the problem is relatively simple to solve. With more structure, or more than two planetary bodies, finding a solution to the problem quickly becomes difficult if not impractical.

One way around this problem is to use permutation-based algorithms. FatiGO uses the step-down max-T procedure of Westafall and Young [21] with 10,000 permutations. Another alternative is to use the p-values for each term as a guideline, and instead attempt to determine how confident one can be in the reported terms with "significant" p-values. GOMine uses two tests to determine this, both based on permutations of the GOI list: a false discovery rate (FDR) and a confidence test (CT). The FDR is calculated by taking the ratio of the mean number of terms observed at the user-defined cutoff significance level in the permutations and the number of terms observed at the cutoff significance level in the microarray data. The CT is simply the count of how many permutations of the GOI had at least as many terms at the user-defined cutoff as the observed microarray data. Combined, these two statistics can give a reasonable overview of the reliability of the results of the term analysis.

Given the degree of structure within both microarray and knowledgebase data sets, and the resulting great difficulty in finding an appropriate multiple test correction, algorithms that permute the microarray data currently appear to be the best way to determine our confidence in the interpretation-analysis results. The downside of this approach is that analysis times will generally increase at a rate somewhat greater than the number of permutations performed. The degree of increase depends primarily on the efficiency of the algorithm that generates the permuted data, which adds new overhead above the interpretation-analysis of each permutated data set. For microarrays, tests involving thousands of permutations will be required at a minimum, increasing run times from mere seconds to minutes or hours. Until such time that an efficient way to untangle the interdependencies of the data sets can be devised, however, permutation tests may be the best method of determining our confidence in statistical inference of the interpretation results.

17.7 CONCLUSIONS

If even a handful of these analysis methods prove fruitful, as is the case already seen, biologists will have a impressive suite of tools to choose from when contemplating how to get a handle on biological meaning of their array results. The better integration of tools and data sources (knowledgebases, annotation, etc.) would make them available to a wider audience. The creation of standards, committee-designed or de facto, may resolve the problem of an overwhelming number of options available to the researcher. New analysis methods may be able to make more sophisticated interpretations of the data through careful multidisciplinary methods combining biological and statistical expertise. Overall, the young domain of microarray, and even "large biological data source," interpretation appears to be off to a healthy start.

So what is the answer to the question "What do I do with this genelist?" Currently, the best answer is that "it depends on the nature of your experiment." Someone who wants to tackle the web of transcriptional regulators will have different needs from the person who wants to explore which metabolic networks appear to be upregulated. Both will have different needs from the researcher who wants to automate hypothesis generation and the experimental process with LIMS. To determine the best way to interpret microarray data, the researcher needs to determine (a) which databases to perform the analysis with and (b) what analysis tools to use. To do this, one must be kept informed of what resources are currently available, through periodic research and regular questioning of others in the same field.

It would certainly be helpful if there were a central listing of databases and analysis tools, presumably through a web-accessible resource. Even more useful would be an "expert system" that helps guide the user through the decision process based on knowledge of the strengths and weaknesses of current resources. The final answer presented here may be an unsatisfactory one. Until such a meta-tool is developed to guide our data analysis, the best way to answer the question is the same one that we can usually rely on in science: we need to do our homework.

ACKNOWLEDGMENTS

The author would like to thank Kei-Hoi Cheung for his help editing this manuscript and Hongyu Zhao, David Tuck, Harmen Bussemaker, and Grier Page for discussions on various aspects of the topic of array interpretation. This work is supported in part by NIH grants T15 LM07056 and P20 LM07253 from the National Library of Medicine (Perry L. Miller).

REFERENCES

1. The Gene Ontology Consortium. Gene Ontology: tool for the unification of biology. *Nature Genetics* 25: 25–29, 2000.
2. K.D. Dahlquist, N. Salomonis, K. Vranizan, S.C. Lawlor, and B.R. Conklin. GenMAPP, a new tool for viewing and analyzing microarray data on biological pathways. *Nature Genetics* 31: 19–20, 2002.
3. M. Kanehisa, S. Goto, S. Kawashima, and A. Nakaya. The KEGG databases at GenomeNet. *Nucleic Acids Research* 30: 42–46, 2002.
4. S. Drăghici, P. Khatri, R.P. Martins, G.C. Ostermeier, and S.A. Krawetz. Global functional profiling of gene expression. *Genomics* 81: 98–104, 2003.
5. T.H. Cormen, C.E. Leiserson, and R.L. Rivest. *Introduction to Algorithms*, 1st ed., Cambridge, MA: MIT Press, 1990.
6. S.W. Doniger, N. Salomonis, K.D. Dahlquist, K. Vranizan, S.C. Lawlor, and B.R. Conklin. MAPPFinder: using gene ontology and GenMAPP to create a global gene-expression profile from microarray data. *Genome Biology* 4: R7, 2003.
7. F. Al-Shahrour, R. Díaz-Uriarte, and J. Dopazo. FatiGO: a web tool for finding significant associations of gene ontology terms to groups of genes. *Bioinformatics*, 20: 578–580, 2003.
8. B.R. Zeeberg, W. Feng, G. Wang, M.D. Wang, A.T. Fojo, M. Sunshine, S. Narasimhan, D.W. Kane, W.C. Reinhold, S. Lababidi, K.J. Bussey, J. Riss, J.C. Barrett, and J.N. Weinstein. GoMiner: a resource for biological interpretation of genomic and proteomic data. *Genome Biology* 4: R28, 2003.
9. M.V. Osier, H. Zhao, and K.-H. Cheung. GOMine — a model for microarray interpretation. *BMC Bioinformatics*, 5: 124, 2004.
10. G. Dennis Jr., B.T. Sherman, D.A. Hosack, J. Yang, W. Gao, H.C. Lane, and R.A. Lempicki. DAVID: database for annotation, visualization, and integrated discovery. *Genome Biology* 4: R60, 2003.

11. K.D. Pruitt and D.R. Maglott. RefSeq and LocusLink: NCBI gene-centered resources. *Nucleic Acids Research* 29: 137–140, 2001.
12. D.A. Benson, I. Karsch-Mizrachi, D.J. Lipman, J. Ostell, and D.L. Wheeler. Genbank. *Nucleic Acids Research* 31: 23–27, 2003.
13. Y. Wang, J.B. Anderson, J. Chen, L.Y. Geer, S. He, D.I. Hurwitz, C.A. Liebert, T. Madeg, G.H. Marchler, A. Marchler-Bauer, A.R. Panchenko, B.A. Shoemaker, J.S. Song, P.A. Thiessen, R.A. Yamashita, and S.H. Bryant. MMDB: Entrez's 3D-structure database. *Nucleic Acids Research* 31: 474–477, 2003.
14. H.J. Bussemaker, H. Li, and E.D. Siggia. Regulatory element detection using correlation with expression. *Nature Genetics* 27: 167–171, 2001.
15. P.E. Hodges, W.E. Payne, and J.I. Garrels. The Yeast Protein Database (YPD): a curated proteome database for *Saccharomyces cerevisiae*. *Nucleic Acids Research* 26: 68–72, 1998.
16. X. Zhou, M.-C.J. Kao, and W.H. Wong. Transitive functional annotation by shortest-path analysis of gene expression data. *Proceedings of the National Academy of Sciences, USA* 99: 12783–12788, 2002.
17. L.F. Wu, T.R. Hughes, A.P. Davierwala, M.D. Robinson, R. Stoughton, and S.J. Altschuler. Large-scale prediction of *Saccharomyces cerevisiae* gene function using overlapping transcriptional clusters. *Nature Genetics* 31: 255–265, 2002.
18. K.H. Cheung, K. White, J. Hagar, M. Gerstein, V. Reinke, K. Nelson, P. Masiar, P. Srivastava, Y. Li, J. Li, J.M. Li, D.B. Allison, M. Snyder, P. Miller, and K. Williams. YMD: a microarray database for large-scale gene expression analysis. In *Proceedings of the American Medical Informatics Association 2002 Annual Symposium*, pp. 140–144, 2002.
19. J. Gollub, C.A. Ball, G. Binkley, J. Demeter, D.B. Finkelstein, J.M. Hebert, T. Hernandez-Boussard, H. Jin, M. Kaloper, J.C. Matese, M. Schroeder, P.O. Brown, D. Botstein, and G. Sherlock. The Stanford Microarray Database: data access and quality assessment tools. *Nucleic Acids Research* 31: 94–96, 2003.
20. S. Dudoit, R.C. Gentleman, and J. Quackenbush. Open source software for the analysis of microarray data. *Biotechniques* (Suppl): 45–51, 2003.
21. P.H. Westafall and S.S. Young. *Resampling-Based Multiple Testing: Examples and Methods for P-value Adjustment*. New York: John Wiley & Sons, 1993.

18 Combining High Dimensional Biological Data to Study Complex Diseases and Quantitative Traits

Grier P. Page and Douglas M. Ruden

CONTENTS

18.1 Introduction 336
18.2 Heritable Changes in Gene Expression 339
18.2.1 eQTLs and pQTLs 340
18.3 Combined HDB Techniques to Identify Candidate or Causal Genes for Complex Diseases and Quantitative Traits 342
18.3.1 Applied Projects — Narrowing the Candidate List 342
18.3.2 Applied Projects — Identification of Causal Genes 345
18.4 Theoretical Papers 346
18.5 Software and Bioinformatics Tools 348
18.6 Issues with Combined High Dimensional Biological Projects 349
18.6.1 The Problem of Errors and Multiplicity Control 349
18.6.2 The Problem of Power 351
18.6.3 The Problem of Intersections 351
18.6.4 The Problem in Incompleteness 352
18.6.4.1 What and When to Sample 352
18.6.4.2 Combining across Species and Tissue Types 353
18.6.4.3 The Problem of Public Data 353
18.6.4.4 The Problem of Causation 353
18.7 Conclusions about Combined HDB Studies 353
References 354

The identification of the genetic and epigenetic factors involved in complex diseases and quantitative traits in humans and nonhumans has not yet met with the hope for success [1–4]. A variety of factors have led to this situation including, but not limited to, lack of power, genetic heterogeneity, allelic heterogeneity, sporadic cases, poorly defined phenotypes, and perhaps an overreliance on reverse genetics or positional cloning. Ten years ago, Francis Collins [5] proposed that the future localization of quantitative traits would be via positional candidates. While this was fine in practice, our knowledge of the functions of genes is at best spotty, thus requiring us to expand the list of positional candidate genes. One manner to refine the list of candidate genes is to use objective quasi-genome wide high dimensional biology (HDB) techniques to get a different point of view on the diseased phenotype. It is hypothesized that, by combing linkage with other HDB techniques, investigators will be more successful in identifying genes for diseases and traits than any single approach. The HDB techniques that have been integrated include many combinations of linkage, association studies, expression microarrays, proteomics, comparative genomics, advanced test crosses, metabolomics, mutagenesis studies, online databases, and, most recently, quantitative epigenetic techniques. This combined approach has been successful in identifying the genes for several traits and it is expected that many more will be identified in the future. In this chapter, we review some of the successful techniques for integrating data, as well as make suggestions for conducting future studies. We also make suggestions for improving these types of studies and encourage researches to take a multipronged approach in understanding complex diseases.

18.1 INTRODUCTION

The first published example of positional cloning was the bithorax complex in *Drosophila*, which occurred in 1983 [6]. The first reported linkage in humans was in 1947 between color blindness and hemophilia [7], but it took till 1989 until the first human gene, cystic fibrosis transmembrane regulator (CFTR), was identified by positional cloning [8]. Its discovery was heralded with much fanfare. In tribute to the evolving power of modern molecular, genomic, and statistical tools, the identification of genes responsible for Mendelian traits has progressed to such an extent that in its recent issue, *Nature Genetics* included the identification of the genes for no fewer than eight different Mendelian conditions. To date, over 1400 genes for 1200 Mendelian traits have been identified [2,9].

The first reporting linkage to a quantitative trait was reported in the mid-1920s. However, it was not until the late 1980s and continuing till today, has there been active reporting of quantitative trait loci (QTL) studies in plants, animals, and humans. Unfortunately, there have been relatively few successes in identification of quantitative trait genes (QTGs) or variations that underlie these traits [10,11].

Why has the identification of causative alleles for complex diseases and quantitative traits been so elusive? To a certain extent, their names reveal the problem; complex diseases are complex and quantitative traits only affect the mean phenotype in a graduated fashion. These traits may be affected by both genetic and environmental factors, as well as gene-by-gene and gene-by-environmental factors. While the traits being studied may aggregate in families, they usually do not segregate in a Mendelian fashion, and even if they did, the late onset of some of these diseases would make family studies difficult. In addition, for quantitative traits such as weight or blood pressure, the values keep changing throughout life, so one is left wondering what the true phenotype is to make matters even more challenging, individual alleles are probably neither necessary nor sufficient to cause the disease; thus, the "disease" alleles may be present in nondiseased population. Furthermore, it is difficult to detect obligate recombination events, and there is undoubtedly extensive allelic and genetic heterogeneity.

Many of the successes in identification of causative genes have been the result of a detailed understanding of the biology of the diseases and traits being studied, or a positional candidate approach. The number of genes in individual organisms lies between 1000 for simple bacteria and up to 30,000 in humans. Because the functions of most genes have been assigned via homology in other organisms rather than directly studied in the organism of question, what is known is undoubtedly incomplete. For example, BRCA1 [12] was cloned in 1994, but despite the efforts of hundreds of investigators, it took until 2001 to prove that BRCA1 can bind DNA [13] and inhibit the nucleolytic activity of the MRE11/RAD50/NBS1 complex, an enzyme implicated in numerous aspects of double-strand break repair. If it takes this much effort to understand the function of BRCA1, one can imagine that even well studied genes may have many yet unknown functions.

Figure 18.1 is an attempt to model the problem using a metabolic pathway. We have highly detailed knowledge of most metabolic pathways from mutational screens

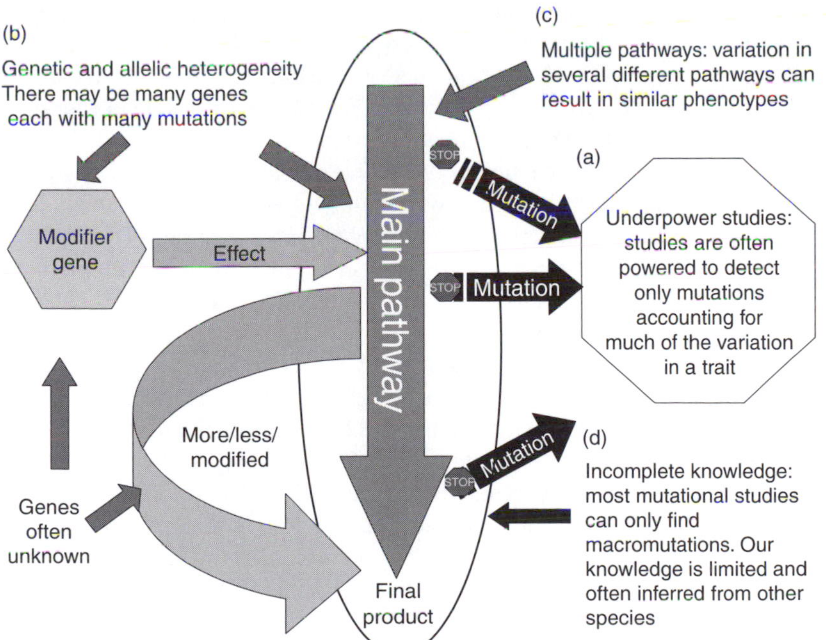

FIGURE 18.1 Why has it been so difficult to find genes for complex diseases and complex traits? There are a variety of reasons for the difficulties including (a) underpowered studies, (b) genetic and allelic heterogeneity, (c) multiple pathways, and (d) incomplete knowledge of the disease mechanisms.

and complementation assays in organisms such as yeast. This has been confirmed in studies of human diseases such as Xeroderma Pigmentosum or Canavan's disease [14]. However, most of these pathways were identified through mutations causing macro changes in phenotypes [15]. As a result, while the genes in the primary pathways are known, the genes in which mutations cause subtle changes are not well known. It is these genes that we believe is likely to harbor most of the variation that predisposes an organism to complex diseases and quantitative traits. In addition, these quantitative trait mutations have, by definition, modest effects upon the phenotype. Therefore, the mutation frequency may drift quite high with only modest selection against deleterious mutations, and in some cases, such as salt-sensitive hypertension in rats [ref], genes may be positively selected for in some environments whereas against in others.

In order to find the unknown genes harboring variation that predisposes an organism to complex diseases, we must have a much broader list of candidate genes to screen for causative genetic variation. The idea to combine HDB techniques to identify novel candidate genes was one of those events that occurred to several different groups simultaneously. While this has been a hot topic on the lecture circuit for several years, Jansen and Nap [16] appear to have been the first to publish the concept in the peer reviewed press. Jansen and Nap coined the term "genetical genomics," which appears not to have yet entered the vernacular, as have the terms "expression profiling" and "marker-based finger printing" of each individual in a population. In the Jansen and Nap [16] paper, they suggest using "genetical genomics" primarily in F2 and F3 intercrosses, in recombinant Inbred lines (RIL) and recombinant Inbred intercross (RIX), in backcross progeny (BC), and in near isogenic lines (NIL). They further suggest that each gene's expression profile be treated as a quantitative trait, which they termed expression QTL (eQTL). They suggest that eQTL can be analyzed via genome wide linkage analysis similar to any other quantitative trait in mapping populations to localize both the *cis*- and *trans*-regulatory factors for the studied gene's expression level. In addition, they suggest that the approach can be used to identify candidate genes as well as to identify novel pathways and epigenetic interactions. The method that they suggest for identifying candidate genes is to combine, via the simple intersection, the information from the multiple sources in order to identify novel candidate genes for complex diseases and traits. A cartoon of the entire process is illustrated in Figure 18.2, a combined microarray QTL study. By examining the intersection of the linkage peaks and the differentially expressed genes, the number of candidate regions or genes may be reduced over a single approach. In addition to microarrays [17,18], the HDB techniques can include proteomics [19], linkage analysis [20–23], association studies [24,25], comparative genome hybridization (CGH) [26], metabolomics [27], ENU mutagenesis [15], imaging, and biochemical system theory [28], and quantitative epigenetic linkage (QEL) studies [29]. As discussed further below, there have recently been a few successes in the identification of candidate quantitative genes and genetic variations using these multipronged HDB approaches.

In the rest of this chapter, we explore some of the studies that have used the combined HDB approach to dissect the genetic etiology of complex diseases and phenotypes. We will also evaluate some of theoretical papers in the area and some of the available software. Finally, we will make some recommendations for the future applications of combined HDB.

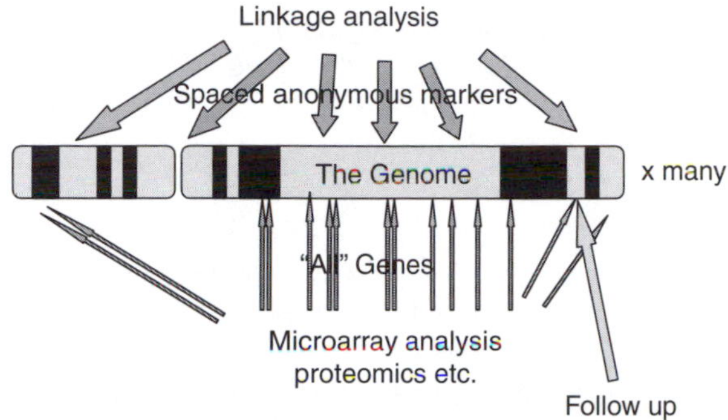

FIGURE 18.2 Premise underlying combined HDB studies. The result of one HDB study say a linkage study identified a portion of the genome as being linked to several regions and a microarray study identified many genes as being differentially expressed. By examining the intersection (or similar function) of these studies we may reduce the number of candidate genes to a manageable level.

18.2 HERITABLE CHANGES IN GENE EXPRESSION

One of the most critical aspects of combined HDB studies discussed above is that all the factors being studied should be heritable. This includes genetic markers, which should behave in a Mendelian fashion, and epigenetic markers that likely will not. It has been known for sometime that genetic variation can cause heritable changes in gene expression. For example, many mutations that lead to disease cause a total elimination of the RNA for the gene. Coding regions of the human genome can contain as much as 0.6% polymorphic sites [30], and these often cause variation in transcription rates and RNA levels [31].

It has long been suspected, but not proven until recently, that genetic variation could heritably and quantitatively affect the amount of RNA generated from each allele of a gene. Yan et al. [32] showed that in several genes there are single nucleotide polymorphisms (SNPs), which lead to differential expression of alleles, and that for at least two of these genes, the differences were heritable in a Mendelian fashion. This study does not actually address if absolute gene expression levels are heritable, but it does indicate that simply examining the raw gene expression level for a gene may not be sufficient to explain differences between samples, as each allele may have slightly different properties.

Cheung et al. [33] studied the heritability of gene expression using lymphoblasts in 35 unrelated Centre d'Etude du Polymorphisme Humain (CEPH) family members using 5000 gene arrays with four chip replicates per person. Five of the most variable genes were verified by Reverse transcription polymerase chain reaction (RT-PCR) in a sample of unrelated CEPH individuals ($n = 49$), the offspring from five CEPH families ($n = 41$), and ten sets of monzygotic twins ($n = 20$). The variability in expression of these genes between these groups was assessed using an F test. The amount of variation between unrelated people is greater than in siblings, and the amount of variation in siblings is greater than in MZ twins. These results suggest that there may be a heritable component to many of the genes' expression levels.

Lo et al. [34] further expanded the study of differentially expressed alleles using the Affymetrix SNP chip. In this study, 1063 of the SNPs on the chip were in transcribed sequences and at least 602 were heterozygous in genomic DNA. All of the genes studied are expressed either in liver or kidney tissues from a set of 7 individuals, as determined with the SNP arrays. 170 Genes showed at least a 4-fold difference in gene expression between the two alleles. Several of the genes were known to be epigenetically imprinted and had mono-allelic expression in specific tissues. A few other genes were in regions of reported imprinting, but were not shown to be imprinted prior to this study. RT-PCR was used to verify some of the preferentially and nonpreferentially expressed alleles, and in 6 of 7 cases tested, the observation was verified. These studies highlight the need to consider expression of a gene beyond a single transcript, and it showed that gene expression levels are both genetically and epigenetically controlled.

18.2.1 eQTLs and pQTLs

If gene expression levels are heritable, one can consider each gene's expression level to be a quantitative trait that can be combined with marker data for a linkage study to identify the loci influence variation in the gene's expression. As mentioned earlier, a locus affecting the expression level of a genes has been named eQTL, and the genetic variation causing differences in gene expression are called expression-level polymorphisms (ELP) [21]. In this chapter, we name a locus affecting protein levels a "pQTL."

The first application eQTL linkage analysis appears to be by Brem et al. [35]. They conducted a genome wide linkage scan on many gene expression levels in budding yeast (*Saccharomyces cerevisiae*). In this study, two strains of budding yeast were mated and 40 haploid segregants (offspring) were grown into colonies and typed for 3312 markers. In addition, the RNA from each of the 40 haploid segregant strains was extracted and microarrayed. The mean heritability of the genes' expression was estimated to be 84%. A total of 1500 genes were differentially expressed between two parental strains of *S. cerevisiae*. Of the differentially expressed genes, 570 of these were linked to one or more loci with $p < 5 \times 10^{-5}$, which in this study, corresponds to a false discovery rate (FDR) of about 10%. It was observed that certain regions of the genome were linked to many different genes' expression levels, and these are called "*trans*-QTL." Many of the linkage clusters included genes that are known to be involved in similar pathways (such as leucine biosynthesis), in a particular cellular organelle (e.g., mitochondria), or under the control of the same transcription factors (e.g., Msn2/4). A good number of genes ($\sim$30%) had their region of maximal linkage close to the genomic position of the genes, indicating that while there is extensive *cis* regulation. Most of the genes with a putative control locus are *trans* acting rather than *cis*. In addition, the *trans*-acting factors are broadly dispersed among many classes of genes [36].

Schadt et al. [37] present extensive work on eQTLs in maize, mouse, and human. This chapter consists of three different studies all aimed at localizing the regions of the genome, which influence the expression of genes. In the murine experiment, 111 F_2 mice from a parental cross of c57Bl/6J and DBA/2J and an $F_1 \times F_1$ intercross were profiled using oligo microarrays. The array had 23,574 genes, but only 7861 were expressed in the tissues that they analyzed. The 7861 expressed genes were used in a genome wide QTL linkage study for each of the genes' expression levels. Significant linkage, that is, with logarithm of the likelihood of the odds (LOD) scores >4.3, was observed in a total of 2123 genes and each linkage "hit" accounted for approximately 25% of the variation in transcript level. However, locus-specific heritability estimates are notoriously high in QTL studies [38]. In addition, the distribution of the eQTLs was not randomly distributed; 74 cM windows had greater than 1% of the eQTLs. In addition, of the genes that could be mapped, 34% had the point of maximal linkage in *cis*; that is, they are *cis*-eQTL. Schadt was able to identify the causative variation for several of the genes. For example, Schadt et al. were able to identify C5 as being under *cis* regulation, replicating the work of Karp [39] (see below). In addition the *cis* effects of several other genes including Alad, St7, and Nnmt were explained.

Furthermore, in the Schadt et al. [66] article, there are additional analyses of a 76-member F_2 pedigree of Maize. Approximately 33% of the genes had a significant eQTL, but over 80% of those eQTLs with a LOD score >7 were *cis*, which may suggest that plants could be under greater *cis* control than yeast and animals. However, some of the genes seem to exhibit genotype-dependent interactions that are similar to epistasis, but the interactions are between the eQTLs of different genes, so it is not strictly epistasis. As a result, any eQTL studies should consider interactions [40,41] and epistasis in their analyses.

Finally, 56 pedigree members from four CEPH pedigrees had their lymphoblastoid cell lines expression analyzed with microarrays. The pedigrees were small for linkage analysis, but 29% of the genes had a heritability that was significantly greater than zero [37]. This study takes advantage of the available *in silico* resources [42] (also see below), for these CEPH pedigrees have been used to develop linkage maps. As a result, many of these pedigrees have been typed for as many as 6000 markers and much of the data is available at Marshfield Center for Medical Genetics http://www.marshfieldclinic.org/research/genetics/.

Lan et al. [43] suggest that, while it is possible to conduct gene mapping for all possible mRNA abundances, it is computationally intensive, subject to high sample variance, and thus underpowered. They suggest the key is to simplify the data. They propose two methods to reduce the dimensionality of the data, principle components analysis and "seeded" hierarchical cluster analysis. They apply their methods to an F_2 *ob/ob* mice derived from BTBR, *ob/ob* and C57Bl/6J, *ob/ob* strains. For this analysis, they used standardized mRNA levels. Several mRNA levels were clustered with murine phenotypes such as insulin, glucose, and body weight. One cluster of genes shows strong correlation with the traits. QTL mapping of the genes in this cluster revealed that several link to the same region of the genome, and they find a large linkage peak by multiple-trait interval mapping (MTM); however, it is expected that genes that are highly correlated will be linked to the same region simply due to the manner in which the mathematics of QTL analysis occurs. In addition, 7 of the 8 genes subject to cluster analysis were analyzed by principal components analysis (PCA), and linkage on the first two PC was conducted. Overlap was found between the PC1 linkage and all three of the hierarchical MTM linkage peaks. The approach of using overlap in linkage among highly correlated genes may be quite interesting due to its ability to "borrow" information among many genes to potentially derive additional power. It is outstanding that the two methods had overlapping linkage peaks. However, it is not clear why only 7 of the 8 genes used in the MTM analysis were used in the PC analysis. Another area for future study is to determine if the null distribution of the LOD score is the same for the MTM, as they are in a single trait interval mapping approach. Therefore, the significance of the reported LOD score is not clear. However, this method may eventually prove to have some advantages over simple single gene eQTL analysis.

The first, and perhaps only, published pQTL study was by Klose et al. [44], who were able to map variation in protein levels in the several regions of the mouse brain in an F_1 backcross. Similar to the findings in expression studies, the regions of maximal linkage to protein levels are often *trans* to the location of the gene in the genome. This was a very ambitious study given the larger amount of work required for proteomic vs. an expression studies. This work highlights some of the difficulties in conducing pQTL studies. For example, even more so than mRNA, each individual protein may be represented by potentially hundreds of variants (and spots) due to numerous posttranslational modifications. We have seen similar results in our proteomics studies [45]. One point that is not in this article, and apparently in no other published studies as well, is that there are no comparisons of the linkage peaks between eQTL and pQTL for the same gene, and for that matter there have not been any published comparisons of the eQTLs of the same gene across tissues, species, or microarray platforms. This may simply be because protein and mRNA expression levels have not yet been studied in the same organism, and the field is young.

18.3 COMBINED HDB TECHNIQUES TO IDENTIFY CANDIDATE OR CAUSAL GENES FOR COMPLEX DISEASES AND QUANTITATIVE TRAITS

The chief use for combined high dimensional approaches so far has been combined microarray and linkage studies to narrow to list of candidate genes for study in a complex disease or trait. A relatively large number of both theoretical and applied papers with varying degrees of success have been written in this area, many of which are describe below. Some of these projects have previously been reviewed by Cheung and Spielman [46].

18.3.1 Applied Projects — Narrowing the Candidate List

Niculescu et al. [47,48] was the first group to apply combined microarray analysis with linkage and cross species mapping to identify putative candidate genes for mania and psychosis. They termed this process "convergent functional genomics." In these studies, they take two slightly different approaches. In the first study [47], rats are either treated with methamphetamine or a control drug dose. Differentially expressed genes in the rat hippocampus were mapped onto the currently mapped human genetic loci associated with bipolar disorder or schizophrenia. This approach was used to identify several novel candidate genes for schizophrenia. Unfortunately, none of the identified genes have yet been found to harbor variation that predisposes people to any psychiatric disorder. This study may be bedeviled by the problem that affects the studies of many complex diseases. The phenotypes were not tightly defined and the animal models may be inappropriate. For example, methamphetamine overdosing in rats may not be a good model for bipolar disorders or schizophrenia, because this type of overdosing causes large morbidity and mortality. Metamphetamine overdosing induces hyperthermia that can essentially "cook" the brain [49]. Therefore, careful consideration should always be paid to the parallels between human diseases and their presumed animal models.

In another article [48], Niculescu et al. suggested examining the genes that are differentially expressed in a variety of brain regions of postmortem biopolar patients, other psychiatric conditions, and controls. They attempted to map the differentially expressed genes onto mapped loci for these disorders. In the article, they do attempt to address the many issues involved in studying postmortem samples; however, they also ignore the fact that they are examining gene expression changes that may be the effect of the psychiatric condition rather than the cause. It is critical in microarray studies to separate cause from effect [50].

Wayne and McIntyre [51] used combined microarray with recombinant inbred *Drosophila* (RIL) to identify candidate genes for ovariole number. A QTL had previously been identified via linkage in *Drosophila* RIL for this trait and deficiencies were developed covering ~74% of the genes in the region. Six of the deficiencies were found that exhibited line by genotype interactions in the correct direction based on the biology. All the genes on the Affymetrix *Drosophila* microarray were classified as either not in QTL, in QTL but not in deficiency, in QTL but on nonsignificant deficiency, or in QTL and on a significant deficiency. A total of 294 genes were on both the array and on a significant deficiency. RNA was extracted from the parental lines of the RIX lines used to localize the QTL. Three replicates of each line were arrayed on the Affymetrix *Drosophila* array. Quite a few genes (21 with a $p < .01$) were differentially expressed, with one gene being significant after a Bonferroni correction for 238 tests. The most significant and four of the other significant genes have virtually no annotation and thus would not have been identified in a conventional candidate gene approach to study ovariole number. Further studies on the most significant genes revealed previously unknown homology (40%) to the human E2IG2 gene, which has been identified in a screen for estrogen targets. The homology between the proteins may suggest similarity between human and *Drosophila* ovulation regulatory mechanisms, and can identify candidates that would have been missed in a tradition candidate gene list. However, *Drosophila* does not have the hormone estrogen, and genetic studies will have to be performed to confirm the significance of these genes on ovariole number.

Eaves et al. [52] used combined microarrays and linkage to study the complex trait of the nonobese diabetic (NOD) model of type I diabetes. NOD is a very complex phenotype in which as many as 20 loci may be responsible for the phenotype. Thus far, the difficult job of identifying the causative genes has been largely fruitless. Seven different congenic strains of single locus NOD mice were microarrayed in each of two tissues, 4-week old female thymuses and 3-month-old spleens. As a result of these studies, more than 400 gene expression differences were observed. In particular, at the

Idd9.1 locus, eight new candidate genes were identified. So far, no confirmed disease genes have been localized.

Tabakoff et al. [53] proposed to combine selective breeding with microarray analyses of brain regions of the selected mice to study complex traits such as alcohol addiction. In the study, four strains were used to study acute function tolerance (AFT) — two that had been selected for high (HAFT-1 and HAFT-2) and two for low (LAFT-1 and LAFT-2). Initially, several high and low lines were established after 20 and 22 generations of selection. After selection, two high lines and two low lines were studied, which included 5 HAFT-1, 5 LAFT-1, 4 HAFT-2, and 4 LAFT-2 individual mice. Some of the strains (HAST-2 and LAFT-2) had been examined in a QTL study [54] of AFT, and several QTL were identified. Differentially expressed genes were identified in both HAFT-1 vs. LAFT-1 comparisons and HAFT-2 vs. LAFT-2 comparisons. These differentially expressed genes were then mapped onto the reported QTL linkage analysis results. Several genes were differentially expressed in both gene expression comparisons and were within a region of linkage. This article has a large number of positives including studying the same strains in linkage and microarray studies. Further, the article has a reasonable sample size for the microarray and replicated microarray studies of the different strains. In addition, some of the genes were verified as being differentially expressed by RT-PCR. However, there were some limitations including the fact that fold change rather than *p*-values were used for assessing significance of change. Also, whole brains were studied, which may limit the resolution of regional specific changes in gene expression and the power of the microarrays. Another limitation is that the microarray results were assumed to be similar to that of linkage analysis, whereas in fact the power was far lower in the microarray study.

Okuda et al. [55] took an approach to identify genes involved in hypertension in Dahl-S rats by carefully phenotyping 102 F_2 male rats from a Dahl-S by Lewis. The F_2 rats were 14 weeks old after 5 weeks of control feeding and 9 weeks of salt loading. The kidneys of the parental strains were 15 weeks old after 10 weeks of control feeding and 5 weeks of salt loading. Three pools of 3 rats from each of the strains were run on the Affymetrix rat U34 array. Several of the genes were highly differentially expressed and polymorphisms were identified in 12 genes. These polymorphisms were tested for co-segregation in the F_2 progeny One gene *pnpo* exhibited significant association with systolic and diastolic blood pressure. Polymorphisms in other genes such as *Comt*, *Sah*, and *Acadsb* were also interesting because possible connection with blood pressure [56]. The *Pnpo* gene may also lie within the range of two reported chromosome 10 QTL for blood pressure [55].

Many complex traits including Geotaxis (gravitaxis) have been selected for extreme values in model organisms. In Toma et al. [57], the RNA from the heads of two selected lines *Hi5* and *Lo*, which were originally selected by Hirsh in the 1950s, were analyzed on cDNA microarrays. The cDNA microarrays, which contained about one-third of the *Drosophila* genome, were used to analyze RNA isolated from duplicate independent subsets of flies. Rather than using the more meaningful *p*-values, genes with a twofold change in expression were considered interesting. Taking advantage of the large collections of *Drosophila* mutants, all of the genes in an extant mutant strain were subjected to Geotaxis testing. The list of genes was further reduced to mutant *D. melanogaster* genes with a neurological defect. A set of control genes were identified that were not differentially expressed, but as well had neurological defects. A total of four differentially expressed and six control genes were verified via qPCR. Each mutation was crossed back onto the same genetic background. Mutants for three (*cry*, female *Pdf*, and *pen*) of the four differentially expressed genes were significantly different from the wild-type strain and none of the six control mutants were different in the geotaxis assays. None of the three genes would have been predicted based upon the previously defined function. Thus, this study successfully identified new candidate genes for harboring variation that is involved in Geotaxis, however, it does not appear that the genetic variation in the *Hi5* or the *Lo* strains has yet been identified to verify that they actually cause quantitative variation in Geotaxis within these strains.

18.3.2 Applied Projects — Identification of Causal Genes

One of the earliest combined microarray and QTL studies that appears to have identified a causative gene was by Aitman et al. [58]. They analyzed spontaneous hypertensive rats (SHRs) and their insulin resistance phenotype by combining four QTL studies (one backcross and three F_2 crosses) [59] of SHR by Wistar Kyoto (WKY) rats. Linkage to rat chromosome 4 was evident in all of the studies. In addition, the authors had a SHR.4 congenic strain, which had the SHR chromosome 4 replaced with a Brown Norway (BN) chromosome 4. The congenic had partial glucose and lipid phenotypes of the SHR, thus indicating a partial phenotypic effect by the SHR chromosome 4. Two expression comparisons were run, SHR/SHR.4 and HSR/BN, using a rat Incyte® cDNA microarray. In both comparisons, cd36 was reduced at least 90% in the SHR strain. Sequencing of cd36 identified many nonsynonymous changes between WKY and SHR, which suggests a gene duplication/deletion event. This was confirmed by genomic sequencing, thereby suggesting that BN and WKY have one functional copy of cd36 and two pseudogenes, whereas SHR only has the pseudogenes [60]. This study has shown the power of combined microarray and QTL studies, and many subsequent studies have followed up on this observation and shown the cd36 has an impact on a variety of phenotypes [58,60,61].

Karp et al. [39] used combined microarray and QTL studies to identify a gene involved in airway hyper responsivity (AHR), a trait important in the asthma. Their studies also highlight some of the issues of using animal models of human diseases. In this study, they phenotyped a wide range of murine strains for their susceptibility to allergen-induced AHR. The whole lungs of A/J (highly responsive), C3H/HeJ (highly resistant), (A/J $\times$ C3H/HeJ) F_1, and phenotypically extreme $F_1 \times$ A/J backcrosses were collected. These samples were run on Affymetrix microarrays. 227 genes showed a greater than threefold change in expression between the parental strains. When the extreme BC progeny were added, a total of 21 genes were differentially expressed. Previous linkage studies in these strains had yielded QTLs on chromosomes 2 and 5. No genes were differentially expressed in the chromosome 2 regions, but C5, which is in the chromosome 5 linkage region and encodes a complement factor used in the cellular immune response, was differentially expressed. C5 was significantly and negatively correlated ($r^2 = -0.66$) with AHR. A 2-bp deletion upstream of the C5 gene had been previously identified in the A/J strain. This mutation causes the mice to be C5 mRNA and protein deficient [62]. This study of C5 has led to a much greater understanding of C5 activity, Th2 pathways, and its role in immunity. While C5 is involved in AHR in these mice strains, there is no evidence for susceptibility polymorphisms in C5 in humans [1,4]. While further studies may find susceptibility polymorphisms in humans, it is an important *caveat* to note that the disease process in humans and animals are likely different in many aspects.

Hitzemann et al. [63] suggest a concrete plan for integrating multiple cross mapping (MCM) with gene expression studies. The MCM paradigm is a new method similar to a diallel cross, but taken to the F_2 in mice, in which all the offsprings are typed for the genotypes and phenotypes of interest. They suggested that by comparing the QTL peaks between this diallel cross and four lab strains, c57Bl/6J, DBA/2J, BALB/cJ, and LP/J, on the phenotype of basal locomoter activity may reduce the spurious linkage results. Using this paradigm, there was a QTL peak on all three B6 intercrosses detected on distal chromosome 1, but no QTL was detected in the other three intercrosses. RNA was removed from whole brain ($n = 6$/strain), dorsomedial striatum ($n = 3$ pools/strain), and central extended amygdala ($n = 3$ pools/strain) of mice from each of the parental strains. The RNA for each mouse was run on the Affymetrix Mu74Av2 array. 59 of the ~136 genes in the 10 Mb region around the distal chromosome 1 QTL were on the array. Five genes from this region were significantly different by analysis of variance (ANOVA) among the groups in whole brain, but only one, Kcnj9, was also significant in the individual brain regions. They then used the *Webqtl.org* (qv) database to study the eQTLs of kcnj9 in RIX strains, and the maximal QTL for kcnj9 is *cis* to kcnj9 gene. A number of polymorphisms were observed between the sequences of kcnj9 between the B6 and D2 strains. Moreover, the A, C, and LP strains are identical to the D2 strain at the kcnj9 locus. One polymorphism at this locus may disrupt a transcription-binding site. Although it has not yet been confirmed, it is possible that one (or more) of these polymorphisms in Kcnj9 may be responsible for variation in basal locomotor activity.

18.4 THEORETICAL PAPERS

A number of papers have been written that address some of the more theoretical aspects of combined high dimensional biological datasets and make suggestion for their application to real data.

The fields of genomics and proteomics have essentially been made possible by the existence of computers, software, and the internet, which allows for the rapid dissemination of ideas and data between groups as well as for the methods to be efficiently implemented. The power of public shared data and the internet for distribution of data were initially proposed by Grupe et al. [42] who suggested that linkage analysis of complex traits can be studied in model organisms with RIL or standard samples such as CEPH families. Such samples only need to be genotyped once and then the organisms can be studied at leisure in many different labs for a variety of phenotypes (including microarrays). This allows biologists to concentrate upon the areas of research that are their expertise rather than developing new techniques, but still allows investigators to take advantage of extensive high dimensional data.

Doerge [21] outlines her vision for how genomic regions that are associated with gene expression could be identified using existing methods and software. In her methods, genes may be considered one at a time, in groups, or as the total in the genome. She proposes that the use of the statistical framework of Sen and Churchill [64] can allow for the dissection of genomic state and location. She suggests that while a single gene can be linked to a region, it is only when many genes are taken together and they all link to a similar region, will it be possible to uncover associated regulatory regions of the genome by statistical analyses. This assertion might have been proven in a couple of applied studies that show that the control region or many genes map to the same or similar locations *trans* to the location of the gene [37,65]. These global *trans* genes are therefore presumably master control genes, such as common transcriptional regulatory factors. She further suggests that the power to look across many genes' expression levels and compare the results may be very important in understanding epistasis in and among genes, for the covariation and relationship among the genes are a key to understanding complex diseases and traits. In addition, Doerge suggests that we need to be cautions for many statistical issues that remain to be solved. These issues include sample size (power), statistical design, adjusting for multiple testing (which would be even more extreme for mapping expression than they are for phenotypes), statistical significance (and assumptions underlying the tests), errors in phenotyping, and handling the deluge of information. Some of these issues will be discussed in greater detail later in this chapter.

A similar vision is presented by Schadt et al. [66], where they describe a new paradigm for the discovery of genes involved in human diseases and traits. The approach involves taking a combination of phenotypes, expression levels, and genotype data to identify a disease associated pattern in human and model organisms. The profile can then be used in finding the genes that are involved in the pattern. All of this information will be combined to identify the key drivers of the disease phenotype, which will then need to be verified and validated to identify the pharmacologically tractable targets [67].

Bunney et al. [68] suggest that microarrays are a good way to identify novel candidate genes for psychiatric disorders and that these studies may be combined with linkage or association studies to further narrow the list of candidate genes. Some including Amin [69] have proposed similar techniques for rheumatic diseases while others have suggested using combined genomics and proteomics to identify targets for cancer research to build better predictive models for cancer [70,71].

Perez-Enciso et al. [72] appear to be the first to attempt to address the power for conducing QTL linkage using gene expression to assist the process. They take an approach that some form of dimensional reduction is needed because of the large number of gene expression values and the low power of most microarray studies. As a result, they consider principle components, canonical analysis, and partial least squares (PLS) approach. They address some complexities such as missing data, which is a real issue in microarray data, especially if much masking is used. The conclusions that they reach are (1) power increases as the number of expression levels underlying the QTL decreases and as the correlation between genes increases; (2) data reduction is necessary to reduce the wide variability in individual gene expression levels; and (3) it is unlikely the cDNA identified by PLS (or other reduction technique) are the actual genes responsible for the QTL. This is true because to the very large number of genes that are studied as well has the high correlation among genes [73], which can result in the incorrect gene being identified. They conclude that association should never be called causation in microarray studies [67]. Furthermore, they acknowledge that one of the real problems with microarrays is that their power is based upon the assumption that the gene expression level is measured without error, whereas the large amount of nonbiological sources of error and technical variability involved in microarrays are likely to make this a critical issue.

Bjornsson et al. have broken new ground by proposing an integrated epigenetic and genetic approach to common human diseases [74]. Their approach uses what they call "common disease genetic epidemiology in the context of both genetic and epigenetic variation," or CDGE for short. In CDGE, the genetic determinants of disease are the classical Mendelian determinants, whereas the epigenic determinants are those nonDNA sequence alterations that arise during the life of an organism. For example, in cancer, there is an increasingly compelling evidence that epigenetic marks, such as chromatin modifications, can influence the progression of cancer. They argue that such epigenetic marks in cancer are only the "tip of the iceberg" and that such marks likely occur in most, if not all, during the late onset of human diseases. In their article, they provide a statistical formalism to quantify the epigenetic and genetic contributions to common diseases. In a related paper, we suggest specific techniques to quantify the epigenetic marks in epigenetic selection experiments, a process that we and others have termed "quantitative epigenetics" [29].

18.5 SOFTWARE AND BIOINFORMATICS TOOLS

A variety of software packages have been developed for the analysis of microarray data [75–77]. For simple combined microarray QTL and eQTL studies, when gene expression values are handled as any other quantitative trait, any QTL linkage software can be used [78,79]. However, some rewriting and automation may be necessary to easily handle the running, interpretation, and displaying of such volumes of data. However, several software packages have also been developed to do combined microarray–QTL studies. ExpressionView® [80] was recently developed for the combined visualization of gene–expression data with QTL data. The software is a Perl Script that can be added to any installation of Ensemble. A set of genes may be submitted to the software with their chromosomal positions, which may then be displayed on a "nomogram" of the human genome. In this software, the overlapping genes and QTLs are automatically identified. In addition, the overlap between multiple QTL studies can also be identified.

WebQtl [65,81] has taken the work of Grupe to the next logical step, by depositing not only marker data for murine inbred and RIX lines, but also extensive phenotypic, expression data on the same lines, often from multiple tissues, in a single location, www.webqtl.org. This service allows one to upload phenotypic data on the RIX lines and run QTL analysis. In addition, one can easily search for correlations between gene expression levels and traits of interest. This tool may become quite useful and is at the forefront what will undoubtedly be developed in subsequent programs. The future will also require additional data mining expertise to extract the valuable nuggets as more and more data is deposited into searchable databases [67,82]. Some have already used the existence of online expression datasets from sites such as the Stanford Microarray Database (SMD) or Gene Expression Omnibus (GEO) for identifying previously unknown patterns between genes [83–85].

18.6 ISSUES WITH COMBINED HIGH DIMENSIONAL BIOLOGICAL PROJECTS

In this section, we address some of the issues that need to be considered when designing, conducting, and analyzing combined high dimensional biological projects. Many of these issues include good experimental design, sample size (power), error rates, errors in phenotypes, and data handling. Many of these are already known issues with some specific high dimensional systems, such as linkage [67] and microarrays [50]. However there are some new issues including problems of incomplete data, as well as problems with intersections, combining disjoint data, determining what is the best tissue/time to sample, combining across species, establishing causation, and relying on previously collected data.

18.6.1 The Problem of Errors and Multiplicity Control

When a "significant" result is obtained in a high dimensional biological experiment, it could be due to one or more of the following reasons:

1. The significant result is actually real.
2. The significant result is a false positive due to random chance.
3. The significant result is due to the tested gene being biologically dependent upon the true effect. For example, it could be due to disequilibrium with the true causative allele in an association study or highly correlated with another gene in a pathway.
4. The significant result is due to some systematic bias in the biology, study, samples, or analysis.

While all of us would like for the significant results to be real, in reality, reasons 2, 3, and 4 are quite likely. Only reason 2 can be reduced by adjusting for multiple testing, but it is often not clear which is the most appropriate adjustment. False positives are often reduced by using high empirical p-values [86], an adjustment such as Bonferroni, or estimating the false discovery rates (FDRs) [87]. Since genes, mRNA levels, and proteins are not independent, it is not clear whether any of these methods are strictly valid. However, it is clear that the p-values will not be valid unless high quality data is analyzed. In other words, a high p-value could simply indicate "garbage in, garbage out."

To achieve high quality, data investigators must identify and reduce or eliminate the biological and nonbiological sources of error in a HDB experiment. In some ways the problem of "biological confounding" is the hardest to deal with. Biological pathways and networks are not completely known, and there may be time-dependent cascades, which can confound microarray studies. In addition, issues such as haplotype blocks, allelic heterogeneity, over dominance, and epistasis can all confound the identification of the true causative alleles in association/linkage [88]. Eliminating biological factors as a source of confounding is an involved process and requires extensive sequencing, microarray, verification, genotyping, and statistical testing. Even after great effort, the identification of truly causative genes, proteins, and polymorphism may still elude us. Some forms of error or bias are not inherent in the phenomena under study but are the result of methodological artifacts, procedural errors, or (hopefully unintentional) investigator biases.

The following is not meant to be an exhaustive list, but rather a delineation of some of the issues and topics that require proper care and attention during the design, conduct, and analysis of HDB studies. There are surprisingly few reviews that highlight the issues that need to be considered in the design and conduct of a study, for linkage exceptions see References 89 and 90 and microarray exceptions see References 50 and 91. Although it may go without saying, a refined understanding of the disease/trait is needed before beginning a study and this knowledge devoted to minimizing the biases and confounding factors even before the study begins. The phenotypes to be studied must be well defined, and heritable. Linkage and association, no matter how significant, have no meaning unless there is a genetic contribution to the trait [92]. Secondly, all biological assays are imprecise. Genotyping [93,94], microarraying [95], proteomics [96], and phenotyping [97,98] often have large error rates. In addition, different image processing techniques can lead to quite different results in microarray [99] and proteomic studies [100]. Relatively, modest levels of error will result not only in significantly diminished power, but also some errors such as null alleles [101] can cause false positive results. Improperly chosen controls can cause both type I and II errors [90]. No matter how much we try, there are always human errors that must be taken into account. Rigorous quality control checks should be implemented in all studies to partially account for these types of human errors. Every statistical test is based upon certain assumptions about the nature of the data, and if these assumptions are violated, the significances generated may not be valid. The assumptions that underpin all statistical tests are generally known, but can be very difficult to detect in small sample sizes. Disturbingly, different software packages can generate different results for essentially the same analysis. Finally, there is always some "cleaning" of data before formal analysis, such as correcting systematic errors (e.g., fixing pedigree errors), or filtering on expression levels or Present/Absent calls. Cleaning and normalization should be conducted to get the most accurate data, not the most "significant" result.

All of these factors can have a have an impact upon the validity of each of the individual HDB studies. Before even considering merging data from two or more types of studies, the validity of the individual HDB studies must be of high quality.

18.6.2 The Problem of Power

One of the common criticisms of microarray and proteomic studies as well as other linkage studies is that they often involve enormous numbers of genes or proteins, and therefore have low power. Among all of the combined microarray–QTL studies, only one, the first done by Brem et al. [35], mentions the power of the study. For a single locus, controlling all of the variation in gene expression (84% heritability), this study had >97% power at $\alpha < 0.01$. However, but if a locus only accounted for one-third of genetic variation, the power was only ~29%. Since only approximately one-third of genes had significant linkage, most genes probably have "oligogenic" effects, if not polygenic control, which makes the consideration of power even more important.

Part of the issue with power in HDB studies, is that, with the exception of linkage, good methods for estimating power have not yet been developed. While power for linkage is well established, many of the assumption used, such as the proportion of variance attributable to a particular locus are too high, resulting in many underpowered unsuccessful studies. Recently, there has been a move to develop methods for the power of microarray studies [102] that are more or less based upon traditional power analyses. However, we believe that rather than traditional power analysis based upon a set α (significance), $(1 - \beta)$ (power), δ (mean difference), and σ (standard deviation), it may be more appropriate to use a nonparametric Bayesian approach that estimates the distribution of all differentially expressed genes rather than any single gene. In the Bayesian approach, one must consider the posterior probability of a true positive (PTP) and the expected discovery rate (EDR) of an experiment [103].

18.6.3 The Problem of Intersections

One of the most commonly used procedures, essentially the only one, for combining data from HDB studies is to take the simple intersection of all "significant" results from the various technologies. This type of approach presupposes that "significant" genes in each type of HDB experiment are equivalent. Unfortunately, this is not usually the case. When combining linkage and microarray studies, for instance, significant linkage results usually have a very conservative p-value, such as a LOD or MLS >3. Furthermore, microarray results are often of genes with a p-value less than .05 or a two-fold change. If there is unequal power, the higher error rate in the lower powered analysis will bring the combined studies down to the least common denominator. Thus, when combining HDB studies, results that are equivalent (have the same FDR) should be combined, so that any issues with the lower powered study will not bias the results.

One assumption of the simple intersection test is that one will enrich for true results. This is true to a certain extent, but it also enriches for false negatives. This is exemplified in Figure 18.3, which illustrates the TP and TN at $\alpha < 0.05$ for either a single or the simple intersection of two similar microarray studies. In this model experiment, there are 13000 genes, with 3000 being differentially expressed between the experimental conditions. As can be seen, while the number of false positives increases, the false negative rates also increases.

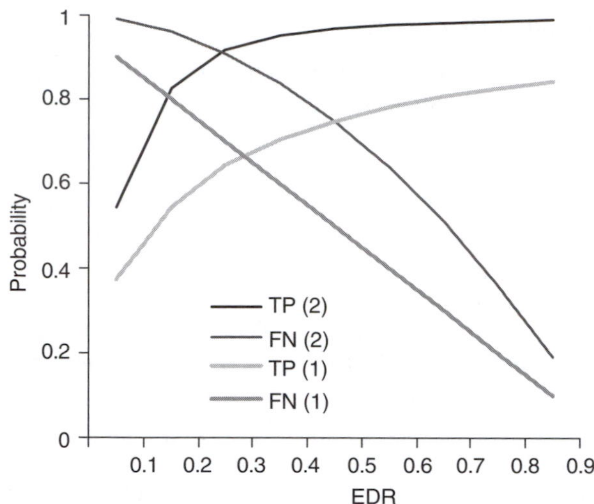

FIGURE 18.3 True positive and false negative rates from one (1) and the intersection of two (2) experiments as a function of EDR or average power. As can be seen, taking the intersection of two or more studies does increase the true positive rate of the results, it also increases the false negative rate.

18.6.4 The Problem in Incompleteness

As was illustrated in several studies above, many if not most arrays, while improving, are incomplete. Moreover, there are no proteomic techniques that can measure all proteins at all molecular weights, PH, salt concentration, and cellular localizations [104]. As a result, despite the assumption that the combined HDB studies are two genome wide studies, the fact is that most of the studies are at least in part incomplete. In addition, while genome wide association studies can probably be conducted, the variability of disequilibrium may make it difficult to determine if one has good information at all locations within the genome [104]. This can partially be fixed by better technologies such as better microarrays, more advanced proteomic technologies, and denser SNPs. However, it should be known that, when conducting combined HDB studies, the results will probably be incomplete, and negative studies may be caused by incomplete data rather than a lack of real effects.

18.6.4.1 What and When to Sample

When combining HDB studies, it is critical that all the data to be combined is appropriate for the question being asked. When studying a psychiatric disorder, it probably does not make more sense to use a muscle microarray or proteomic study that it would to use a linkage study of diabetes. But that is often what investigators do when they use convenience samples such as public data (see Section 18.6.4.3) or cross species studies (see Section 18.6.4.2). In addition, is one if looking for the causes of obesity it may not make sense to sample already obese individuals, which may only reveal the effects of obesity. Moreover, it can take time for a treatment to have an effect on gene expression, with bacteria and yeast responding much more quickly. Both the timing and selection of tissues for studies must be carefully considered and the biology of the disease and system being studied should guide the selection of tissues and time points.

18.6.4.2 Combining across Species and Tissue Types

Animal models are often used as models of human disease. However, while the diseases may be similar in phenotype, they may be different in genetic etiology. Several of the published combined HBD studies take human linkage data and murine microarray data, but there are several issues with this approach. For example, mutations in *leptin* in mice cause obesity, but there are apparently very few pathololgical mutations in the human *Leptin* gene. The murine model of human panic disorder is a very heavy dose of methamphetamine, and this type of overdose is not what causes human panic disorder. Thus, while investigators can combine the HDB data from multiple species, they must be aware that the intersection across species will not necessarily be informative.

18.6.4.3 The Problem of Public Data

Grupe et al. suggest that a great increase in productivity and throughput can be achieved by sharing of HDB data across many labs through public databases. Several databases, such as GEO and Array Express, have been established. GEO has greater than 14,000 deposited arrays of information, and many journals require that data be made public in order to be published. This has led to a proliferation of the public repositories, but there are no safeguards on the quality of the data in these databases. Some is undoubtedly of high quality, but others are certainly not. Thus when using public data for combined HDB studies, one should verify that whether the data is of sufficient quality so that when the data is combined the results are accurate.

18.6.4.4 The Problem of Causation

It is essentially impossible to establish causation statistically because while a preponderance of evidence may be accumulated [67] it does not provide causation. For that matter it is even difficult to establish direction of effects in microarray and proteomic studies without a time series, or genetic mutants [66]. For example, in linkage analysis polymorphisms cause phenotypes, not the other way round, so the direction is known. But in expression studies, if two or more genes change it is not possible to determine which is the cause and which is the effect without further studies.

18.7 CONCLUSIONS ABOUT COMBINED HDB STUDIES

It is apparent that the use of combined HDB studies has been and will continue to be valuable for the identification of genes involved in complex diseases and quantitative traits. Several studies have used this approach to identify many *cis* and *trans* eQTLs as well as genes that are involved in Asthma [39] and insulin action [60]. Software has been developed to aid in the combined HDB studies as the volume of data especially in eQTL and pQTL studies is prodigious and new methods are still being developed for displaying the data.

However, most HDB techniques and their statistical analysis are still in development and are only valid if the studies are designed well, conducted with the highest possible standards, in samples that provide high power. In addition, the studies are liable to errors from a wide variety of sources, including technical and biological, those need to be addressed during the conduct of these studies. There is still quite a bit of room for the development of new techniques to improve the utility of combined HDB studies in the studies of complex diseases and quantitative traits.

So far, most combined HDB studies have focused upon combining two types of HDB data. But it is apparent that more power would be obtained from combining more than two types of data. In addition, the large number of public resources for microarrays [105] and QTL studies (http://www.jax.org/staff/churchill/labsite/datasets/index.html), make it possible to have combined studies which may be very powerful if high quality data sets are combined. This is exemplified by meta analysis of linkage [106] and QTL studies that can reveal areas for more intensive study. Eventually, we may reach a point where investigators will be able to draw on HDB data from the many existing sources and supplement this data with their own experiments in order to test hypotheses or identify genes for complex diseases and traits.

REFERENCES

1. Ioannidis J.P., Trikalinos T.A., Nertzani E.E., and Contopoulos-Ioannidis D.G. Genetic associations in large versus small studies: an empirical assessment. *Lancet* 2003; 567–571.
2. Ioannidis J.P. Genetic associations: false or true? *Trends Mol. Med.* 2003; 135–138.
3. Hirschhorn J.N., Lohmueller K., Byrne E., and Hirschhorn K. A comprehensive review of genetic association studies. *Genet. Med.* 2002; 45–61.
4. Lohmueller K.E., Pearce C.L., Pike M., Lander E.S., and Hirschhorn J.N. Meta-analysis of genetic association studies supports a contribution of common variants to susceptibility to common disease. *Nat. Genet.* 2003; 177–182.
5. Collins F.S. Positional cloning moves from perditional to traditional. *Nat. Genet.* 1995; 347–350.
6. Bender W., Spierer P., and Hogness D.S. Chromosomal walking and jumping to isolate DNA from the Ace and rosy loci and the bithorax complex in *Drosophila melanogaster*. *J. Mol. Biol.* 1983; 17–33.
7. Haldane J.B., and Smith C.A.B. A new estimate of the linkage between the genes for colour-blindness and haemophilia in man. *Ann. Eugen.* 1947; 10–31.
8. Riordan J.R., Rommens J.M., Kerem B., Alon N., Rozmahel R., Grzelczak Z., Zielenski J., Lok S., Plavsic N., and Chou J.L. Identification of the cystic fibrosis gene: cloning and characterization of complementary DNA. *Science* 1989; 1066–1073.
9. Hirschhorn J.N., and Altshuler D. Once and again-issues surrounding replication in genetic association studies. *J. Clin. Endocrinol. Metab.* 2002; 4438–4441.

10. Frary A., Nesbitt T.C., Grandillo S., Knaap E., Cong B., Liu J., Meller J., Elber R., Alpert K.B., and Tanksley S.D. FW2.2: a quantitative trait locus key to the evolution of tomato fruit size. *Science* 2000; 85–88.
11. Hanis C., Boerwinkle E., Chakraborty R., Ellsworth D.L., Concannon P., Stirling B., Morrison V.A., Wapelhorst B., Spielman R.S., Gogolin-Ewens K.J., Shepherd J.M., Williams S.R., Risch N., Hinds D., Iwassaki N., Ogata M., Omori Y., Petzold C., Rietzsch H, Schröder H.E., Schulze J., Cox N.J., Menzel S., Boriaj V.V., Chen X., Lim L.R., Lindner T., Mereu L.E., Wang Y.Q., Xiang K., Yamagata K., and Bell G.I. A geneome-wide search for human non-insulin-dependent (type 2) diabetes genes reveals a major susceptibility locus on chromosome 2. *Nat. Genet.* 1996; 161–166.
12. Miki Y., and Swenson J. A strong candidate for breast and ovarian cancer susceptibility gene BRCA1. *Science* 1994; 66–71.
13. Paull T.T., Cortez D., Bowers B., Elledge S.J., and Gellert M. Direct DNA binding by Brca1. *Proc. Natl Acad. Sci., USA* 2001; 6086–6091.
14. Kaul R., Gao G.P., Balamurugan K., and Matalon R. Cloning of the human aspartoacylase cDNA and a common missense mutation in Canavan disease. *Nat. Genet.* 1993; 118–123.
15. Balling R. ENU mutagenesis: analyzing gene function in mice. *Annu. Rev. Genom. Hum. Genet.* 2001; 463–492.
16. Jansen R.C., and Nap J.P. Genetical genomics: the added value from segregation. *Trends Genet.* 2001; 388–391.
17. Eisen M.B., Spellman P.T., Brown P.O., and Botstein D. Cluster analysis and display of genome-wide expression patterns. *Proc. Natl Acad. Sci., USA* 1998; 14863–14868.
18. Lee C.K., Weindruch R., and Prolla T.A. Gene-expression profile of the ageing brain in mice. *Nat. Genet.* 2000; 294–297.
19. Patterson S.D., and Aebersold R.H. Proteomics: the first decade and beyond. *Nat. Genet.* 2003; 311–323.
20. Morton N.E. Sequential tests for the detection of linkage. *Am. J. Hum. Genet.* 1955; 277–318.
21. Doerge R.W. Mapping and analysis of quantitative trait loci in experimental populations. *Nat. Rev. Genet.* 2002; 43–52.
22. Mackay T.F. The genetic architecture of quantitative traits. *Annu. Rev. Genet.* 2001; 303–339.
23. Lander E.S., and Schork N.J. Genetic dissection of complex traits. *Science* 1994; 2037–2048.
24. Allison D.B. Transmission-disequilibrium test for quantitative traits. *Am. J. Hum. Genet.* 1997; 676–690.
25. Page G.P., and Amos C.I. Comparison of linkage-disequilibrium methods for localization of genes influencing quantitative traits in humans. *Am. J. Hum. Genet.* 1999; 1194–1205.
26. Pollack J.R., Perou C.M., Alizadeh A.A., Eisen M.B., Pergamenschikov A., Williams CF, Jeffrey S.S., Botstein D., and Brown P.O. Genome-wide analysis of DNA copy-number changes using cDNA microarrays. *Nat. Genet.* 1999; 41–46.
27. Taylor J., King R.D., Altmann T., and Fiehn O. Application of metabolomics to plant genotype discrimination using statistics and machine learning. *Bioinformatics* 2002; S241–S248.
28. Voit E.O., and Radivoyevitch T. Biochemical systems analysis of genome-wide expression data. *Bioinformatics* 2000; 1023–1037.

29. Garfinkel M., Sollars V., Lu X., and Ruden D.M. Multigenerational section and detection of altered histone acetylation and methylation patterns. In: Tollefsbol T., ed. *Methods in Molecular Biology, Vol. 287: Epigenetic Protocols*. Totowa, NJ: Humana Press, Inc., 2004; pp. 151–168.
30. Stephens J.C., Schneider J.A., Tanguay D.A., Choi J., Acharya T., Stanley S.E., Jiang R., Messer C.J., Chew A., Han J.H., Duan J., Carr J.L., Lee M.S., Koshy B., Kumar A.M., Zhang G., Newell W.R., Windemuth A., Xu C., Kalbfleisch T.S., Shaner S.L., Arnold K, Schulz V., Drysdale C.M., Nandabalan K., Judson R.S., Ruano G., and Vovis G.F. Haplotype variation and linkage disequilibrium in 313 human genes. *Science* 2001; 489–493.
31. Antonarakis S.E., Irkin S.H., Cheng T.C., Scott A.F., Sexton J.P., Trusko S.P., Charache S., Kazazian H.H., Jr. beta-Thalassemia in American Blacks: novel mutations in the "TATA" box and an acceptor splice site. *Proc. Natl Acad. Sci., USA* 1984; 1154–1158.
32. Yan H., Yuan W., Velculescu V.E., Vogelstein B., and Kinzler K.W. Allelic variation in human gene expression. *Science* 2002; 1143.
33. Cheung V.G., Conlin L.K., Weber T.M., Arcaro M., Jen K.Y., Morley M., and Spielman R.S. Natural variation in human gene expression assessed in lymphoblastoid cells. *Nat. Genet.* 2003; 422–425.
34. Lo H.S., Wang Z., Hu Y., Yang H.H., Gere S., Buetow K.H., and Lee M.P. Allelic variation in gene expression is common in the human genome. *Genome Res.* 2003; 1855–1862.
35. Brem R.B., Yvert G., Clinton R., and Kruglyak L. Genetic dissection of transcriptional regulation in budding yeast. *Science* 2002; 752–755.
36. Yvert G., Brem R.B., Whittle J., Akey J.M., Foss E., Smith E.N., Mackelprang R., and Kruglyak L. Trans-acting regulatory variation in *Saccharomyces cerevisiae* and the role of transcription factors. *Nat. Genet.* 2003; 57–64.
37. Schadt E.E., Monks S.A., Drake T.A., Lusis A.J., Che N., Colinayo V., Ruff T.G., Milligan S.B., Lamb J.R., Cavet G., Linsley P.S., Mao M., Stoughton R.B., and Friend S.H. Genetics of gene expression surveyed in maize, mouse and man. *Nature* 2003; 297–302.
38. Allison D.B., Fernandez J.R., Heo M., Zhu S.K., Etzel C., Beasley T.M., and Amos C.I. Bias in estimates of quantitative-trait-locus effect in genome scans: demonstration of the phenomenon and a method-of-moments procedure for reducing bias. *Am. J. Hum. Genet.* 2002; 575–585.
39. Karp C.L., Grupe A., Schadt E., Ewart S.L., Keane-Moore M., Cuomo P.J., Kohl J., Wahl L., Kuperman D., Germer S., Aud D., Peltz G., and Wills-Karp M. Identification of complement factor 5 as a susceptibility locus for experimental allergic asthma. *Nat. Immunol.* 2000; 221–226.
40. Yi N., George V., and Allison D.B. Stochastic search variable selection for identifying multiple quantitative trait loci. *Genetics* 2003; 1129–1138.
41. Yi N., Xu S., and Allison D.B. Bayesian model choice and search strategies for mapping interacting quantitative trait loci. *Genetics* 2003; 867–883.
42. Grupe A., Germer S., Usuka J., Aud D., Belknap J.K., Klein R.F., Ahluwalia M.K., Higuchi R., and Peltz G. In silico mapping of complex disease-related traits in mice. *Science* 2001; 1915–1918.
43. Lan H., Stoehr J.P., Nadler S.T., Schueler K.L., Yandell B.S., and Attie A.D. Dimension reduction for mapping mRNA abundance as quantitative traits. *Genetics* 2003; 1607–1614.
44. Klose J., Nock C., Herrmann M., Stuhler K., Marcus K., Bluggel M., Krause E., Schalkwyk L.C., Rastan S., Brown S.D., Bussow K., Himmelbauer H., and

Lehrach H. Genetic analysis of the mouse brain proteome. *Nat. Genet.* 2002; 385–393.

45. Venkatraman A., Landar A., Davis A.J., Ulasova E., Page G., Murphy M.P., Darley-Usmar V.D., and Bailey S.M. Oxidative modification of hepatic mitochondria protein thiols: effect of chronic alcohol consumption. *Am. J. Physiol. Gastrointest. Liver Physiol.* 2003; 286: G521–G527.
46. Cheung V.G., and Spielman R.S. The genetics of variation in gene expression. *Nat. Genet.* 2002; 522–525.
47. Niculescu A.B., Segal D.S., Kuczenski R., Barrett T., Hauger R.L., and Kelsoe J.R. Identifying a series of candidate genes for mania and psychosis: a convergent functional genomics approach. *Physiol. Genomics* 2000; 83–91.
48. Niculescu A.B., III, and Kelsoe J.R. Convergent functional genomics: application to bipolar disorder. *Ann. Med.* 2001; 263–271.
49. Brown P.L., Wise R.A., and Kiyatkin E.A. Brain hyperthermia is induced by methamphetamine and exacerbated by social interaction. *J. Neurosci.* 2003; 3924–3929.
50. Page G.P., Edwards J.W., Barnes S., Weindruch R., and Allison D.B. A design and statistical perspective on microarray gene expression studies in nutrition: the need for playful creativity and scientific hard-mindedness. *Nutrition* 2003; 997–1000.
51. Wayne M.L., and McIntyre L.M. Combining mapping and arraying: an approach to candidate gene identification. *Proc. Natl Acad. Sci., USA* 2002; 99: 14903–14906.
52. Eaves I.A., Wicker L.S., Ghandour G., Lyons P.A., Peterson L.B., Todd J.A., and Glynne R.J. Combining mouse congenic strains and microarray gene expression analyses to study a complex trait: the NOD model of type 1 diabetes. *Genome Res.* 2002; 232–243.
53. Tabakoff B., Bhave S.V., and Hoffman P.L. Selective breeding, quantitative trait locus analysis, and gene arrays identify candidate genes for complex drug-related behaviors. *J. Neurosci.* 2003; 4491–4498.
54. Kirstein S.L., Davidson K.L., Ehringer M.A., Sikela J.M., Erwin V.G., and Tabakoff B. Quantitative trait loci affecting initial sensitivity and acute functional tolerance to ethanol-induced ataxia and brain cAMP signaling in BXD recombinant inbred mice. *J. Pharmacol. Exp. Ther.* 2002; 1238–1245.
55. Okuda T., Sumiya T., Iwai N., and Miyata T. Pyridoxine $5'$-phosphate oxidase is a candidate gene responsible for hypertension in Dahl-S rats. *Biochem. Biophys. Res. Commun.* 2004; 647–653.
56. Jordan J., Lipp A., Tank J., Schroder C., Stoffels M., Franke G., Diedrich A., Arnold G., Goldstein D.S., Sharma A.M., and Luft F.C. Catechol-o-methyltransferase and blood pressure in humans. *Circulation* 2002; 460–465.
57. Toma D.P., White K.P., Hirsch J., and Greenspan R.J. Identification of genes involved in *Drosophila melanogaster* geotaxis, a complex behavioral trait. *Nat. Genet.* 2002; 31: 349–353.
58. Aitman T.J., Glazier A.M., Wallace C.A., Cooper L.D., Norsworthy P.J., Wahid F.N., al Majali K.M., Trembling P.M., Mann C.J., Shoulders C.C., Graf D., St Lezin E., Kurtz T.W., Kren V., Pravenec M., Ibrahimi A., Abumrad N.A., Stanton L.W., and Scott J. Identification of Cd36 (Fat) as an insulin-resistance gene causing defective fatty acid and glucose metabolism in hypertensive rats. *Nat. Genet.* 1999; 76–83.
59. Aitman T.J., Gotoda T., Ebans A.L., Imrie H., Heath K.E., Trembling P.M., Truman H., Wallace C.A., Rahman A., Doré C., Flint J., Kren V., Zidek V., Kurtz T.W., Pravenex M., and Scott J. Quantitative trait loci for cellular defects in glucose and fatty acid metabolism in hypertensive rats. *Nat. Genet.* 1997; 197–201.

60. Glazier A.M., Scott J., and Aitman T.J. Molecular basis of the Cd36 chromosomal deletion underlying SHR defects in insulin action and fatty acid metabolism. *Mamm. Genome* 2002; 108–113.
61. Collison M., Glazier A.M., Graham D., Morton J.J., Dominiczak M.H., Aitman T.J., Connell J.M., Gould G.W., and Dominiczak A.F. Cd36 and molecular mechanisms of insulin resistance in the stroke-prone spontaneously hypertensive rat. *Diabetes* 2000; 2222–2226.
62. Wetsel R.A., Fleischer D.T., and Haviland D.L. Deficiency of the murine fifth complement component (C5). A 2-base pair gene deletion in a 5′-exon. *J. Biol. Chem.* 1990; 2435–2440.
63. Hitzemann R., Malmanger B., Reed C., Lawler M., Hitzemann B., Coulombe S., Buck K., Rademacher B., Walter N., Polyakov Y., Sikela J., Gensler B., Burgers S., Williams R.W., Manly K., Flint J., and Talbot C. A strategy for the integration of QTL, gene expression, and sequence analyses. *Mamm. Genome* 2003; 733–747.
64. Sen S., and Churchill G.A. A statistical framework for quantitative trait mapping. *Genetics* 2001; 371–387.
65. Chesler E.J., Wang J., Lu L., Manly K.F., and Williams R. Genetic correlates of gene expression in recombinant inbred strains: a relational model to explore for neurobehavioral phenotypes. *Neuroinformatics* 2003; 343–357.
66. Schadt E.E., Monks S.A., and Friend S.H. A new paradigm for drug discovery: integrating clinical, genetic, genomic and molecular phenotype data to identify drug targets. *Biochem. Soc. Trans.* 2003; 437–443.
67. Page G.P., George V., Go R.C., Page P.Z., and Allison D.B. "Are we there yet?": Deciding when one has demonstrated specific genetic causation in complex diseases and quantitative traits. *Am. J. Hum. Genet.* 2003; 711–719.
68. Bunney W.E., Bunney B.G., Vawter M.P., Tomita H., Li J., Evans S.J., Choudary P.V., Myers R.M., Jones E.G., Watson S.J., and Akil H. Microarray technology: a review of new strategies to discover candidate vulnerability genes in psychiatric disorders. *Am. J. Psychiat.* 2003; 657–666.
69. Amin A.R. A need for a 'whole-istic functional genomics' approach in complex human diseases: arthritis. *Arthritis Res. Ther.* 2003; 76–79.
70. Hanash S.M., Madoz-Gurpide J., and Misek D.E. Identification of novel targets for cancer therapy using expression proteomics. *Leukemia* 2002; 478–485.
71. Hanash S.M., Bobek M.P., Rickman D.S., Williams T., Rouillard J.M., Kuick R., Puravs E. Integrating cancer genomics and proteomics in the post-genome era. *Proteomics* 2002; 69–75.
72. Perez-Enciso M., Toro M.A., Tenenhaus M., and Gianola D. Combining gene expression and molecular marker information for mapping complex trait genes: a simulation study. *Genetics* 2003; 1597–1606.
73. Lander E.S. Array of hope. *Nat. Genet.* 1999; 3–5.
74. Bjornsson H., Fallin M., and Feinberg A. An integrated epigenetic and genetic approach to common human disease. *Trends Genet.* 2004; 20: 350–358.
75. Tusher V.G., Tibshirani R., and Chu G. Significance analysis of microarrays applied to the ionizing radiation response. *Proc. Natl Acad. Sci. USA* 2001; 5116–5121.
76. Baldi P., and Long A.D. A Bayesian framework for the analysis of microarray expression data: regularized *t*-test and statistical inferences of gene changes. *Bioinformatics* 2001; 509–519.
77. Huber W., and Gentleman R. Matchprobes: a bioconductor package for the sequence-matching of microarray probe elements. *Bioinformatics* 2004; 20: 1651–1652.

78. Broman K.W., Wu H., Sen S., and Churchill G.A. R/qtl: QTL mapping in experimental crosses. *Bioinformatics* 2003; 889–890.
79. Manly K.F., Cudmore R.H., Jr., and Meer J.M. Map Manager QTX, cross-platform software for genetic mapping. *Mamm. Genome* 2001; 930–932.
80. Ibrahim S., Fisher C., El-Alaily H., Soliman H., and Anwar A. Improvement of the nutritional quality of Egyptian and Sudanese sorghum grains by the addition of phosphates. *Br. Poult. Sci.* 1988; 721–728.
81. Wang J., Manly K., and Williams R. WebQTL: web-based complex trait analysis. *Neuroinformatics* 2003; 299–308.
82. Brazma A., Hingamp P., Quackenbush J., Sherlock G., Spellman P., Stoeckert C., Aach J., Ansorge W., Ball C.A., Causton H.C., Gaasterland T., Glenisson P., Holstege F.C., Kim I.F., Markowitz V., Matese J.C., Parkinson H., Robinson A., Sarkans U., Schulze-Kremer S., Stewart J., Taylor R., Vilo J., and Vingron M. Minimum information about a microarray experiment (MIAME)-toward standards for microarray data. *Nat. Genet.* 2001; 365–371.
83. Qian J., Lin J., Luscombe N.M., Yu H., and Gerstein M. Prediction of regulatory networks: genome-wide identification of transcription factor targets from gene expression data. *Bioinformatics* 2003; 1917–1926.
84. Segal E., Yelensky R., and Koller D. Genome-wide discovery of transcriptional modules from DNA sequence and gene expression. *Bioinformatics* 2003; I273–I282.
85. Stuart J.M., Segal E., Koller D., and Kim S.K. A gene-coexpression network for global discovery of conserved genetic modules. *Science* 2003; 249–255.
86. Risch N., and Merikangas K. The future of genetic studies of complex human diseases. *Science* 1996; 1516–1517.
87. Benjamini Y., and Hochberg Y. Controlling the false discovery rate: a practical and powerful approach to multiple testing. *J. R. Stat. Soc. Ser. B (Methodol.)*, 1995; 289–300.
88. Culverhouse R., Suarez B.K., Lin J., and Reich T. A perspective on epistasis: limits of models displaying no main effect. *Am. J. Hum. Genet.* 2002; 461–471.
89. Terwilliger J.D., and Goring H.H. Gene mapping in the 20th and 21st centuries: statistical methods, data analysis, and experimental design. *Hum. Biol.* 2000; 63–132.
90. Ellsworth D.L., and Manolio T.A. The emerging importance of genetics in epidemiologic research II. Issues in study design and gene mapping. *Ann. Epidemiol.* 1999; 75–90.
91. Yang Y.H., and Speed T. Design issues for cDNA microarray experiments. *Nat. Rev. Genet.* 2002; 579–588.
92. Ott J. *Analysis of Human Genetic Linkage*. Johns Hopkins Press. Baltimore, MD, 1991.
93. Gordon D., Levenstien M.A., Finch S.J., and Ott J. Errors and linkage disequilibrium interact multiplicatively when computing sample sizes for genetic case-control association studies. *Pac. Symp. Biocomput.* 2003; 490–501.
94. Gordon D., Finch S.J., Nothnagel M., and Ott J. Power and sample size calculations for case-control genetic association tests when errors are present: application to single nucleotide polymorphisms. *Hum. Hered.* 2002; 22–33.
95. Coombes K.R., Highsmith W.E., Krogmann T.A., Baggerly K.A., Stivers D.N., and Abruzzo L.V. Identifying and quantifying sources of variation in microarray data using high-density cDNA membrane arrays. *J. Comput. Biol.* 2002; 655–669.

96. Kim H., Page G.P., and Barnes S. Proteomics and mass spectrometry in nutrition research. *Nutrition* 2004; 155–165.
97. Rice J., Saccone N., and Rasmusssen E. Definition of the phenotype. In: Rao D., and Province M., eds. *Genetic Dissection of Complex Traits*. Academic Press, San Diego, NJ 2001; pp. 69–76.
98. Egan S., Nathan P., and Lumley M. Diagnostic concordance of ICD-10 personality and comorbid disorders: a comparison of standard clinical assessment and structured interviews in a clinical setting. *Aust. N. Z. J. Psychat.* 2003; 484–491.
99. Irizarry R.A., Hobbs B., Collin F., Beazer-Barclay Y.D., Antonellis K.J., Scherf U, and Speed T.P. Exploration, normalization, and summaries of high density oligonucleotide array probe level data. *Biostatistics* 2003; 249–264.
100. Rosengren A.T., Salmi J.M., Aittokallio T., Westerholm J., Lahesmaa R., Nyman T.A., and Nevalainen O.S. Comparison of PDQuest and Progenesis software packages in the analysis of two-dimensional electrophoresis gels. *Proteomics* 2003; 1936–1946.
101. Ewen K.R., Bahlo M., Treloar S.A., Levinson D.F., Mowry B., Barlow J.W., and Foote S.J. Identification and analysis of error types in high-throughput genotyping. *Am. J. Hum. Genet.* 2000; 727–736.
102. Pan W., Lin J., and Le C.T. How many replicates of arrays are required to detect gene expression changes in microarray experiments? A mixture model approach. *Genome Biol.* 13: 325–338.
103. Gadbury G., Page G., Edwards J., Weindruch R., Permana P.A., Mountz J., Allison D.B. Power analysis and sample size estimation in the age of high dimensional biology: a parametric bootstrap approach and examples from microarray research. *Stat. Meth. Med. Res.* 75: 161–173.
104. Freeman W.M., and Hemby S.E. Proteomics for protein expression profiling in neuroscience. *Neurochem. Res.* 2004; 1065–1081.
105. Edgar R., Domrachev M., and Lash A.E. Gene expression omnibus: NCBI gene expression and hybridization array data repository. *Nucleic Acids Res.* 2002; 207–210.
106. Marazita M.L., Murray J.C., Lidral A.C., Arcos-Burgos M., Cooper M.E., Goldstein T, Maher B.S., Daack-Hirsch S., Schultz R., Mansilla M.A., Field L.L., Liu Y.E., Prescott N., Malcolm S., Winter R., Ray A., Moreno L., Valencia C., Neiswanger K., Wyszynski D.F., Bailey-Wilson J.E., Albacha-Hejazi H., Beaty T.H., McIntosh I., Hetmanski J.B., Tuncbilek G., Edwards M., Harkin L., Scott R., and Roddick L.G. Meta-analysis of 13 genome scans reveals multiple cleft lip/palate genes with novel loci on 9q21 and 2q32–35. *Am. J. Hum. Genet.* 2004; 161–173.

Index

A

Affymetrix, 4, 30, 36
  normalization methods, 18
Agglomerative
  clusters, 139, *see also* bottom-up clusters
  fusion, hierarchical, 169
  hierarchical clustering, 153
  linkage methods, 140
  tree, dendrograms, 142
Airway hyper responsivity, 345–346
Algorithm,
  GCVSS, 18
  normalization, 24
Allegory of the cave, 59
American Statistical Association, 39
Among-arrays variance, 12, 13
Among-subjects variance, 13
Analysis of variance, *see* ANOVA
Analyzing data, parametric linear models, 239
ANOVA,
  fixed effects linear models, 234
  fixed linear model, 204–205
  Kerr et al., 83
  microarray experiments, 246
  test of contrasts, 85
ANOVA-based normalization methods, 21
Arabidopsis, carbohydrate storage in, 42
Arabidopsis Functional Genome Consortium, 33
ARE, 254
Array,
  design, 30, 31
  effect, variance of, 14
  effects, 13, 124
    matrix form, 114
  quality, 30
  variability, 241
    fixed effects linear models, 236
Array-specific variance, 12
Array-wide metrics, 36
  quality metrics, 41
Artifacts, sources of, 160
Assessment tools, clusters, 148
Asymptotic relative efficiency, *see* ARE
Auto antibodies, 3
Auto antigen, 3
  1152-feature array, 3
  microarrays, 3
Auto immune serum, 3

B

B-statistics, 279
  testable hypotheses, 282
  hierarchical models, 273
Background fluorescence, 32
Balanced permutation method, null distribution, 270
Banner plots, 143
Basic Local Alignment Search Tool, *see* BLAST
Bayes,
  law, 268, 271
  theorem, 301
Bayesian
  approaches, 300
  FDR, 301
  inference, 267
  information criteria, *see* BIC
  posterior probabilities, 80
Behrens–Fisher problem, 257
Bernoulli, 70
  component, gamma–gamma models, 275
  trial, 88
Best,
  linear unbiased estimator, *see* BLUE
  linear unbiased predictor, *see* BLUP
Beta (2, 2) prior distribution, 275
Between-gene pair intensity, hypotheses testing, 190–191
B_g statistic, 281
BH, 298, 300
  rule, 301
Bias, 23, 24, 33, 44
  reduction of, 13–14
BIC, 151
Bilogical confounding, 350
Bimodal distribution, 42, 153
Binary variables, distance, 136
Bio Analyzer, 36
Bioconductor, 314
Bioinformatic quality, 40

Biological,
  averaging, 99, 103
  function, 309
  pathway, 291
  process, 309, 324
BIOML, 311
Bio-ontologies, 310
Biosciences domain, semantic heterogeneities, 306
Bivariate normal distribution, 15
BLAST, 309
Blocking, 24, 112
BLUE, 121, 123
  fixed linear model, 205
  mixed linear model, 207
BLUP, mixed linear model, 207
Bonferroni, 295
  adjustment,
    two-sample case, 248
    probe-level modeling, 213
  correction, 39–40
    multiple testing, 292
  cutoff, 40
    probe-level modeling, 210–213
  procedure, multiple testing, 293
  technique, 80
  threshold, 46
Bootstrap,
  methods, cluster analysis, 171
  techniques, 261
  with surgery, 300
Bootstrapping, hypotheses testing, 242
Bottom-up clusters, 139, *see also* agglomerative clusters
By spot statistical analysis, 23

C

Calculate pool, *see* CP
Canberra distance, 136–138
Cardinality, hypotheses testing, 189
Cascades, 350
CAST, 146
Categorical variables, distance, 136
Causation, 353
cdf, 186
cDNA,
  arrays, 32, 33
  microarray, 4
  microarray data, normalizing, 120
  mixed-effects model, 21
  spotted arrays, 11, 299
Cellular component, 309, 324
Central limit theorem, 65
CFTR, 336
Chebby checker method, 259–260
Chebyshev's inequality, nonparametric methods, 248
Choreography engine, 316
Circular reasoning, 64
Class discovery, 132
Classification, 61–62
Cliff's δ statistics, 257–259, 262
Cluster Affinity Search Technique, *see* CAST
Cluster,
  analysis,
    computer packages, 155
    typical steps, 163
  significant, 139
  solutions, detecting stable, 171–173
  stability,
    cDNA chip, 162
    reliability and valididty, 160
    replicability, 161
Clustering,
  algorithm, 160, 167
  approaches, comparison of, 148
Cluster-specific discrepancy rates, 150
Combined modes, deduction and induction, 69
Combining HDB studies, 352–353
Compatibility, 314
Complete
  linkage,
    agglomerative algorithm, 140
    cluster analysis, 169
  null hypothesis, 294
    nonparametric methods, 247–248
    permutation-based nonparametric methods, 253
Compression, 314
Confidence interval, multiplicative model, 202
Consensus trees, 151
Cluster analysis, 171, 172
Contamination, scenarios, 103
Continuous measures,
  distance; similarity, 135–136
Contrast in ANOVA, 82
Control array, 31
Controlling procedures, resampling-based, 299
Convex-combination method, 151
Cophenetic correlation, 170
Correlation matrices, random, 68
Co-segregation, 344
CP, 100
Critical metrics, 34
cRNA, 36
Cross hybridization, 32
CT, 331
Cumulative
  distribution function, *see* cdf
  errors, 35

Curated
knowledge systems, 322
knowledgebase, 327
Cyclic loess, single-channel arrays, 20
Cystic fibrosis transmembrane regulator, *see* CFTR
Cytochrome, 41
Cytokines, 3, 4

D

D chips, 48
DA, 323
DAG, 323
Data,
analysis, semantic heterogeneities, 306
dissemination, 306
integrity, 30
mining, 290, 328
modeling, semantic heterogeneities, 306
normalization,
semantic heterogeneities, 306
and cleaning, 165
quality, 30
set types, 24
simulation, 25
Database,
for annotation, visualization, and integrated discovery, *see* DAVID
integration, 329
DAVID, 327, 329
Degrees of freedom, 105
multiplicative model, 202
mixed effects models, 237
probe-level modeling, 213
unequal-variance two-group t-test, 229
Dendrograms, 138, 139, 143, 148
cluster analysis, 168
Differential expression, posterior probability of, 269
Differently expressed genes, identification, 107
Dimension reduction, 193
UMSA, 181
Direct acyclic graph, *see* DAG
Directly associated, *see* DA
Distance,
measure, in cluster analysis, 135
metrics, 186
Distribution of p-values, 84
Divisive clusters, 139, *see also* top-down clusters,
DNA,
microarray, 2
oligonucleotide microarrays, 4
Dominance,
matrix, Cliff's δ statistics, 258
double, 193
single, 193
triple, 192
Donuts, 37
Drosophila, 96
Dye effects, matrix form, 114

E

EB and SAM, 261–262
ecdf, 186
EDR, 63, 78–80, 84
mixture model, 87
role of sample size on, 88–90
ELISA, 3
EM,
algorithm, 275
lognormal–normal models, 276
Empirical
Bayes,
approach, 107
method, 253
methodology, 248
BLUP, mixed linear model, 207
cumulative distribution function, *see* ecdf
risk minimization, *see* ERM
Environmental sensitivity, 35
Epigenetic, 348
factors, 335
markers, 339
Epistasis, 341
eQTL, 342, 348, 354
eQTLs and pQTLs, 340
Equal-variance two-group t-test, microarray experiments, 227
Equivalence
criterion, 96, 104
formula, 96
ERM, 180
Error,
component, random, 164
probabilities prediction, 61
probabilities, types 1 and 2 variance, fixed effects linear models, 233
type, 2, 60
Errors,
type 1, 66
and multiplicity control, 349–350
EST, 184
Estimands, 61
Estimating, null distribution, 269
Estimation, 61

Estimators, 295
  unbiased, 98
  generalized least squares, 120, 124
Euclidean distance, 136–138, 143
  cluster analysis, 167, 172
  hypotheses testing, 187
Evaluation of normalization methods, 24–25
Evolutionary genetics, 208–214
Excessive normalization, 38
Exchangeability, gamma–gamma models, 274
Exchangeable random variables, 268–269
Expected,
  discovery rate, *see* EDR
  maximization algorithm, *see* EM algorithm
Experimental errors, hypotheses testing, 190
Expressed sequence tags (ESTs), 4
Expressed sequence tags, *see* EST
Expression,
  indexes, 219
  levels, distribution of, 103
  profiles, 34, 42
  QTL, 338
  values, comparison of, 291
eXtensible Markup Language, *see* XML

F

F statistic, 208
F test, 13
False,
  discovery rates, *see* FDR
  negative rate, 63
  positive rate, 63
  positives, 60, 349
FatiGO, 325–326, 327
FDA, 125
FDR, 38, 60, 80, 108, 248, 294, 340
  Bayesian interpretation of, 294
  control of the positive, 295
  control of, 298
  SAM, 231
  uncertainty, 82
Fisher's,
  exact test, 326
  formula, nonparametric bootstrap, 250
Fishing expedition, 165
Fixed,
  linear model, 200, 216
    hypotheses, 209
  model expression index, *see* FMEI
Fligner–Policello test, 257, 258
FMEI, 204, 214
  and MMEI, probe-level modeling, 210
Fold-change, 61, 78
  probe-level modeling, 216
  analysis, 226
  criterion, 106–108
Food and Drug Administration (FDA), 70
Forensic statisticians, 59
Four factor main effects, fixed effects linear models, 234
Frame-based ontology, 310
F-ratio,
  ANOVA, 246
  two-sample case, 248
Frequentist,
  approach, 268
  framework, 300
  test, 6
  testing, 66
  paradigm, 60
    hypotheses testing, 273
F-statistic, 83
Fudge factor, SAM, 231
FWER, 294, 295
  control of, 295, 298
  and FDR, comparison, 293

G

GADPH rates, 45
Gamma,
  –gamma models, 274, 278
  distribution, 277
Gap statistic, 150–151
Gauss–Markov,
  assumption, 65
  theorem, 121, 123
GCOS, 37
GCVSS, 18
Gel electrophoresis, 36
Gen MAPP, 325
Gene Chip, 30, 37
Gene,
  expression, 5, 25
    data simulators, 68
    design, 113
    estimating, 97
    fold-change analysis, 225
    precision of estimator, 105
    quantification, 100
    unbiased estimate of, 99
  identifiers, 323
  Ontology Next Generation, *see* GONG
  Ontology, *see* GO
  profiles, level, scatter, and shape, 166
  reassignments, hypotheses testing, 190
  region heterogeneity, hypotheses testing, 187

sequence diversity analysis, 185
sharing, 150
variability, mixed effects models, 237
Generalized,
cross-validation smoothing spline, *see* GCVSS
Latin square, 112–113, 115–117
least squares, mixed linear model, 206
two-level hierarchical models, 273
Genes
contiguous, 190
marker, 165
over- or underexpressed, 189
positional candidate, 336
reduction–oxidation, 41
selection of, 164
signature, 193
up- or downregulated, 120, 326
Gene-specific,
mean, prior distribution, 285
scatter, 248
variances,
equal-variance two-group *t*-test, 228
hierarchical linear model, 284
Genetic,
etiology, 353
markers, 339
networks, 159
Genetical genomics, 338
Genomic heterogeneity, 184
hypotheses testing, 188
Global,
error, measure of, 293
metrics, 36, 45
normalization, 14
p-value, 295, 297
test, 292
GO, 41
annotation, 47
Miner, 326
Slim, 309
GODB, 323, 328
GOMine, 326
GONG, 310
Greedy algorithm, 170
Gridable covariance structure, 68
Gridding outliers, 36
Grouping, parametric linear models, 239

H

Hematopoietic differentiation, 144
Heritable, 350
Hierarchical,
Bayesian methods, 273
clustering algorithms, 160
clusters, 148
cluster analysis, lung cancer, 132–134
models,
parameterization of variances, 280
sharing of information, 278
probability model, 282
Holm's procedure, 299
Homogeneity of variance, 66
fixed effects linear models, 234
Homology, 337
Homoscedastic variation, ANOVA, 246
Homoscedasticity, 65
rank-based statistics, 255
Housekeeping gene, 14–18
differentials, 120
unweighted average, 123
weighted average, 122–123
Human Latin square, 200
Hyperprior distribution, hierarchical linear model, 284
Hypotheses,
gene sequence diversity analysis, 187
testing,
between condition pair, 191
fixed effects linear models, 234
tests, probe-level modeling, 217

I

Image quality, 37
in situ synthesis, 32
in vitro transcription, *see* IVT
Inclusive association, 323
Incomplete normalization, 38
Incompleteness, 352
Inference principle, 193
Inferential paradigm, 60
Information,
borrowing, gamma–gamma models, 275
overload, 322
Inner product matrix, hypotheses testing, 192
Integrated epigenetic and genetic approach, 348
Integration, 328
Intensities, mixed linear model, 206
Intensity,
functional, hypotheses testing, 192
responses, equal-variance two-group *t*-test, 228
vector, hypotheses testing, 192
Interaction, fixed effects linear models, 233
Intercross, 340
Interesting genes, 326
Interferon-γ, 4
Interleukins, 4
International Life Sciences Institute, ILSI, 44

Interoperability, 308
Interquartile range, *see* IQR
Interrelated web servers, 313
Intersection, 351
Ionizing radiation, contrasts, 270
IQR, 19
Isotype subclass, 3
IVT, 36

J

Jacob Bernoulli, 70
Jensen's inequality, 105

K

k-Medoids, 144
KEGG, 310, 323, 327
k-means cluster analysis, 143, 170
Knowledgebases, 323, 330
Kolmogorov–Smirnov
  statistics, 259
  test, 261
KW, 262
Kyoto Encyclopedia of Genes and Genomes, *see* KEGG

L

Labeling, 36
Laboratory Information Management Systems, *see* LIMS
LDA, 178
Leave out one cross-validation, *see* LOOCV
Likelihood function, 285
LIMS, 322, 331
Linear discriminant analysis, *see* LDA
Linear normalization methods, 17–18, 20–22
Locally weighted regression, *see* loess
Loess, 12, 17, 18, 23
  regression, 14
  orthonormal basis, 20
  probe-level modeling, 209
Loess-based normalization methods, 19
Log transformation, 33
  ANOVA, 246
Log_2-transformed expression, 11
Logarithmic transformation, 11–12
Logistic regression, 271
Lognormal–normal model, 276, 278
Log-ratio means, 280
Log-transformed response, fixed linear model, 204–205
Long oligo microarray, 4
LOOCV, 47–49
Loss function, 302

M

MAGE-ML, 308
Manhattan distance, 136–138
Mann–Whitney,
  U statistic, 258
  U test, 252
MAPP, 325
Marginal null data distribution, gamma–gamma models, 275
MAS, 39, 43–44, 198
Master control genes, 347
Matrix form, mixed linear model, 206
Maximum,
  likelihood estimators, lognormal–normal models, 276
  likelihood-based methods, mixed linear model, 207
MBEI, 202, 204, 214
MCM, 346
MDS, 147
Mean,
  and variance,
    gene-specific, 280
    log-ratio, 281
  difference, permutation-based nonparametric methods, 252
Mendelian, 336
Metabolic pathway, 337
Metadata, 316
Meta-methodolgy, 59
Meta-methods, 70
Method of moment estimators, 282
MGED, 307, 308
MGED-OM, 307
MIAME, 38, 307
Microarray,
  analysis suite, *see* MAS
  analysis, 5
  data,
    simulators, 68
    standard-based, 307
    postanalysis interpretation, 322
  experiments, diagnostic techniques, 241
  gene expression,
    database, *see* MGED
    markup language, *see* MAGE-ML
    object model, *see* MGED-OM
  pathway profiles, *see* MAPP
Mismatch, *see* MM
Mixed,
  linear model, 216
    hypotheses, 209

models, 38
Wolfinger *et al*, 83
Mixed-effects normalization, 21
Mixture model, 153, 301
MM, 198, 202
MMDB, 327
MMEI, 208, 214
Model-based expression index, *see* MBEI
Model-view controller, 329
Modified t-statistic, 82
Molecular,
function, 309, 324
modeling database, *see* MMDB
Monte Carlo,
hypotheses testing, 188
two-sample case, 248
mRNA,
levels, biological averaging, 100
pooling, 95–97
MTM, 341
Multi-dimensional scaling, *see* MDS
Multi-modalities, 153
Multiple,
comparison, 80, 290
cross mappings, *see* MCM
testing, 292
Multiple-trait interval mapping, *see* MTM
Multiplicative model, 198, 215
hypotheses, 209
Li–Wong's, 202
probe-level modeling, 200
Multiplicity, 292
Multivariate normal distribution, 87
MvA plot, 19–20, 23

N

National Institute of Health, *see* NIH
Negative log-transformed p-values, 48
Neighbor-joining evolutionary tree, 208
Net Affix, 32
Neural networks, 147
Newton–Raphson, mixed linear model, 207
NIH, 327
Nimble Gen, 4
Nonellipsoidal clusters, 141
Nonlinear,
classifiers, UMSA, 181
statistics, 18
Nonnegligible eigenvalues, SVD, 183
Nonparametric,
Bayesian approach, 351
bootstrap, 249–250
closeness measure, hypotheses testing, 192
estimator, 187
inference, 185, 247
statistics, 66
Normal,
–normal models, 274
distribution, 106
cluster analysis, 171
consensus trees, 151
fixed effects linear models, 234
Normality, hypotheses testing, 241
Normalization,
methods, cyclic loess, 20
global and local, 17–20
outlier-resistant, 14
subtraction of global means, 15
what is, 10
Northern, 41
Nucleolytic, 337
Null,
distribution, 269
permutation estimate of, 300
permutation-based nonparametric methods, 254
hypotheses, 60, 62
hypotheses testing, 191
multiple testing, 294
null distribution, 269
permutation-based nonparametric methods, 252, 254
rank-based statistics, 255–257
small sample sizes, 80
spike-in study, 68
two-sample case, 248
WRS, 258
scores,
ionizing radiation, 270
null distribution, 269
variance, 83
Number of significant genes, comparison, 240

O

Object-oriented, relational model vs., 306
OD ratio, 34
Off-the- shelf optimization, lognormal–normal models, 276
Oligonucleotide arrays, high-density, 11
Omic,
sciences, 70
technologies, 58
One-channel systems, 300
Open Biological Ontologies, 308
Operons, 68
Optical density ratio, *see* OD ratio
Optimal soft-margin classifier, UMSA, 181
Orthogonalization, 35

Outliers, 14, 39, 42, 45
analysis, 47
arrays, 38
automatic rejection, 165
cluster analysis, 165
pooled design, 102
Overall mean, 18
Over-analysis, 38

P

Pairwise similarity, 166
Parametric,
assumption, null distribution, 271
equal-variance *t*-test, 250–251
F-ratio, 246
linear models, fixed-effects models, 232–236
assumptions, 223
comparison, 239–241
fold-change by standard deviation, 240
mixed effects models, 236–239
variations of, 224
Parsing paradigms, 311
Partial least squares approximation, 347–348
Pattern identification analysis, 322
PCA, 38, 46, 147, 341–342
PCA/SVD, UMSA, 181
PCAm, 48–49
PCR, 4, 37
Pearson
correlation, 135, 136–138
cluster analysis, 167
v. Shalala, 70
PEML, 311–312
Perfect match, *see* PM
Permutation and bootstrap, nonparametric methods, 248
Permutation-based algorithms, 331
Permutation-based tests, limitation, 261
Pharmacologically tractable targets, 347
Pilot study, 34, 50
Plasmode, 68–69
Plato, 59
Plug-and-play analysis, 329
PM, 198, 202
Poisson regression, 271
Polyhierarchical tree, 324–325
Polymorphisms, 344, 346
Pooled designs, contamination, 100–105
Pooling, 33
cost, 99
Popular-level reality, 59
Positional cloning, 336
Posterior,
distribution
hierarchical linear model, 283
semiparametric model, 278
probability, Bayes approach, 272
Post-translational modification, 3
Power, 63, 298, 351
pQTL, 342
Prediction, 61
Predictors, 291
Primate data, 200
Principal components analysis, *see* PCA
Priors,
distribution,
hierarchical linear model, 284
hierarchical normal–normal, 285
probability,
Bayes approach, 272
nondifferential expression, 272
Bayesian approach, 274
hierarchical models, 274
Pro Peak, UMSA, 182, 184
Probabilities of expression, hierarchical linear model, 284
Probe,
affinity, 30
patterns, Affymetrix, 201
sets, 22, 31, 42, 50,
level metrics, 45
expression profile of, 214
Probe-level modeling,
fixed linear model, 204
mixed linear model, 205–208
models, 213
primate example, 208
Protein microarray, 2
immunologic targets, 3
Proteome-wide covariance structure, 67
Proteomic tools, 2
Proteomics Experiment Markup Language, *see* PEML
Pseudo-null distribution, 66
Psychometric test theory, 163
Pub Med, 58
Public data, 353
p-value, 39–40, 46
p-value plot, 84
Cliffs δ statistics, 258
equal-variance two-group *t*-test, 228
fold change by ANOVA, 235
fold-change by mixed model, 238
fold-change by *t*-test, 230
hypotheses testing, 191
permutation-based nonparametric methods, 251
p-value of the, 293
unequal-variance two-group *t*-test, 229
Chebby checker, 259
discrete distribution of, 86

FWER corrected, 295
mixture modeling, 272
multiple testing, 292
multiple testing, 293
probe-level modeling, 210–213
validity, 66

Q

Q statistics, multiplicative model, 202
qq plots, 20, 103–104
qRT-PCR, 100
QTG, 336
QTL, 336, 343, 345, 348, 354
linkage, 347
Quality control metrics, 35
Quantile,
–quantile plots, *see* qq plots
normalization methods, 20
normalized, 44
Quantitative,
epigenetics, 348
trait genes, *see* QTG
traits, 336, 338, 348
loci, *see* QTL
Query languages, 306
q-value, 295

R

Random,
array variability, mixed effects models, 237
effects, mixed linear model, 205
error vector, ANOVA, 246
variables, mixed linear model, 206
variation, 12
Randomization, 24, 35, 112
tests, 86
Rank sum,
ANOVA, 254
test, Mann–Whitney–Wilcoxon, 184
Rank-based
nonparametric approach, 254
statistics, hypotheses, 256
tests, limitation, 261
Ranking, permutation-based nonparametric methods, 254
Recursive partitioning, *see* RP
Reduced normal equation, 114
Reference genes, 17
Referent distribution,
nonparametric methods, 247
permutation-based nonparametric methods, 252
Regulons, 143
Relevance networks, 147
REML, 207
Replicate arrays, 78
Replication, 24, 112
Resample, bootstrap, 251
Resampling approach, multiplicative model, 202
Residual, 47
maximum likelihood, *see* REML
Response, microarray experiments, 225
Reverse transcription polymerase chain reaction, *see* RT-PCR
RMA, 43–44, 48
Robust multi-array analysis, *see* RMA
RP, 178
RT-PCR, 41, 120, 339, 344

S

SAM, 231, 253
hypotheses testing, 189
SAS, 47
Satterthwaite
degrees of freedom, 213
-type methods, 208
unequal-variance two-group *t*-test, 229
SC, 148–149
Scatter-plot
matrix, probe-level modeling,
bivariate normality, 199, 209–212
Search for proof, 64
Security, 313
Self-organizing maps, *see* SOMs
Semantic heterogeneities, 306
Semiparametric model, 276
Shepard plot, 170
SHR, 345
Shrinkage estimators, 302
Significance,
cutoff, 295
testing, 13
Significant,
genes, 351
p-values, 331
Silhouette coefficient, *see* SC
Similarity metrics, 166–167
Simulation
results, probe-level modeling, 217
studies, probe-level modeling, 215
Single linkage,
agglomerative algorithm, 140
cluster analysis, 169
Single nucleotide polymorphisms, *see* SNPs
Single-channel arrays, 20–22
Singular value decomposition, *see* SVD
Smart storing rules, 295

SMD, 314
SNPs, 339, 340
SOAP, 312
Software sources, 300
SOM, 144
SOS, learning function, 145
Sources,
 of error, nonbiological, 348
 of variation, mixed linear model, 205
Spearman correlation, 135
Spherical clusters, 141
Spike-in,
 controls, 35, 40, 45, 50
 data sets, 68
Splinter cluster, 142
Split-plot design, 22
Spontaneous hypertensive rats, *see* SHR
Stability,
 of clusters, 147
 p-values, 47
Standard error, bootstrap, 251
Statistical,
 inference,
  permutation-based nonparametric methods, 253
 issues, 347
 methodology, 60
 models, probe-level modeling, 217
Stochastic error, Affymetrix, 200
Strong control, 294, 295
Strict association, 323
Student's *t*-test, 65
Subset normalization, 22
Supervised,
 component analysis, 182–184
 learning, 180
Support vector machines, *see* SVM
SVD, 38, 47
 hypotheses testing, 189, 192
 UMSA, 183
SVD and UMSA, 183
SVM, 178–181
Syntactic heterogeneities, 306
Systematic bias, 112

T

t-statistics, 13
Target ligands, 3
t-distribution, 300
Technical variability, 348
Test statistic,
 hypotheses testing, 191
 mixed effects models, 237
Thresholding estimators, 302
Tissue microarray, 2
TN, 78–80
 mixture model, 87
 role of sample size on, 88–90
Tool operability, 307
Top-down clusters, 139, *see also* divisive clusters
TP, 78–80
 mixture model, 87
 role of sample size on, 88–90
Traditional power analysis, 351
Transcriptional regulatory factor, 347
Transcriptome-wide covariance structure, 67
Translational medicine, 5
True,
 differential expression, 268
 fold change, multiplicative model, 202
 negative rate, 63
 positive rate, 63
t-statistic, 299
 unequal-variance two-group *t*-test, 229
t-test, 13
 permutation-based nonparametric methods, 253
 probe-level modeling, 216
 two-sample case, 248
 of null hypothesis, 85
 combining information, 278
t-type statistic, 83
Tukey, 289, 291
Tukey's biweight function, 198
Tumor necrosis factor, 4
Tumor necrosis factor-α, 85
Two-channel arrays, gene expression ratios, 279
Two-color,
 arrays, 32, 37
 systems, 11
Two-dimensional measure, hypotheses testing, 192
Two-group,
 comparisons, microarray experiments, 225
 t-test, multiplicative model, 202
Two-sided,
 hypothesis, 227
 p-value, 326
Type 1 error, 80
 combined modes, 69
 equal-variance two-group *t*-test, 228
 probe-level modeling, 216
 rank-based statistics, 255
 two-sample case, 248
 mixed effects models, 238
Type 1 error rate, ANOVA, 246
Type 2 error, 80
Type I contamination, 101–103
Type II contamination, 101–103

U

UDDI, 312
UML, 308
UMSA, 179
  ERM, 180
Uncentered correlation, 136–138
Unequal variance
  t-test, 253
  two-group *t*-test, 229–231
Unified,
  maximum separability analysis, *see* UMSA
  modeling language, *see* UML
Unigene ID, 32
Univariate tests, ANOVA, 246
Universal Description, Discovery, and Integration, *see* UDDI
Unrooted directed graph, 325
Unsupervised dimension reduction, UMSA, 181
U-statistics, 184, 186
Utility of pooling, 105
UV absorbance ratios, 36, 41

V

Variability, 12, *see also* Variance; Variation
  array-specific, 15
  biological, 34
  biological, 97–99
  of disequilibrium, 352
  pool-to-pool, 97
Variance, 10, 39, 44, *see also* Variability; Variation
  mixed effects models, 237
  of biased estimator, 101
  outliers, 66
  reduction of, 13–14
  residual, 15
  scaling for heterogeneity, 19
  sources of, 163
  stabilizing transformation, 33, 38
  type models, 12
Variation, technical, 97–99
Visualization, 329
Volcano plots, probe-level modeling, 210

W

WADP, 150–152
Ward's method, cluster analysis, 169
Weak control, 294
Web Service Description Language, *see* WSDL
Web services choreography, 313
Weibull distribution, 271
Weighted average discrepant pairs, *see* WADP
Welsh correction, 87
Whole-genomic sequences, 4
Wilcoxon,
  rank test, 299
  rank-sum
    statistic, 270–271
    test, 254
  test, 83
Within-group heterogeneity, hypotheses testing, 187
WSDL, 312
WWII inductees, 95

X

XML, 311, 329
  -based standards, 310

Y

Yale Microarray Database, *see* YMD
YMD, 330
YPD, 328

Z

z-score rank ordered list, 327